Research activity involving algae in the classes Chrysophyceae and Synurophyceae ('chrysophytes') has increased dramatically over the last decade. These beautiful and delicate organisms are pivotal for studies of protistan evolution, studies of food web dynamics in oligotrophic freshwater ecosystems and assessment of environmental degradation resulting from eutrophication and acid rain. They also represent excellent model cellular systems for studying processes inherent in basic metabolism, biomineralization, endo- and exocytosis and macro-assembly of cell surface layers.

This book gives a broad overview of current research, emphasizing the phylogeny, ecology and development of these organisms. Each chapter also contains reviews of the literature and presents ideas for future research.

Phycologists, paleoecologists, limnologists and plankton ecologists will find this a mine of valuable information.

CHRYSOPHYTE ALGAE

CHRYSOPHYTE ALGAE

Ecology, phylogeny and development

Edited by

CRAIG D. SANDGREN

*Associate Professor, Biological Sciences, and Research
Associate, Center for Great Lakes Studies, University of
Wisconsin – Milwaukee, Wisconsin, USA*

JOHN P. SMOL

*Professor, Department of Biology, Queen's University,
Kingston, Ontario, Canada*

and

JØRGEN KRISTIANSEN

*Associate Professor, Botanical Institute, Department of
Phycology, University of Copenhagen, Copenhagen, Denmark*

CAMBRIDGE
UNIVERSITY PRESS

CAMBRIDGE UNIVERSITY PRESS
Cambridge, New York, Melbourne, Madrid, Cape Town, Singapore, São Paulo, Delhi

Cambridge University Press
The Edinburgh Building, Cambridge CB2 8RU, UK

Published in the United States of America by Cambridge University Press, New York

www.cambridge.org
Information on this title: www.cambridge.org/9780521102414

First published 1995
Reprinted 1996
This digitally printed version 2009

A catalogue record for this publication is available from the British Library

Library of Congress Cataloguing in Publication data

Chrysophyte algae: ecology, phylogeny, and development / edited by
Craig D. Sandgren, John P. Smol & Jørgen Kristiansen.
        p.   cm.
Papers presented at the Third International Chrysophyte Symposium,
held Aug. 12–16, 1991 at Queen's University, Kingston, Ontario,
Canada.
Includes index.
ISBN 0 521 46260 6 (hc)
1. Chrysophyceae – Congresses.   I. Sandgren, Craig D.   II. Smol,
J. P. (John P.)   III. Kristiansen, Jørgen.   IV. International
Chrysophyte Symposium (3rd: 1991: Queen's University)
QK569.C62C47   1995
589.4'87 – dc20   94-11379   CIP

ISBN 978-0-521-46260-0 hardback
ISBN 978-0-521-10241-4 paperback

Contents

Preface

Research activity involving algae in the algal classes Chrysophyceae and Synurophyceae (collectively, 'chrysophytes') has increased dramatically over the last decade. Chrysophyte algae are primarily freshwater flagellate organisms that typically constitute a dominant or subdominant portion of the phytoplankton biomass in lakes with moderate to low productivity. They have long been under-studied by aquatic ecologists because of difficulties in culturing and preserving chrysophytes, and because of the tendency to study eutrophic, human-impacted lake systems in which chrysophytes are relatively rare. However, the critical importance of chrysophytes in carbon flux through oligotrophic, freshwater food webs has been the subject of considerable recent research. The mixotrophic nutritional capacity of most chrysophytes provides them with a distinct advantage in nutrient-poor lakes. Chrysophyte-rich lakes are often poorly buffered and are thus sensitive to anthropogenic acidification. Because of the extreme interest in lake acidification in both North America and Europe, siliceous chrysophyte microfossils (resting cysts, cell scale layers) have rapidly become primary paleolimnological research tools. Chryso-phyte microfossil research has, in turn, stimulated considerable renewed interest in their biogeography and in environmental factors influencing seasonal and spatial distribution patterns. This activity has revealed fundamental gaps in our knowledge of chrysophyte nutritional physiology and basic metabolism. Chrysophyte algae also continue to occupy a central position with regard to evolutionary relationships among groups of chromophyte, chlorophyll a and c containing algae (diatoms, chryso-phytes, brown algae, dinoflagellates, etc.) and also in linking chromophytes to several non-photosynthetic protistan groups (amoebae, heliozoans, choanoflagellates, oomycete molds). These relationships have been the subject of considerable ultrastructural phylogenetic research over the last

15 years, and chrysophytes continue to be pivotal in current molecular phylogenetic studies of photosynthetic protists.

It is this diverse range of recent research activities concerning chrysophyte algae that stimulates and justifies the need for special international symposia. The First International Chrysophyte Symposium (1st ICS), which primarily grew out of the recognized need for focussed phylogenetic discussions concerning these important algae, was held in Fargo, North Dakota, following the annual AIBS meeting in 1983. The 2nd ICS was held in Berlin in 1987, following the last International Botanical Congress. The Third International Chrysophyte Symposium (3rd ICS), which led to the publication of this volume, was held on the beautiful, traditional campus of Queen's University in Kingston, Ontario, Canada, August 12–16, 1991. The 3rd ICS was one of many special events held on the campus during the celebration of Queen's University Sesquicentennial Anniversary. Dr Craig D. Sandgren served as Program Chairman for the 3rd ICS and Chair of the International Organizing Committee. Dr John P. Smol of the Biology Department of Queen's University served as Chair of the Local Organizing Committee. In addition, many scientists at Queen's University gave freely of their time to make the symposium a success. Foremost among these volunteers were Katharine Duff, Barbara Zeeb and John Glew, who all assisted with many aspects of the symposium. Funding for some of the symposium's activities was provided by a conference grant from the National Sciences and Research Council of Canada. In addition, we would like to acknowledge support from Queen's University: the Principal's Sesquicentennial Fund, the Office of Research Services, and the Department of Biology. Other contributions included those from Olympus Microscopes, Hitachi Scientific Instruments, and The College of Letters and Sciences, University of Wisconsin – Milwaukee.

The symposium focussed on current trends in research concerning chrysophyte algae. It continued a tradition of strong emphasis on evolutionary and systematic biology, but also included special emphasis on ecological and paleolimnological studies. The goals of the International Organizing Committee were: (1) to summarize the current status of chrysophyte research in these areas for the general scientific community; (2) to stimulate research efforts in specific topics where critical new information is required; and (3) to promote chrysophyte algae as convenient cellular systems for studying basic questions in cell biology such as basic metabolism, biomineralization, endo- and exocytosis, and extracellular macro-assembly. To these ends, the scientific program included three and a half days of invited plenary lectures, workshops, contributed paper sessions and posters. Each of the full meeting days emphasized one of the following themes: 'Phylogeny, Systematics and Evolution', 'Development, Physiology and Nutrition', 'Ecology, Paleoecology and Reproduction'. Plenary speakers were asked to review the literature in their particular area and present ideas for future research.

This volume contains chapters contributed by the plenary speakers plus several original papers submitted by other symposium participants. The book is arranged into subsections according to the symposium themes. We hope this work accomplishes the goals of summarizing current understanding of chrysophyte algae and of stimulating new research by both chrysophyte specialists and the general scientific community.

September 1993

Craig D. Sandgren
John P. Smol
Jørgen Kristiansen

Contributors

David A. N. Barker
School of Biological Sciences, University of Birmingham, Birmingham B15 2TT, UK

Martin E. Boraas
Department of Biological Sciences, University of Wisconsin – Milwaukee, PO Box 413, USA

Rose Ann Cattolico
Department of Botany, University of Washington, Seattle, WA 98195, USA

Gertrud Cronberg
Institute of Ecology/Limnology, University of Lund, Box 65, S-221 00 Lund, Sweden

Terrence P. Delaney
Department of Molecular Genetics, CIBA-GEIGY Corporation, 3054 Cornwallis Road, Research Triangle Park, NC 27709, USA

Pertti Eloranta
Department of Limnology and Environmental Protection/Limnology (E-House, Viiki), FIN-00014, University of Helsinki, Finland

Antje Gutowski
Institut für Meeresbotanik, Universität Bremen, PO Box 330 440, D-28334 Bremen, Germany

Linda K. Hardison
Department of Botany, University of Washington, Seattle, WA 98195, USA

Dale A. Holen
Penn State University, Worthington, Scranton Campus, 120 Ridge View Drive, Dunmore, PA 18512, USA

Anthony Kontoulis
School of Botany, University of Melbourne, Parkville 3052, Victoria, Australia

Jørgen Kristiansen
Department of Mycology and Phycology, Botanical Institute, University of Copenhagen, Øster Farimagsgade 2 D, DK-1353 Copenhagen K, Denmark

Barry S. C. Leadbeater
School of Biological Sciences, University of Birmingham, Birmingham B15 2TT, UK

Martha Ludwig
School of Botany, University of Melbourne, Parkville 3052, Victoria, Australia

Brigitte Martin-Wagenmann
Institut für Systematische Botanik und Pflanzengeographie, Freie Universität Berlin, Altensteinstrasse 6, D-14195 Berlin, Germany

Øjvind Moestrup
Department of Mycology and Phycology, Botanical Institute, University of Copenhagen, Øster Farimagsgade 2 D, DK-1353 Copenhagen K, Denmark

Kenneth H. Nicholls
Limnology Section, Water Resources Branch, Ontario Ministry of the Environment, 125 Resources Road, PO Box 213, Etobicoke, Ontario M9P 3V6, Canada

Charles J. O'Kelly
Bigelow Laboratory for Ocean Sciences, McKown Point, West Boothbay Harbor, ME 04575, USA

Hans R. Preisig
Institut für Systematische Botanik, Universität Zürich, Zollikerstrasse 107, CH-8008 Zürich, Switzerland

John A. Raven
Department of Biological Sciences, University of Dundee, Dundee DD1 4HN, UK

Leela S. Saha
University Department of Botany, Bhagalpur University, Bhagalpur (Bihar) 812 007, India

Craig D. Sandgren
Department of Biological Sciences, University of Wisconsin – Milwaukee, PO Box 413, Milwaukee, WI 53201, USA

Peter A. Siver
Department of Botany, Connecticut College, New London, CT 06320, USA

John P. Smol
Paleoecological Environmental Assessment and Research Laboratory, Department of Biology, Queen's University, Kingston, Ontario K7L 3N6, Canada

William E. Walton
Center for Great Lakes Studies, University of Wisconsin – Milwaukee, 600 E. Greenfield Avenue, Milwaukee, WI 53204, USA

Richard Wetherbee
School of Botany, University of Melbourne, Parkville 3052, Victoria, Australia

Daniel E. Wujek
Department of Biology, Central Michigan University, Mount Pleasant, MI 48859, USA

1

History of chrysophyte research: origin and development of concepts and ideas

JØRGEN KRISTIANSEN

This account covers a period of about 200 years – from the early beginnings of light microscopy until electron microscopy became established in the middle of the twentieth century. It takes its beginning in the middle of the eighteenth century, in rural Denmark at idyllic Frederiksdal, a manor house 20 km north of Copenhagen. In 1750, Countess Catharina Schulin had lost her husband Count Johan Siegesmund Schulin, who had held a high position in the Danish government administration. She was left alone with two children, a large estate, and an enormous staff of servants. She sought comfort in religion and summoned young theology students from the University to preach for her (Anker 1943).

Among these was 20-year-old Otto Friedrich Müller (Fig. 1.1). He won the friendship of the countess, and after 3 years he was appointed as a private tutor to her young son. He took this position very seriously, so that the young count at the age of 10 years and 9 months was able to pass the entrance examination to the University. Later, Müller accompanied the young count on the European tour that was traditional for young noblemen.

Müller had leisure time during his appointment and the beautiful surroundings of Frederiksdal stimulated his interest in natural history. Countess Schulin was interested not only in religion, but also in many aspects of culture and science. She was extremely pleased with Müller's service and encouraged and supported him in his scientific studies. She bought books on natural history, collections of natural objects such as sea shells (which are still to be seen on the estate) and even furnished him with both a simple and a compound microscope. As the latter was not corrected for chromatic aberration, he preferred the simple one for most of his studies. These microscopes were of the Watkin's model, but made by a Copenhagen instrument maker (Anker 1943).

Figure 1.1. O.F. Müller (1730–1784).

The simple microscope (with only one lens) had been known for many years, but it was not put to serious scientific use until the beginning of the seventeenth century. The invention of the compound microscope is generally ascribed to Hans Janssen, as early as 1590. However, these microscopes had been used primarily by amateurs, and no significant scientific progress using microscopes had taken place during the 100 years since Robert Hooke (1665) and Leeuwenhoek (1673) made their epoch-making discoveries (see Dobell 1960). The microscope had been used mainly as a toy, as is apparent from, for example, Ledermüller (1763). The study of microscopic organisms was almost considered outside real science and not thought appropriate for serious scientists. At this time it was even doubted whether these microscopic 'beings' were real living organisms.

Müller found in the microscopic world a neglected but also rewarding field, and his studies became of profound importance for the development of algal science. He wrote in a letter to Linnaeus: 'Fog and rain was my promenade weather, mosses and fungi the company I sought, worms and infusoria my nightly pleasure' (Schiødte 1870–1). Müller was an admirer of the great Linnaeus, who had just published his *Systema Naturae* (first edition 1835). But he was in opposition to Linnaeus' division of the microscopic organisms into only two genera – *Volvox* and *Chaos* – because he found a much greater diversity in the microscopic world. In the numerous ponds and other small waterbodies around Frederiksdal he found many interesting organisms to study, and he published his results in several books. Most famous among these is, perhaps, *Animalcula Infusoria* from 1786. The designation *Animalcula Infusoria* was quite modern at the time, created by Wrisberg 1765.

Müller was, of course, not aware that in this work he described the first chrysophyte. But, on the 50 plates, among hundreds of drawings of what were later to be recognized as ciliates, flagellates, desmids, diatoms, rotifers, etc., three organisms on plates III and IV are of interest here.

The first chrysophyte described is *Volvox vegetans*, later known as *Anthophysa vegetans*. This colorless, colonial, stalked organism is easily recognized from the drawings. *Volvox* comprised all organisms with motile globular colonies, according to Linnaeus' concept.

The next is the colonial organism called *Volvox uva*. Ehrenberg subsequently (1838) renamed it *Uvella virescens* and pointed out its possible identity with his *Synura uvella*. The third chrysophyte is *Enchelis punctifera*, which Ehrenberg later renamed *Microglena punctifera*. The identity with the organism of that name today is somewhat questionable, and even now the status of the genus *Microglena* is uncertain (Kristiansen 1988).

While Müller was describing his numerous organisms and classifying them in the Linnean system under the class of Vermes, order Infusoria, some similar organisms were at the same time being studied from quite another viewpoint. Priestley, in his early studies on photosynthesis, wrote in 1773 that a green deposit in water gave 'a very pure air in sun-light', but he had no suspicion that his green deposit was in fact vegetable matter (Sachs 1906). It was for some time subsequently known as 'Priestley's matter' and not thought to be of any particular structure; neither was it associated with the microscopic studies of Müller and others.

At about the same time, in France during the French Revolution, the botanist M. Villars published a local flora of the province of Dauphiné (1789). It contained almost exclusively phanerogams, but also a few fungi, lichens, and algae. On the very last plate, as the very last figure, there is

a primitive drawing denoted as *Conferva foetida*, the description of which
in the text (p. 1010) leaves no doubt of its identity as what is today known
as the palmelloid chrysophyte *Hydrurus foetidus* ('filamentis crassis,
vermiformibus, fluitantibus et gelatinosis').

Müller and Villars did not know each other, and even if they had they
would never have suspected that they had got hold of two ends of the
same case. But in their scant descriptions and primitive drawings lay
hidden the start of the study of chrysophytes, providing both structural
details and the first hints of the range of thallus organization types, from
the unicellular flagellate to the branched filament.

Another Frenchman, Vaucher (1803 – year XI of the French Revolu-
tionary calendar), redescribed Villars' *Conferva foetida*. He did not
consider it among the true Conferves, but referred it to the genus *Ulva*
because of its gelatinous consistency, as *Ulva foetida*. His remarks about
its characteristic smell are noteworthy: 'L'odeur qu'elle répand est
très-forte, et ressemble aux odeurs animales et surtout à celle des corps
qui commencent à entrer en putréfaction'. He gave reasonable illustrations at both high and low magnification.

However, the position of *Hydrurus* in a taxonomic system was still a
phycological nightmare. In 1824 the Swedish botanist C.A. Agardh (then
professor in Lund, but for the last 25 years of his life he was a bishop:
Fig. 1.2) published the first comprehensive survey of the Algae: *Systema
Algarum*. It contained a sound classification scheme, and even an attempt
at a phylogeny, with diatoms at the bottom and fucoids on the top. He
placed *Hydrurus* together with other algae of gelatinous habit in the
second order Nostochinae, and introduced the name *Hydrurus* (*vaucherii*).
But except for the diatoms, no unicellular forms were classified with the
Algae. They were still considered *Infusoria* and treated as such by the
zoologists.

During this period – as ever since – there were difficulties in defining
algae. Early phycology was haunted by the principle of spontaneous
generation. As late as 1833 this principle was advocated by Kützing,
who thought it valid for simple forms such as *Palmella* and *Protococcus*,
from which more elaborate forms might evolve which were able to
propagate themselves by spores (cf. Sachs 1906). Another problem lay in
defining the boundary between the lower animals and plants. The
difficulty was preliminarily solved by classifying all organisms capable of
independent movement with the animals; thus whole groups of 'algae'
(e.g. Volvocinea (green flagellates), Bacillariae (diatoms)) were claimed by
the zoologists. Logically, when formation of swarmers (zoospores) was

Figure 1.2. Upper left: C.A. Agardh (1785–1859). Upper right: C.G. Ehrenberg (1795–1876). Lower left: F. Dujardin (1801–1860). Lower right: M. Perty (1804–1884)..

seen for the first time in a filamentous alga, the phenomenon was described as a change of a plant into an animal! In this way, Trentepohl (1807) explained the escape of zoospores in *Vaucheria*. As Sachs (1906) expressed it: 'The remarkable thing was not that such views were entertained, but that the majority of botanists combined with them a belief in the constancy of species.'

The most impressive treatment of the *Infusoria* in the first half of the nineteenth century was given by the German scientist Christian Gottfried Ehrenberg (Fig. 1.2), who lived from 1795 to 1876 and worked as a professor in Berlin (Laue 1895). He was a very versatile person who studied and published on many very different subjects within biology. During his many travels through Africa and Asia, often with Alexander von Humboldt, he obtained a wide knowledge of the diversity of life. His first publications were about higher plants and fungi, but he also published on many subjects within zoology, including different types of monkeys and coral reefs, and he was the first to demonstrate the neural filaments in the animal body. However, his greatest reputation was as the founder of the science of micropaleontology, and he studied microfossils in sediments from all over the world.

Here, his awe-inspiring works on *Infusoria* are of most interest. They comprised a whole series of contributions starting in 1829 and culminating in the magnificent folio work *Infusionsthierchen als vollkommene Organismen* (1838) where, on 532 pages and 69 colored plates, he described the structure and appearance of a great many microscopic organisms (many of them discovered by himself) and classified them in a logical system.

In the family Kugelthiere (Volvocina) we find several chrysophytes: *Synura uvella* (Traubenartige Strahlenkugel) and *Uroglena volvox* (Wälzende Strahlenauge), and also the problematic *Syncrypta volvox*. This last species has been included in almost every textbook and popular book on algae since Ehrenberg's time, but it may actually represent a stage of *Synura sphagnicola* (Kristiansen 1988). Even Ehrenberg mentioned the similarity between *Syncrypta* and *Synura*. Other chrysophytes – *Dinobryon sertularia* (Wirbel-Moosthierchen) and *Epiyxis* – are placed in another family, Dinobryina, for loricate organisms.

All these chrysophytes were classified in the class Polygastrica, Magenthiere ('stomach-animals'). This peculiar designation reflects Ehrenberg's main theory on the structure of the Infusoria. He was in strong opposition to the prevalent view that these organisms could be considered only as shapeless bodies of slime and jelly originated by spontaneous generation. On the contrary, he argued that they were highly organized creatures with stomach, intestine, sexual organs, eyes, etc. Such organs were evident in the rotifers, and he argued that they must also be present in the other, smaller *Infusoria* where they were less readily apparent.

Ehrenberg's ideas were received with strong opposition. The coupe de grâce had already been given in 1841 by Felix Dujardin (Fig. 1.2) in his work *Histoire naturelle des Zoophytes: Infusoires*, supported by Siebold who in 1845 defined the Protozoa as unicellular (Locy 1915). Dujardin was professor in zoology and botany in Rennes in France and he possessed a highly critical mind. He opposed Ehrenberg's theories that these small animals should have diverse organs, but instead maintained that they were similar to the cells of multicellular organisms and, like these, were filled by a structureless slime which he termed 'sarcode'.

As for the chrysophytes, Dujardin operated with the same groups as Ehrenberg, with *Uroglena*, *Synura* and *Syncrypta* in the family Volcocina, the loricate *Dinobryon* and *Epipyxis* in the Dinobryina, and *Anthophysa* in the Monadina. Dujardin illustrated his work with beautiful plates, and he writes that he had the best microscope of his time: compound, achromatic, × 300–440 magnification, and enabling him to study the true form and structure of the organisms 'au lieu de la diviner à travers un contour diffus et nébuleux'. As for the plant versus animal problem, he had almost the same views as Ehrenberg, but he excepted diatoms (including desmids) from animals; these were plants to him.

Maximilian Perty (Fig. 1.2) was professor in Bern in Switzerland from 1834. He was a very gifted person with many interests, as can be seen from his autobiography (1879). But from his youth, *Infusoria* were his favorite subject for study. He was a faithful follower of Dujardin's cellular concept of the *Infusoria*, and he laughed at Ehrenberg's suggestion (ironically the only one at least partially accepted by posterity) that the red stigma in most flagellates was an eye: 'What should such simple creatures want to see?' His taxonomy is similar to Dujardin's. The *Infusoria* fall in two groups, Ciliata and Phytozoidea (flagellates), and his chrysophytes are distributed among several families in the same way.

Perty was a very clever microscopist, and with his Ploessl microscope he had equipment which in magnification and optical quality (and combined with his skillful eye) rivalled modern microscopes. He was able to see flagella and *Mallomonas* bristles. In his famous work from 1852, *Zur Kenntnis kleinster Lebensformen*, he described the genus *Mallomonas* with the type species *M. acaroides*, which he also called *M. ploesslii* in honor of his microscope manufacturer. He saw the bristles and the crenulated structure of the cell surface, but not individual scales; they were not seen until 1893 by Seligo.

Also following in Dujardin's footsteps we find Fresenius, who in 1858 published some observations on *Mallomonas* (that the flagellum is the locomotion organelle, and that the bristles are moved only passively) and also on *Anthophysa*.

This early period of fragmented chrysophyte research culminated in the monumental flagellate work *Der Organismus der Flagellaten* by Stein 1878, the designation *flagellates*' having been created by Cohn 25 years previously (1853). This work marks the beginning of the next era in which chrysophytes are distinguished as a separate taxonomic category. Stein lived between 1818 and 1885 and spent most of his working life as a professor in Berlin. His flagellate work represents an important step in chrysophyte research. Microscope quality, observation ability, and skill in drawing and printing had now reached such proficiency that many of his illustrations are of publishable quality even today; indeed many of them are still widely used in textbooks. However, still more important was his realization that the brown flagellates, which previous authors had scattered widely among other flagellates, should be combined into one family. He demonstrated that *Synura*, *Uroglena*, and *Syncrypta* belonged together (and not with the green Volvocina with which they had hitherto been classified) because they had the 'brown chlorophyll' ('diatomin') in common, located in two longitudinal bands in the cell, one of which contained a red eyespot. He also found that many non-colonial flagellates placed in the family Monadida appeared to have the same structure (*Microglena*, *Chrysomonas*, *Hymenomonas*, *Chrysopyxis*, *Stylochrysis*). Therefore, he united all these forms in the one family Chrysomonadina, where he also placed *Anthophysa*. He also felt tempted to incorporate *Dinobryon* (family Dinobryina) into this new family because of the essentially identical cell structure, but ultimately kept this genus separate because of its lorica.

He was of the opinion that the members of the family Chrysomonadina thus defined were typical animals, because many of them were known to engulf solid food. But they were also closely related to the other flagellate groups and he felt that they in many ways served as a link between them.

Just a few years after Stein's book, the equally monumental work on *Infusoria* by Saville-Kent (1880–1) was published. He was influenced by Stein and used many of his illustrations. As regards chrysophytes, he renamed them Chrysomonadidae, and included *Dinobryon* in that family.

About this time, some concepts regarding larger patterns in chrysophyte relationships had begun to emerge. Heralded by studies such as Trentepohl's (1807) discovery of zoospores in *Vaucheria*, the existence of different morphological stages during a life history became more and more accepted. An important contribution in this connection was a small paper by Cienkowsky from Odessa (1870). He recognized the relationship between flagellates and Palmellacea (among these *Hydrurus*): that many palmella-like organisms also can form swarmers similar to certain flagellates, and that on the other hand many flagellates can form palmelloid stages. Here we find the first vague ideas about 'organization levels' – a theme which Pascher 40 years later demonstrated clearly and convincingly, for both the chrysophytes and other algae. Cienkowski (1870) was also the first to describe the endogenous cyst formation characteristic of chrysophytes. Such ideas were further substantiated in Woronin's (Fig. 1.3) work on *Chromophyton* in 1880. He described how *Chromophyton* changed between flagellate and palmella stages and cysts. He thought it showed similarities to the green algal genus *Palmella*, but the brown color spoke against such a position. He was on the brink of establishing the organization levels as later described by Pascher (1914), but had not the necessary courage.

Woronin's work provoked an immediate reaction. Wille (1885 and some earlier papers) was so impressed by the fact that *Chromophyton* swarmers looked like *Dinobryon* cells that he postulated that both *Chromophyton* swarmers and palmella might belong in the *Dinobryon* life cycle. This was of course not correct, and he was much ridiculed by Fisch (1885), but his ideas nevertheless contained a core of truth.

The study of *Hydrurus* was repeatedly taken up, and one of the most important contributions for the understanding of chrysophyte relationships was the paper by Rostafinski (Fig. 1.3): 'L'*Hydrurus* et ses affinitées' (1882). He described parts of its life history and found similarity with *Chromophyton* in cell construction and palmella formation. So he united these genera in a new class called Syngeneticae, from which he proposed diatoms and brown algae could be derived. These ideas won adherents, even if they were severely criticized by Lagerheim (Fig. 1.3) who, incidentally, discovered the cysts of *Hydrurus* (1887).

The questions regarding the plant or animal nature of chrysophyte flagellates persisted and were accentuated by the fact that representatives from both kingdoms (as constructed in the nineteenth century) now were united in the same class. Should this class then be included in the Algae, as Rostafinski had proposed, or within the Protozoa?

Figure 1.3. Upper left: M. Woronin (1838–1903). Lower left: J.T. Rostafinski (1850–1928). Upper right: N.G. Lagerheim (1860–1926). Lower right: O. Kirchner (1851–1925).

Schmitz (1882) frankly proposed that organisms with brown chloro-plasts ('chromatophores' as he termed these organelles) were as much algae as those with green chloroplasts. The decisive step was taken by Hansgirg (1884 and followed up in 1886) in his *Prodromus der Algenflora*

Figure 1.4. Left: G. Klebs (1857–1918). Right: E. Lemmermann (1867–1915).

von Böhmen. Here he included Rostafinski's Syngeneticae as an order within the class Phaeophyceae, with its two families Chromophytonea and Hydrurea. Further, he included other organisms with brown chloroplasts, such as the members of the Chrysomonadina and Phaeothamnieae, although he hesitated about this step because Klebs (1882) had found Rostafinski's system too artificial.

The other perception that all flagellates should be located within the Protozoa also had its proponents, both among botanists such as Kirchner (Fig. 1.3; 1878, 1891) and especially among protozoologists such as Fromentel (1874) and Bütschli (1888) – a view continued in many textbooks to the present day (Lee *et al.* 1985). Such a view requires that chrysophyte taxonomy should be governed by the zoological code of nomenclature.

One of the most brilliant scientists involved in chrysophyte studies at the end of the nineteenth century was Georg Klebs (Fig. 1.4). He started as an assistant with de Bary in Strassburg, and later became professor in Tübingen and then in Basel (Küster 1918). He continued Stein's research on flagellates and in 1892 he published the famous *Flagellatenstudien*. He was originally critical of Rostafinski's ideas, but in 1892 he changed his views and agreed that *Hydrurus* should be joined with the brown flagellates because he was now convinced of the importance of silicified cysts as a common character shared by many organisms with brown chloroplasts. He established a system of five flagellate groups, one

Figure 1.5. Left: G. Senn (1875–1945). Right: A. Pascher (1881–1945)..

of which, the Chromomonadinae, contained the families Chrysomonadina and Cryptomonadina. He defined the Chrysomonadina in accordance with Stein but now with *Dinobryon* included, and he divided them in the following groups: Chrysomonadina nuda (e.g. *Chromulina*), Chryso- monadina loricata (e.g. *Dinobryon*) and Chrysomonadina membranata (e.g. *Mallomonas*). *Hydrurus* was placed in an appendix.

One of Georg Klebs' students was Gustav Senn (Fig. 1.5). At the age of only 25, he wrote for Engler & Prantl's *Die natürlichen Pflanzenfamilien* (1900) the chapters on flagellates, among these the order Chrysomona- dineae. He used the number of flagella for subdivision into families: one flagellum, Chromulinaceae; two (almost) equal, Hymenomonadaceae; and two unequal, Ochromonadaceae. Already 24 genera were recog- nized. *Hydrurus* was classified with the Chromulinaceae on the basis of reproductive cell morphology. With this work, so unlike any of its predecessors, we move into the twentieth century. A taxonomy had been established which, in its main ideas, was fundamental for Pascher and later for Bourrelly. Indeed, the chrysophyte volume in the new edition of Pascher's *Süsswasserflora* (Starmach 1985) used a further development of this taxonomic system from 1900. Paradoxically, this was Senn's last contribution to phycology. During his 30-year professorship in Basel, Switzerland, he studied mainly physiology of higher plants and devoted much time to university administration. He died in 1945, just as he retired from the university (Becherer 1945).

In the beginning of the twentieth century, knowledge of chrysophyte species and their distribution had been greatly enlarged by Ernst Lemmerman (Fig. 1.4). He began, as many other phycologists of his day, as a teacher, but later obtained a museum post (Bitter 1919). His treatment of the chrysophytes in *Kryptogamenflora der Mark Brandenburg* (1910) is an excellent compilation of the existing knowledge to date. It contains 27 genera with a total of 78 species, arranged according to Senn's system.

The year 1945 marked the death of Senn, and also of his heir and successor in phycological taxonomy, Adolf Pascher. Pascher (Fig. 1.5) was one of the greatest contributors to the study of chrysophytes. He started his scientific career in Prague in 1899, and he remained there throughout his life, from 1927 as professor at the German Karl's University. He studied all groups of freshwater algae and published more than a hundred scientific papers on their structure, taxonomy and phylogeny. However, during the 1930s he devoted more and more of his time to administration and to political activity, which caused him to take his own life in May 1945 (Geitler 1946; Pascher 1953; Pohl 1955).

Pascher described numerous new species (103) and genera (43), and made detailed studies in taxonomy and life histories. But he went a step further and synthesized his observations and ideas in his epoch-making system describing the parallel morphological evolution among the different classes of algae. He became renowned as editor of the famous *Süsswasser-flora* series, of which a new (third) edition is currently in press.

The chrysophytes were among Pascher's favorite algae. His first chrysophyte paper (1909) followed Senn's system and relied heavily on Lemmermann's authority. In his next works (1910, 1912) the ideas about parallel evolution began to emerge, and in his treatment of the Chrysomonadina in the *Süsswasserflora* (1913), the organization levels were preliminarily introduced and used within the framework of Senn's system. He operated with three main groups: Euchrysomonadinae (Chrysomonadales, Isochrysidales, Ochromonadales), Rhizochrysidinae, and Chrysocapsinae. This flora contained 124 species; by comparison the second edition (Starmach 1985) describes 966 species.

Pascher's system became established in principle in his 1914 paper where he suggested that every group of flagellates was ancestral to, and continued by, a more or less complete series of more complex organization levels. This system was discussed in more detail in 1931 and arranged within the Plant Kingdom. The brown flagellates (Chrysomonadina) were

Table 1.1. *Two important classification schemes for the class Chrysophyceae based on light microscopy*

Organization level	Pascher (1910, 1914, 1925, 1931)	Bourrelly (1957, as modified 1965)			
		Acontochrysophycidae	Heterochrysophycidae	Ochromonadales	Isochrysophycidae
Flagellate	Chromulinales Ochromonadales Isochrysidales	NONE	Chromulinales	Ochromonadales	Isochrysidales Prymnesiales
Rhizopodial	Rhizochrysidales	Rhizochrysidales	–	–	NONE
Palmelloid	Chrysocapsales	Chrysosaccales	NONE	NONE	NONE
Coccoid	Chrysosphaerales	Stichogloeales	Chrysosphaerineae	Chrysapionineae	NONE
Filamentous/parenchymatic	Chrysotrichales	Phaeoplacales	Thallochrysidineae	Phaeothamnineae	NONE

thus considered ancestral to a series containing rhizopodial, capsal (palmelloid), coccal (walled), trichal (filamentous) and parenchymatic levels; this whole series Pascher named Chrysophyceae (Table 1.1). The modern algal class Chrysophyceae was thus established in recognizable form in 1914.

Pascher's system was widely accepted: it was, for example, adopted by West & Fritsch (1927) in *A Treatise on the British Freshwater Algae*, the first edition of which (West 1904) had been based on Hansgirg. Pascher published many more papers on aspects of this morphological classification system: for example in 1925 on 'algal-like' chrysophytes, on various other organization levels, about transitions from one level to another, about species having the potential to express themselves on several levels. This is also the focus of one of his last papers (1943) – how *Dinobryon* can also occur in rhizopodial and palmelloid stages – which harks back to Wille's ideas 60 years previously.

It is a great loss to chrysophyte science that Pascher never completed a chrysophyte monograph intended for *Rabenhorst's Kryptogamenflora*. The manuscript remained unfinished (Pohl 1955) and was subsequently lost.

During the first half of the twentieth century, Pascher was the central figure in chrysophyte taxonomy. His morphological classification system was widely used by phycologists: compare Huber-Pestalozzi's chrysophyte volume in *Die Binnengewässer* (1941) and Fritsch's chapter in *Manual of Phycology* (1951). Still there was great activity in other chrysophyte research during this period, mainly on structural details, on revisions of smaller groups, or on descriptions of new species; but all within the framework of Pascher's system.

A few of the most prominent researchers from that period will be briefly characterized.

Scherffel made many original observations on chrysophyte biology and life histories (1911, 1927) while working in Budapest.

J. Boye Petersen (Fig. 1.6) was curator in the Copenhagen Botanical Museum 1920–51 and was professor in thallophytes from 1951 to 1958. He studied primarily soil algae, but also chrysophytes. One of his important early works was a small paper (1918) in which he demonstrated by staining technique the heterokont flagella of the chrysophytes. In the same paper he showed that the 'envelope' of *Synura* consists of silica scales. He resumed these investigations almost 40 years later with electron microscopy (Nygaard 1962).

A.A. Korshikov (Fig. 1.6) worked in Kharkov in Ukraine and published

Figure 1.6. Left: J. Boye Petersen (1887–1961). Right: A.A. Korshikov (1889–1945).

several papers on chrysophytes. One of the most important is his study on *Synura* (1927), in which he showed that this genus includes several species which can be distinguished by the shape of their silica scales. He died in 1945. His research was continued by Matvienko, who published both a Russian (1954) and a Ukrainian (1965) chrysophyte flora.

Krieger (Fig. 1.7), who worked throughout his life in Berlin, is best known for his comprehensive studies on desmids and on the fine structure of diatom valves. However, in 1930 he published an excellent critical monograph on the genera *Mallomonas* and *Dinobryon*, which was widely used (Bethge 1955).

W. Conrad (Fig. 1.7), had a position at the Natural History Museum in Brussels beginning in 1929. He described many species (more than 75) of chrysophytes and published critical monographs on several genera (e.g. on *Mallomonas* 1933) and on chrysophytes with three flagella (now classified as Haptophyceae/Prymnesiophyceae). Subsequently, he increasingly studied ecological topics, especially relating to brackish environments (Kufferath 1944).

Heinrichs Skuja (Fig. 1.7) was born in Latvia, and worked in the University of Riga. In 1944, he was forced to leave his country and settled in Uppsala, Sweden, where he was given an extraordinary professorship and gathered many students around him. He studied the biology and occurrence of a great many chrysophytes and described numerous (at least 70) new species. He was the last of the great masters who illustrated their works with their own drawings, which were of unrivalled mastership. It

Figure 1.7. Above: W. Krieger (1886–1954). Lower left: W. Conrad (1888–1943). Lower right: H. Skuja (1892–1972).

is said that he began the drawing of a plate of illustrations in the upper left corner and he completed it in the lower right corner with his signature HS (Thomasson 1973).

Nygaard (born 1903) introduced the use of chrysophyte cysts in paleolimnology in his 1956 Gribsø paper (Kristiansen 1986).

Of course there were attempts to revise or surplant Pascher's system, but not with real success. Some systems, however, are historically interesting. Lameere (1929, 1932), for instance, included foraminifers, heliozoans and radiolarians in the chrysophytes. Hollande (1952) tried to combine Pascher's system with Senn's, thus emphasizing the importance of flagellar number. Copeland (1956) worked from somewhat different principles. In his class Heterokonta he also included some lower fungi, and his chrysophytes consisted of two orders based on flagellar number: Ochromonadea (two flagella) and Silicoflagellata (one). His system was not accepted, but some of his ideas on gross classification pointed towards the future.

Hollande's system was the basis for the great work by Bourrelly: *Recherches sur les Chrysophycées* (1957). On the basis of thorough studies of cell structure, levels of thallus organization and life histories, Bourrelly tried to replace Pascher's system with a more phylogenetically sound classification. He considered the forms without flagellate stages the primitive condition (although he admitted that in some cases flagellate stages might have been lost), and he derived the whole class from a hypothetically non-flagellate ancestor. He distinguished two main groups of orders: those without a cell wall (the monadoid and rhizopodial levels) and those with a wall (the filamentous and coccoid levels). The palmelloid order Chrysosaccales was considered to have a somewhat intermediate position. On each level the taxonomic orders were distinguished on flagellar number. Inside each order he arranged the families in a proposed phylogenetic scheme, illustrated by diagrams.

In 1965, Bourrelly modified his system and divided the Chrysophyceae into three subclasses on the basis of flagellar presence and number: Acontochrysophycidae without flagella in any stages; Heterochryso-phycidae, with one flagellum (Chromulinales) or two unequal flagella (Ochromonadales); and Isochrysophycidae with two equal flagella. Within these subclasses and orders, classification is based on organization levels and also other features (Table 1.1). Starmach used Bourrelly's system in his flora (in Polish 1981, rewritten in German for the *Süsswasserflora* in 1985), but it proved difficult to arrange the system for practical identifica-tion purposes. On the other hand, this flora is our most modern survey on freshwater chrysophytes, with its total of 966 species (inclusive of 40 Haptophyceae/Prymnesiophyceae and 84 Craspedophyceae).

Bourrelly's classification system for chrysophytes represented the culmination and the triumph of the light microscopy era, and at the same time marked the end of it. By the early 1960s electron microscopy was rapidly taking over the scene and has become the standard tool for providing details of cell structure. But the first electron microscopic study including a chrysophyte had already been published (Brown 1945). The use of electron microscopy has provided incredibly detailed knowledge of cell structure, showing the immense complexity of cytoplasm and organellyes; in some ways Ehrenberg's ideas about the complexity of the flagellate cell have been rehabilitated. This, together with an increased use of biochemical and molecular-biological approaches, has resulted in the development of new taxonomic methods, and in new and contrasting interpretations of phylogenetic relationships among algae. Our ideas about delimitation and classification of Chrysophyceae and chrysophytes continue to change, but these developments are still taking place and go beyond the historical context of this paper. For the present status of chrysophyte taxonomy, see Preisig (this volume).

I shall end with the paradox that the class Chrysophyceae was not formally described and typified until 1976 by David Hibberd, one of the recent great electron microscopists within the chrysophytes. This was 62 years after the name of the class was proposed, and almost 200 years after the first members of the class were described.

References

Agardh, C.A. 1824. *Systema Algarum*, pp. 1–312 Lund.

Anker, J. 1943. Otto Friedrich Müller. *Act. Hist. Sci. Mat. Med.* **II**, 1–317.

Becherer, A. 1945. Gustav Senn 1875–1945. *Verh. Schweiz. Naturf. Ges.* **125**: 376–80.

Bethge, H. 1955. Willi Krieger 1886–1954. *Ber. Dtsch. Bot. Ges.* **68a**: 141–3.

Bitter, G. 1919. Ernst Lemmermann. *Abh. Naturw. Ver. Bremen* **24**: 273–91.

Bourrelly, P. 1957. Recherches sur les Chrysophycées. *Rev. Algol.* Mém.-Hors-Sér., pp. 1–412.

Bourrelly, P. 1965. La classification des Chrysophycées: ses problèmes. *Rev. Algol.* **1**: 56–60.

Brown, H.P. 1945. On the structure and mechanics of the protozoan flagellum. *Ohio J. Sci.* **45**: 247–301.

Bütschli, O. 1878. Beiträge zur Kenntnis der Flagellaten und einiger verwandten Organismen. *Z. Wiss. Zool.* **30**: 205–81, pl. 11–15.

Cienkowski, L. 1870. Ueber Palmellaceen und einige Flagellaten. *Arch. Mikrosk. Anat.* **6**: 421–36, pl. XXIII–XXIV.

Cohn, F. 1853. Beiträge zur Entwicklungsgeschichte der Infusorien. *Z. Wiss. Zool.* **4**: 253–81.

Conrad, W. 1933. Revision du genre *Mallomonas* Perty (1852) incl. *Pseudo-Mallomonas* Chodat (1920). *Mém. Mus. R. Hist. Nat. Belg.* **56**: 1–82.

Copeland, H.F. 1956. *The Classification of Lower Organisms.* Palo Alto, Calif. pp. 1–302.

Dobell, C. 1960. *Antony van Leeuwenhoek and his 'Little Animals'.* New York. (1st edn 1932) pp. 1–435.

Dujardin, F. 1841. *Histoire naturelle des zoophytes. Infusoires.* Paris.

Ehrenberg, C.G. 1838. *Die Infusionthierchen als vollkommene Organismen I-II.* Leipzig. pp. 1–547, pl. 1–64.

Fisch, C. 1885. Untersuchungen über Flagellaten und verwandte Organismen. *Z. Wiss. Zool.* **42**: 47–125, pl. 1–4.

Fresenius, G. 1858. Beiträge zur Kenntnis mikroskopischer Organismen. *Abh. Senckenb. Naturf. Ges.* **2**: 1–34, pl. X–XII. Frankfurt a.M.

Fritsch, F.E. 1951. Chrysophyta. In: Smith, G.M. (ed.) *Manual of Phycology.* Waltham, Mass. pp. 83–104.

Fromentel, E. de. 1874. *Études sur les Microzoaires ou Infusoires proprement dits.* Paris. 363pp., 30 pl.

Geitler, L. 1946. Adolf Pascher. *Naturwissenschaften* **33**: 320.

Hansgirg, A. 1884. Bemerkungen zur Systematik einiger Süsswasseralgen. *Österr. Bot. Z.* **34**: 389–94.

Hansgirg, A. 1886. *Prodromus der Algenflora von Böhmen.* I–II. Prague, pp. 1–288, 1–268.

Hibberd, D.J. 1976. The ultrastructure and taxonomy of the Chrysophyceae and Prymnesiophyceae (Haptophyceae): a survey with some new observations on the ultrastructure of the Chrysophyceae. *Bot. J. Linn. Soc.* **72**: 55–80.

Hollande, A. 1952. Classe des Chrysomonadines. In: Grassé, P. (ed.) *Traité de Zoologie* **I**: 451–570.

Huber-Pestalozzi, G. 1941. Chrysophyceae, Farblose Flagellaten, Heterokonta. Huber-Pestalozzi, G. (ed.) Das Phytoplankton des Süsswassers, vol. 2(1), pp. 1–365. In: Thienemann, A (ed.) *Die Binnengewässer* 14. Stuttgart.

Kirchner, O. 1878. Algen. In: Cohn, F. (ed.) *Kryptogamenflora von Schlesien II.* Breslau. pp. 1–284.

Kirchner, O. 1891. *Die mikroskopische Pflanzenwelt des Süsswassers.* Hamburg, pp. 1–60, pl. 1–5.

Klebs, G. 1882. Referat über die Arbeit von Rostafinski: Hydrurus. *Bot. Zeitung* 1882.

Klebs, G. 1892. Flagellatenstudien. I–II. *Z. Wiss. Zool.* **55**: 265–351, pl. 13–16, 353–444, pl. 17–18.

Korshikov, A.A. 1929. Studies on the chrysomonads. *Arch. Protistenk.* **67**: 253–90.

Krieger, W. 1930. Untersuchungen über Plankton-Chrysomonaden. *Bot. Arch.* **29**: 258–329.

Kristiansen, J. 1986. Gunnar Nygaard: a guiding influence on paleolimnological research. In: Smol, J. *et al.* (eds.) Diatoms and lake acidity. *Dev. Hydrobiol.* **29**: 1–3.

Kristiansen, J. 1988. The problem of 'Enigmatic Chrysophytes'. *Arch. Protistenk.* **135**: 9–15.

Kufferath, H. 1944. A la mémoire de Walter Conrad, protistologiste (1888–1943). *Bull. Mus. R. Hist. Nat. Belg.* **20**(1): 1–16.

Küster, E. 1918. Georg Klebs 1857–1918. *Ber. Dtsch. Bot. Ges.* **36**: 90–116.

Lagerheim, G. 1887. Zur Entwicklungsgeschichte des Hydrurus. *Ber. Dtsch. Bot. Ges.* **6**: 75–84.

Lameere, A. 1929, 1932. *Précis de Zoologie* I. Brussels. pp. 1–395.

Laue, M. 1895. *Christian Gottfried Ehrenberg. Ein Vertreter deutscher Naturforschung im neunzehenten Jahrhundert*, 1795–1876. Berlin. pp. 11–287.

Ledermüller, M.F. 1760. *Mikroskopische Gemuths- und Augen-Ergötzung.* Erstes Funfzig. Nürnberg.

Lee, J.L., Hutner, S.H. & Bovee, E.C. 1985. *An Illustrated guide to the Protozoa.* Society of Protozoologists, Kansas. pp. 1–629.

Lemmermann, E. 1910. *Kryptogamenflora der Mark Brandenburg III.* Leipzig. pp. 1–712.

Locy, W.A. 1915. *Die Biologie und ihre Schöpfer.* Gustav Fischer, Jena. pp. 1–415.

Matvienko, A.M. 1954. Zolotistye Vodorosil. *Opredelitelj Presnovodnych Vodoroslej SSSR* **3**: 1–188.

Matvienko, A.M. 1965. Zolotisti vodorosti-Chrysophyta. *Viznatsnik prisnovodnich vodorostej Ukrainskoi* RSR III (1): 1–367.

Matvienko, A.M. & Shvaps, M.F. 1989. *Aleksandr Arkadenits Korshikov: professor Kharkovskogo Unniversiteta.* Tsentralnaja Nautsnaja Bibliteka, Kharkov. pp. 1–43.

Müller, O.F. 1786. *Animalcula infusoria fluviatilia et marina.* Hafnia. 376pp., 50 pl.

Nygaard, G. 1956. Ancient and recent flora of diatoms and Chrysophyceae in Lake Gribsø. *Fol. Limnol. Scand.* **8**: 32–94.

Nygaard, G. 1962. Johannes Boye Petersen (1887–1961). *Rev. Algol. N.S.* **6**: 83–6, pl. 8.

Pascher, A. 1909. Ein kleiner Beitrag zur Kenntnis der Chrysomonadinen Böhmens. *Lotos* **57**: 148–54.

Pascher, A. 1910. Chrysomonaden aus dem Hirschberger Grossteich. *Int. Rev. Ges. Hydrobiol. Hydrogr. Monogr. Abh.* **1**: 1–66, pl. I–III.

Pascher, A. 1912. Über Rhizopoden und Palmellastadien bei Flagellaten (Chrysomonaden), nebst einem Übersicht über die braunen Flagellaten. *Arch. Protistenk.* **25**: 153–200, Tf. 9.

Pascher, A. 1913. Chrysomonadinae. In: Pascher, A. (ed.) *Die Süsswasserflora Duetschlands, Österreichs u. d. Schweiz* **2**: 7–95.

Pascher, A. 1914. Über Flagellaten und Algen. *Ber. Dtsch. Bot. Ges.* **34**: 136–60.

Pascher, A. 1925. Die braune Algenreihe der Chrysophyceen. *Arch. Protistenk.* **52**: 469–564, pl. 15.

Pascher, A. 1931. Systematische Übersicht über die mit Flagellaten in Zusammenhang stehenden Algenreihen und Versuch einer Einreihung dieser Algenstämme in die Stämme des Pflanzenreiches. *Beih. Bot. Centralbl.* **48**(2): 317–32.

Pascher, A. 1943. Zur Kenntnis verschiedener Ausbildungen der planktonischen Dinobryen. *Int. Rev. Ges. Hydrobiol. Hydrogr.* **43**: 110–23.

Pascher, A. 1953. Lebenslauf und Arbeitsverzeichnis. *Arch. Protistenk.* **98**: III–XXXII.

Perty, M. 1852. *Zur Kenntnis kleinster Lebensformen.* Bern. 228pp. 17 pl.

Perty, M. 1879. *Erinnerungen aus dem Leben eines Natur- und Seelenforschers des neunzehnten Jahrhunderts.* Leipzig & Heidelberg. 486pp.

Petersen, J.B. 1918. Om Synura uvella Stein og nogle andre Chrysomonadiner. *Vid. Medd. Dansk. Naturh. Foren.* **69**: 345–57, pl. V.

Pohl, F. 1955. Adolf Pascher 1881–1945. *Ber. Dtsch. Bot. Ges.* **68a**: 117–20.

Rostafinski, J.T. 1882. L'Hydrurus et ses affinités. *Ann. Sci. Nat. Bot.* **14**: 4–25, pl. 1.

Sachs, J. von 1906. *History of Botany* (1530–1860). Oxford. pp. 1–568.

Saville-Kent, W. 1880–1. *A Manual of the Infusoria* I–II. London. 913pp. 51 pl.

Scherffel, A. 1911. Beitrag zur Kenntnis der Chrysomonadinen. *Arch. Protistenk.* **22**: 299–344, pl. 16.

Scherffel, A. 1922. Beitrag zur Kenntnis der Chrysomonadimen: II. *Arch. Protistenk.* **57**: 331–61, pl. 15.

Schiødte, J.C. 1870–1. Af Linné's Brevveksling: Aktstykker til Naturstudiets Historie i Danmark. *Naturhist. Tidsskr.* **3**(7): 233–521.

Schmitz, F. 1882. Die Chromatophoren der Algen. *Verh. naturhist. Ver. Preuss. Rheinlande Westfalens* **40**: 1–180, pl. 1.

Seligo, A. 1893. Ueber einige Flagellaten des Süsswasserplankton. *Festgabe Westpreussischen Fischereiverein hundertundfünfzigjährigen Jubiläum Naturforsch. Ges. Danzig.* pp. 1–8, Tf. 1.

Senn, G. 1900. Flagellata. In: Engler, A. & Prantl, K. (eds.) *Die natürlichen Pflanzenfamilien* IIa. Leipzig. pp. 93–102.

Starmach, K. 1981. Chrysophyceae: Złotowiciowce. In: Starmach, K. *et al.* (eds.) *Flora Slodkowodna Polski* **5**: 1–175.

Starmach, K. 1985. Chrysophyceae und Haptophyceae. In: Ettl, H. *et al.* (eds.) *Süsswasserflora von Mitteleuropa*, 1. Gustav Fischer, Jena. pp. 1–515.

Stein, F. Ritter von, 1878. *Der Organismus der Infusionsthiere III: Der Organismus der Flagellaten I.* Leipzig. 150pp., 24 pl.

Thomasson, K. 1973. Prof. Dr. Heinrichs Leonhards Skuja 8.IX.1892–19. VIII.1972. *Int. Rev. Ges. Hydrobiol.* **58**: 441–2.

Trentepohl, J.F. 1807. Beobachtungen über die Fortpflanzung der Ectospermen des Herrn Vaucher, insbesonderheit der Conferva bullosa Linn. nebst einigen Bemerkungen über die Oscillatorien. In: Roth, A.W. (ed.) *Botanische Bemerkungen und Berichtigungen.* Leipzig. pp. 181–216, 1 pl.

Vaucher, J.-P. 1803. *Historie des conferves d'eau douce.* Geneva. pp. 1–285, pl. I–XVII.

Villars, M. 1789. *Histoire des Plantes de Dauphiné*, vols. I–III + pl. Grenoble, Lyon & Paris.

West, G.S. 1904. *A Treatise on the British Freshwater Algae.* Cambridge. pp. 1–372.

West, G.S. & Fritsch, F.E. 1927. *A Treatise on the British Freshwater Algae.* Cambridge. pp. 1–534.

Wille, N. 1885. Ueber Chromulina-Arten als Palmellastadium bei Flagellaten. *Bot. Centralbl.* **23**: 1–6.

Woronin, M. 1880. Chromophyton Rosanoffii. *Bot. Zeitung* **38**: 626–31, 642–4, pl. 9.

Wrisberg, H.A. 1765. *Observationum de Animalculis Infusorii Satura.* Goettingen. pp. 1–110, pl. 1–2.

Part I
Phylogeny, systematics and evolution

2

Evolution of plastid genomes: inferences from discordant molecular phylogenies

TERRENCE P. DELANEY, LINDA K. HARDISON
& ROSE ANN CATTOLICO

Introduction

Substantial morphological, biochemical and genetic diversity exists among photosynthetic eukaryotes. Some of the most conspicuous differences involve plastid-based characters. These features, such as the photosynthetic pigment complement, have provided a basis for defining plant and algal taxa. Christensen (1962) classified algae into three divisions according to their pigment composition. In this system, the division chlorophyta includes plants and algae that possess chlorophylls *a* and *b*, the Chromophyta are algae that predominantly contain chlorophylls *a* and *c*, while the Rhodophyta are defined in part as those algae having primarily chlorophyll *a* and accessory phycobilin pigments. However, autotrophic eukaryotes evolved in a stepwise manner with the mitochondrial and plastid organelles originating from formerly free-living prokaryotic endosymbionts (Gray & Doolittle 1982). In examining the evolution of these organisms, one must consider the evolution of not only the host, but also its organelles with their prokaryotic ancestors.

By examining cytosolic and nuclear features exclusive of plastid characters, one can infer relationships between extant plants and algae. Such studies have lent support to the validity of the Rhodophyta and Chlorophyta as true phylogenetic assemblages (natural taxa). For example, the complete absence of actual or vestigial flagella in rhodophytes (Gabrielson *et al.* 1985) and conserved features of karyokinesis and flagellar root structure in chlorophytes (Mattox & Stewart 1984) support the evolutionary cohesiveness of these groups. Furthermore, comparison of nuclear-encoded ribosomal RNA (rRNA) supports the model of a monophyletic origin for green and red algae (Perasso *et al.* 1989; Chapman & Buchheim 1991).

In contrast to the chlorophytes and rhodophytes, the division chromophyta appears to be an artificial assemblage, encompassing such divergent groups as the dinoflagellates, the brown and the golden brown algae. More recent treatments favor a classification of the Chromophyta based on a variety of morphological characters including the possession of thrust-reversing flagella with tripartite tubular mastigonemes (Cavalier-Smith 1989), the transitional helix in the flagellum (Preisig 1989) and tubular mitochondrial cristae (reviewed by Round 1989). According to these studies, the Chromophyta are better defined by characters other than pigmentation, since certain chlorophyll-*c*-containing algae are only distantly related to the main chromophyte group (e.g. dinoflagellates and cryptophytes; Gunderson *et al.* 1987). In addition, certain colorless protists exhibit morphological features similar to those found in the Chromophyta, suggesting their evolutionary affinity with the group (Beakes 1989). Examination of nuclear-encoded rRNA shows that certain organisms lie outside the algal mainstream. These include *Euglena gracilis*, which branched off the eukaryotic lineage very early, and the dinoflagellate *Prorocentrum micans*, which is found to be more closely related to the ciliates than to other algae (Gunderson *et al.* 1987).

Plastids from diverse representatives of the plant kingdom show many similarities in their biochemical and genetic traits. For example, the plastids found in all algal taxa utilize water as the primary electron donor in photosynthesis. Among prokaryotes, only the cyanobacteria use water as a source of reductant (reviewed by Olson & Pierson 1987). Unifying features are also found in the organization and structure of plastid genomes from Chlorophyta, Rhodophyta and Chromophyta. These include similarities in genome size and structure, with circular genomes ranging from 95 to 300 kilobase pairs (kbp), and the conservation of certain gene sets (Delaney & Cattolico 1989; Palmer 1985; Valentin *et al.* 1992).

Despite these similarities, important differences between algal plastids have led to the separation of these organisms into high-level taxonomic groupings. The molecular data examined in a number of plastid genomes support the taxonomy proposed by Christensen, which divides algae into three divisions. In addition, these data support classification systems based on biochemical and morphological criteria. For example, homologous genes encoding components of the phycobilisome are found in the plastid genomes of Rhodophyta and in *Cyanophora paradox* (Shivji 1991; Bryant *et al.* 1985), suggesting a relationship between plastids in these organisms. Moreover, at least 14 genes that are nuclear located in the Chlorophyta appear to be plastid encoded in both chromophytes and rhodophytes (reviewed by Valentin *et al.* 1992). These genes include proteases, ATPase components, transport polypeptides and the small subunit of the Calvin cycle enzyme ribulose-1,5-bisphosphate carboxylase (Rubisco).

While these data are informative, important problems have yet to be resolved. Do the differences observed among extant plastids reflect multiple symbiotic origins of the plastids, or evolutionary divergence from a single common ancestral plastid stock? Furthermore, has plastid evolution occurred along a simple linear pathway, or have other events such as lateral gene transfer or organelle fusion contributed to the diversity observed among modern plastids? These issues have great importance in understanding both the co-evolution of the plastid and its host, and the evolution of the metabolic processes that attended the acquisition of plastids from formerly free-living endosymbionts.

The careful comparison of homologous molecular sequences in related organisms is a powerful method for addressing questions of evolutionary relationships. To be useful, the molecular sequences used in such analyses must be distributed throughout the group of organisms or organelles being studied. In addition, the molecules must be sufficiently conserved to observe informative changes against a minimal background of multiple mutations at a site. Finally, it is critical when inferring phylogenies that one compare only those portions of the molecule that are homologous, in order to achieve a meaningful alignment among species. The statistical treatment of this kind of data has been reviewed and will not be discussed here (Olsen 1987; Felsenstein 1988).

The application of molecular sequences data to evolutionary biology is especially useful in organisms that show little morphological differentiation. For example, examination of 16S rRNA sequences led to a substantial reorganization of bacterial classification to better reflect the evolution of this group (Woese 1987). In these studies, 16S ribosomal RNA could be used as a molecular chronometer for all bacteria because it is highly conserved and ubiquitous. Interest in 16S rRNA has led to the accumulation of a substantial database of molecular sequence data from all forms of life. The presence of rRNA in plastids of all plants and algae and in their prokaryotic ancestor(s) makes this molecule especially useful for studies of plastid evolution.

To provide independent indicators of plastid relationships, two other molecules have been examined from plastids and certain prokaryotic autotrophs. These sequences were derived from the genes (*rbc*L and *rbc*S) encoding the large and small subunit proteins of Rubisco, respectively. The *rbc*L gene is plastid localized in all plants and algae examined.

However, the coding site of the *rbc*S gene varies, being nuclear localized in the Chlorophyta, but plastid encoded in all chromophytic and rhodo-phytic algae examined to date (Reith & Cattolico 1986; Douglas & Durnford 1989; Hwang & Tabita 1989; Valentin & Zetsche 1989, 1990*a*). In addition to differences in *rbc*S coding location, the primary structure of chromophyte and rhodophyte Rubisco genes is substantially different from those found in Chlorophyta and cyanobacteria (Boczar *et al.* 1989; Valentin & Zetsche 1989; 1990*b*; Douglas *et al.* 1990). These observations make the phylogeny of these genes especially significant.

In this chapter the evolutionary histories of plastids from diverse representatives of terrestrial plants and algae are discussed. Evolutionary arguments are based on phylogenies derived from genes encoding plastid 16S ribosomal RNA, and the Rubisco small and large subunit proteins. An unexpected result of this analysis is that different genes encoded on the same organellar genome may exhibit different evolutionary histories, suggesting the possibility of novel modes of gene acquisition in the evolution of plastid genomes.

Results and discussion

The plastid 16S rRNA sequence has been determined from the marine chromophyte *Olisthodiscus luteus* (Delaney & Cattolico 1991). The gene sequence showed that the rRNA was 1541 nucleotides in length, and could be folded into a secondary structure similar to that obtained from many sources (Fig. 2.1). The position and size of the base-paired regions in 16S rRNA are highly conserved in evolution (Gutell *et al.* 1985), enabling the majority of the model to be aligned with 16S rRNA sequences from other organisms. Figure 2.2 shows a sequence alignment from five organisms, and illustrates the areas of high conservation interspersed with divergent regions. Only regions that can be aligned with little ambiguity are considered in our phylogenetic analyses. These regions are over- and underlined in Fig. 2.2 and comprise 1338 nucleotides. Sequence data from 17 species are used in the 16S rRNA phylogeny presented (Table 2.1).

The aligned sequences were analyzed using a distance matrix approach, where the pairwise similarity values were determined between each of the sequences. These values were used to calculate the inferred evolutionary distance between each pair of sequences. This calculation takes into account the inevitable silent multiple mutations at a site, in which a later mutation restores the site to its original identity (Jukes & Cantor 1969). The resulting data are presented in a matrix of evolutionary distance

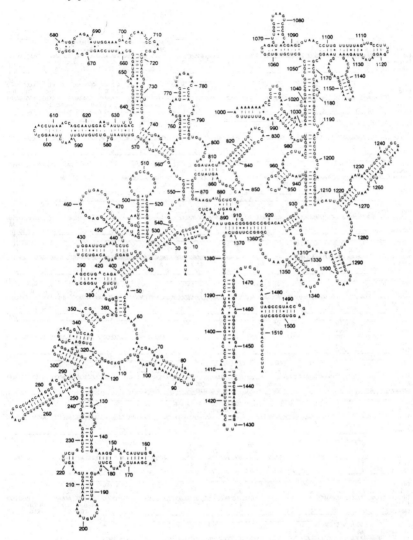

Figure 2.1. Secondary structure model of plastid 16S rRNA from the chromophyte *Olisthodiscus luteus* (Delaney & Cattolico 1991). Putative helical regions are indicated by tick marks or dots for canonical and non-canonical base-pairs, respectively. (Reprinted by permission of Oxford University Press.)

values (Table 2.2). The distance matrix is then examined using a least squares optimization algorithm (Fitch & Margoliash 1967) to identify a phylogenetic tree with a topology and scale that best fits the data (Fig. 2.3). The bootstrap statistical test (Felsenstein 1988) was conducted to

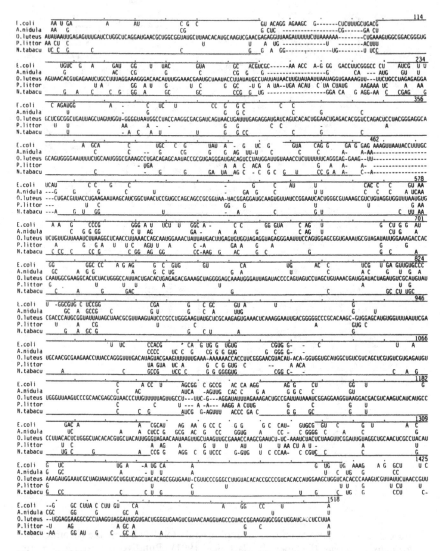

Figure 2.2. Representative alignment of 16S rRNA sequences. The *Olisthodiscus luteus* sequence is listed in full, while the other sequences are shown only where they differ from that in *O. luteus*, and at their termini. The regions that were used for phylogenetic analysis are indicated by lines above and below the sequence alignment (1338 nucleotides in total). The species included are *Escherichia coli*, *Anacystis nidulans* (*Synechococcus* 6301), *O. luteus*, *Pylaiella littoralis* and *Nicotiana tabacum*.

Table 2.1. *Organisms and gene sequences considered in this study*

Organism	Classification	Sequences examined[a]	References[b]
Olisthodiscus luteus	Chromophyta	16S, SS, LS	1, 2, 3
Pylaiella littoralis	Chromophyta	16S, LS	4, 5
Ochromonas danica	Chromophyta	16S	6
Cylindrotheca sp. strain N1	Chromophyta	LS	7
Cryptomonas φ	Cryptophyta	SS, LS	8, 9
Porphyridium aerugineum	Rhodophyta	SS, LS	10
Euglena gracilis	Chlorophyta	16S, SS, LS	11, 12, 13
Chlamydomonas reinhardii	Chlorophyta	16S, SS, LS	14, 15, 16
Chlorella ellipsoidea	Chlorophyta	16S, LS	17, 18
Marchantia polymorpha	Chlorophyta	16S, LS	19
Nicotiana tabacum	Chlorophyta	16S, SS, LS	20, 21, 22
Zea mays	Chlorophyta	16S, SS, LS	23, 24, 25
Cyanophora paradoxa	Glaucophyta	16S, SS	26, 27
Pseudoanabaena PCC 6903	Cyanobacteria	16S	26
Prochlorothrix hollandica	Cyanobacteria	16S	28
Synechococcus PCC 6301	Cyanobacteria	16S, SS, LS	29, 30, 31
Anabaena PCC 7120/7122	Cyanobacteria	SS, LS	32, 33
Escherichia coli	γ-Proteobacteria	16S	34
Alcaligenes eutrophus	β-Proteobacteria	SS, LS	35
Agrobacterium tumefaciens	α-Proteobacteria	16S	36
Chromatium vinosum	γ-Proteobacteria	LS	37
Rhodospirillum rubrum	α-Proteobacteria	LS	38

[a] 16S, 16S ribosomal RNA nucleotide sequence; SS and LS, small and large subunits of Rubisco amino acid sequences, respectively.
[b] References: 1, Delaney & Cattolico (1991); 2, Boczar *et al.* (1989); 3, Hardison *et al.* (1992); 4, Markowicz *et al.* (1988); 5, Assali *et al.* (1990); 6, Witt and Stackebrandt (1988); 7, Hwang & Tabita (1991); 8, Douglas & Durnford (1989); 9, Douglas *et al.* (1990); 10, Valentin & Zetsche (1989); 11, Graf *et al.* (1982); 12, Sailland *et al.* (1986); 13, Gingrich & Hallick (1985); 14, Dron *et al.* (1982*a*); 15, Goldschmidt-Clermont & Rahire (1986); 16, Dron *et al.* (1982*b*); 17, Yamada (1988); 18, Yoshinaga *et al.* (1988); 19, Ohyama *et al.* (1986); 20, Tohdoh & Sugiura (1982); 21, Mazur & Chui (1985); 22, Shinozaki & Sugiura (1983*a*); 23, Schwarz & Kössel (1980); 24, Matsuoka *et al.* (1987); 25, McIntosh *et al.* (1980); 26, Giovannoni *et al.* (1988); 27, Starnes *et al.* (1985); 28, Turner *et al.* (1989); 29, Tomioka & Sugiura (1983); 30, Shinozaki & Sugiura (1983*b*); 31, Reichelt & Delaney (1983); 32, Nierzwicki-Bauer *et al.* (1984); 33, Curtis & Haselkorn (1983); 34, Brosius *et al.* (1978); 35, Anderson & Caton (1987); 36, Yang *et al.* (1985); 37, Viale *et al.* (1989); 38, Nargang *et al.* (1984).

evaluate the reliability of the phylogeny. This statistical method constructs alternate sequence data sets by randomly sampling from the original alignment of sequences; distance matrices and phylogenies are calculated for each data set. These phylogenies are examined to assess which

Table 2.2. *Distance matrix of 16S rRNA sequences*

	Olis	Pyla	Ochr	Eugl	Chla	Chlo	Marc	Nico	Zea	Cyan	Pseu	Proc	Syne	Agro	Esch
Olis	–	90.5	87.6	85.2	83.2	85.1	86.9	86.2	85.2	86.2	85.6	84.0	87.1	78.5	78.8
Pyla	102	–	86.8	85.4	82.2	84.0	85.7	84.7	83.6	85.4	84.0	82.6	85.6	77.7	77.2
Ochr	135	145	–	83.0	82.2	83.6	84.1	83.4	82.6	83.7	85.0	82.6	84.5	76.7	77.8
Eugl	165	162	193	–	82.6	83.8	85.3	84.3	83.4	82.4	82.4	81.5	83.9	76.7	77.7
Chla	191	203	203	198	–	85.5	88.1	87.2	86.8	83.3	84.9	82.9	85.1	78.0	79.2
Chlo	166	180	185	182	161	–	89.9	88.7	88.2	84.9	85.5	84.2	84.7	77.8	78.1
Marc	144	158	178	163	129	109	–	97.3	96.0	88.0	87.4	87.0	87.5	79.6	79.6
Nico	153	172	187	176	140	122	28	–	97.5	87.6	87.3	86.8	87.1	79.6	78.9
Zea	165	185	198	187	146	128	41	26	–	86.5	85.7	85.5	86.0	78.3	77.9
Cyan	152	162	184	201	188	169	131	136	149	–	90.6	87.5	88.2	76.2	75.1
Pseu	159	179	168	201	169	162	138	139	159	101	–	90.4	90.7	83.5	81.7
Proc	180	198	198	212	195	178	143	145	161	137	102	–	91.4	77.4	78.2
Syne	141	160	174	181	167	171	137	142	156	129	100	92	–	79.5	80.6
Agro	253	265	280	279	261	263	238	238	256	287	186	269	239	–	82.6
Esch	249	272	264	265	244	258	238	248	262	302	210	258	225	199	–

The upper right sector displays the percent nucleotide identity between aligned pairs of sequences. The lower left sector shows the inferred evolutionary distance between species pairs (average number of substitutions per site × 1000). Abbreviations correspond to the first four letters of organisms listed in Table 2.1.

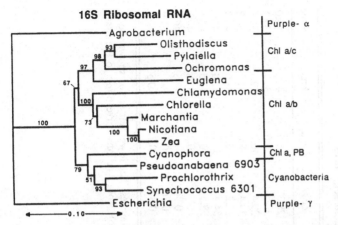

16S Ribosomal RNA

Figure 2.3. Phylogeny derived from 16S rRNA sequences. The scale shows the inferred nucleotide substitutions per site. Numbers adjacent to the branches indicate the frequency occurrence of the cluster to the right in 100 bootstrap replications. On the right are listed the prokaryotic taxa, or the pigment complement for each of the plastid types examined. Chl, chlorophyll.

branches on the tree resist rearrangement and are the most reliable. The phylogenetic tree presented in Fig. 2.3 represents the consensus of 100 such bootstrapped trees, with the frequency of occurrence of each branch indicated.

Several points can be made from the 16S rRNA based phylogeny. Each of the plastid 16S rRNA genes appears to be related to one another and to those of the cyanobacteria examined, while these groups are only distantly related to the proteobacterial genes included in this tree. The proteobacteria (Stackebrandt *et al.* 1988) were formerly referred to as purple bacteria (Woese 1987). This result suggests an evolutionary origin of plastids from a cyanobacterium or a close relative as opposed to a proteobacterial ancestor. Within the plastids, subdivisions are apparent between the Chlorophyta, Cyanophora (Rhodophyta-like) and Chromophyta.

Interestingly, the 16S rRNA from *Euglena gracilis* seems to be more closely related to genes from Chromophyta than other Chlorophyta. Earlier work using ultrastructural data led Gibbs (1978) to suggest an origin for the *Euglena* plastid distinct from that of other chlorophyte plastids. Since then, molecular sequence data have shown the nuclear genome of this alga to be far removed from that of other algae (Sogin *et al.* 1986), while plastid sequence data discussed here show a relatively

Table 2.3. *Distance matrix of Rubisco small subunit amino acid sequences*

	Olis	Alca	Cryp	Porp	Chla	Eugl	Nico	Zea	Cyan	Anab	Syne
Olis	–	55.4	62.0	62.0	39.1	35.9	30.4	28.3	35.9	43.5	38.0
Alca	0.60	–	52.2	56.5	42.4	38.0	37.0	34.8	37.0	39.1	39.1
Cryp	0.49	0.67	–	70.7	45.7	37.0	32.6	29.4	40.2	42.4	39.1
Porp	0.49	0.58	0.35	–	38.0	33.7	32.6	29.4	38.0	41.3	34.8
Chla	0.97	0.89	0.81	1.00	–	62.0	51.1	52.2	52.2	48.9	48.9
Eugl	1.07	1.00	1.03	1.14	0.49	–	50.0	54.4	39.1	42.4	43.5
Nico	1.25	1.03	1.17	1.17	0.69	0.71	–	68.5	40.2	43.5	43.5
Zea	1.34	1.10	1.29	1.29	0.67	0.62	0.38	–	40.2	47.8	45.7
Cyan	1.07	1.03	0.94	1.00	0.67	0.97	0.94	0.94	–	57.6	46.7
Anab	0.86	0.97	0.89	0.91	0.73	0.89	0.86	0.76	0.56	–	60.9
Syne	1.00	0.97	0.97	1.10	0.73	0.86	0.86	0.81	0.78	0.50	–

The upper right sector displays the percent amino acid identity between aligned pairs of sequences. The lower left sector shows the inferred evolutionary distance between species pairs (average number of substitutions per site). Abbreviations correspond to the first four letters of organisms listed in Table 2.1.

Table 2.4. *Distance matrix of Rubisco large subunit amino acid sequences*

	Olis	Alca	Pyla	Cyli	Cryp	Porp	Chla	Eugl	Chlo	Marc	Nico	Zea	Anab	Syne	Chro	Rhod
Olis	—	73.6	87.5	86.4	82.8	85.9	57.5	56.4	57.3	59.1	59.5	56.9	58.4	58.6	55.3	28.7
Alca	0.31	—	72.9	72.3	71.8	74.0	57.5	56.4	57.9	57.9	57.9	56.1	59.0	58.8	55.9	31.8
Pyla	0.13	0.32	—	84.4	83.0	84.8	55.5	56.2	56.2	57.8	58.0	55.7	56.4	56.9	54.6	28.7
Cyli	0.15	0.33	0.17	—	81.5	86.4	56.7	56.7	57.0	58.1	58.5	55.8	58.3	58.7	55.6	28.8
Cryp	0.19	0.33	0.19	0.21	—	85.5	55.3	54.9	54.9	56.2	56.4	54.6	56.0	56.9	53.1	29.1
Porp	0.15	0.30	0.17	0.15	0.16	—	56.0	55.7	56.4	57.3	58.2	56.4	58.0	58.9	54.2	28.4
Chla	0.56	0.56	0.60	0.58	0.60	0.59	—	92.4	94.0	90.4	89.3	86.6	84.6	82.4	74.6	30.0
Eugl	0.58	0.58	0.59	0.58	0.61	0.60	0.08	—	89.8	87.8	86.6	84.0	82.0	80.4	72.6	30.2
Chlo	0.57	0.56	0.59	0.57	0.61	0.58	0.06	0.11	—	90.4	88.4	86.4	84.9	83.5	74.2	30.0
Marc	0.54	0.56	0.56	0.55	0.58	0.57	0.10	0.13	0.10	—	91.8	88.6	85.1	82.6	73.7	30.0
Nico	0.53	0.56	0.55	0.55	0.58	0.55	0.12	0.14	0.12	0.09	—	90.2	84.0	81.3	72.4	30.0
Zea	0.58	0.59	0.60	0.59	0.62	0.58	0.15	0.18	0.15	0.12	0.10	—	80.2	79.1	70.6	30.5
Anab	0.55	0.54	0.58	0.55	0.59	0.55	0.16	0.20	0.16	0.16	0.18	0.22	—	84.4	76.4	31.1
Syne	0.54	0.54	0.58	0.54	0.58	0.54	0.17	0.22	0.18	0.19	0.21	0.24	0.17	—	76.0	31.1
Chro	0.60	0.59	0.62	0.60	0.65	0.63	0.19	0.32	0.30	0.31	0.33	0.35	0.27	0.28	—	30.2
Rhod	1.32	1.20	1.32	1.31	1.30	1.33	1.27	1.26	1.27	1.28	1.28	1.25	1.23	1.23	1.26	—

See the notes to Table 2.3 for additional information.

       Chrysophyte algae

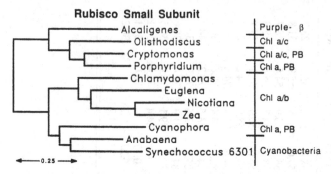

Figure 2.4. Phylogeny derived from Rubisco small subunit sequences. The scale shows the inferred amino acid substitution per site. On the right are listed the prokaryotic taxa, or the pigment complement for each of the eukaryotic taxa examined. Chl, chlorophyll.

close relationship of the *Euglena* plastid genome to those of Chromophyta (Douglas & Turner 1991; Markowicz & Loiseaux-de Göer 1991; this work). The great disparity between the phylogenetic placement of *Euglena* nuclear and chloroplast genes provides support for Gibbs' hypothesis of an ancestor for the *Euglena* plastid that is distinct from plastids of other chlorophytes.

A separate phylogenetic analysis was conducted using protein sequence data derived from the Rubisco large and small subunit genes from algae, vascular plants, cyanobacteria, and chemoautotrophic bacteria (Table 2.1). These data were analyzed using methodologies similar to those described for rRNA. The aligned amino acid sequences were used to generate separate phylogenies based on either Rubisco small or large subunit proteins (Tables 2.3, 2.4; Figs. 2.4, 2.5). To test the reliability of these phylogenies, the data were re-examined using several different methods of analysis (Doolittle & Feng 1991; Feng & Doolittle 1991). The concordant phylogenies that resulted from each of the analytical methods provide support for the trees presented (data not shown).

The phylogenies derived from Rubisco small (Fig. 2.4) and large subunit proteins (Fig. 2.5) are nearly identical despite the different levels of sequence conservation for these genes (Tables 2.3, 2.4). In each tree, one of the major branches encompasses cyanobacteria, chlorophytic plastids and, in the case of the large subunit, the γ-proteobacterium *Chromatium vinosum*. A second major branch leads to a group containing the β-proteobacterium *Alcaligenes eutrophus* and plastids from chromophytic and rhodophytic algae (summarized in Table 2.5). The large subunit tree

Rubisco Large Subunit

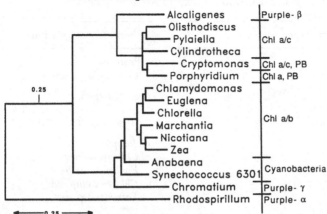

Figure 2.5. Phylogeny derived from Rubisco large subunit sequences. The scale shows the inferred amino acid substitution per site. The branch leading to *Rhodospirillum rubrum* is shortened by 0.25 units as indicated by the tick mark. On the right are listed the prokaryotic taxa, or the pigment complement for each of the eukaryotic taxa examined. Chl, chlorophyll.

has an additional branch leading to the α-proteobacterial gene from *Rhodospirillum rubrum*. This bacterium has a unique Rubisco with only two large subunits, in contrast to the more common organization of eight small and eight large polypeptides that comprise the Rubisco holoenzyme. This feature may explain the extreme divergence of the *Rhodospirillum* large subunit from other large subunits, since its functional organization may be under less stringent evolutionary constraint. Extremely divergent sequences can artifactually appear to root a phylogenetic tree, since such sequences exhibit low similarity values with all other sequences considered.

The segregation of plastid Rubiscos into two groups, which are similar to distinct proteobacteria (designated β-proteobacteria-type and γ-proteobacteria-type; Table 2.5) contrasts sharply with data from 16S rRNA. Phylogenetic analysis of rRNA genes shows all algal plastids to cluster within a single plastid–cyanobacterial radiation, which is distinct from the proteobacteria (Fig. 2.3). Thus to define the evolutionary origins of algal plastids exclusively on any single gene phylogeny can be misleading.

Chlorophytic plastids and cyanobacteria have Rubisco- and 16S rRNA-based phylogenies that concur. However, the phylogenies of these genes in rhodophytic and chromophytic plastids suggest that the plastid Rubisco and 16S rRNA genes have been derived from separate evolutionary

Table 2.5. *Two major classes of Rubisco genes*

β-proteobacteria-type	γ-proteobacteria-type

β-Proteobacteria
  Alcaligenes eutrophus
	γ-Proteobacteria
	  Chromatium vinosum

Chromophyta
  Olisthodiscus luteus
  Pylaiella littoralis
  Fucus serratus
  Cylindrotheca
	Cyanobacteria
	  Synechococcus PCC 6301
	  Anabaena PCC 7120/7122
	
	Chlorophyta
	  Chlamydomonas reinhardii

Cryptophyta
  Cryptomonas φ
	Euglena gracilis
	  Chlorella ellipsoidea
	  Terrestrial plants

Rhodophyta
  Porphyridium aerugineum
	Glaucophyta
	  Cyanophora paradoxa

In the left-hand column are those genes showing greatest similarity to those from the β-proteobacterium *Alcaligenes eutrophus*. In the right-hand column are those genes with the greatest similarity to the γ-proteobacterium *Chromatium vinosum*.

pathways (i.e. that they originate from different prokaryotic taxa). Several mechanisms for the evolution of plastid genomes may be hypothesized that can accommodate the discordant phylogenies for Rubisco and rRNA genes. These mechanisms must account not only for the organism that served as the original endosymbiont, but also the potential exchange of DNA between this prokaryote and other free-living organisms, plasmids and/or viruses. Similarly, genetic exchange involving a captive organelle could have occurred after endosymbiosis. Speculation with so many variables is difficult, and the following schemes are presented to evoke discussion.

The presence of two classes of Rubisco genes in algal plastids may have resulted from the divergence of an ancestral prokaryote into lineages having distinct α-, β-, and γ-type Rubisco genes. Transfer of these genes from β- and γ-proteobacteria into the prokaryotic ancestor(s) of plastids (e.g. cyanobacteria-like cells) or into established organelles themselves would give rise to rhodophytic/chromophytic or chlorophytic plastids with the gene signatures that we have reported.

Alternatively, one can postulate the existence of Rubisco prior to the divergence of cyanobacteria and proteobacteria. Lineages descending from the common ancestor to these organisms would cluster in Rubisco-based phylogenies (i.e. γ-proteobacteria, cyanobacteria and chlorophytic plastids), while lineages that experienced higher rates of evolution for these genes would lie outside the main group. Faster rates of Rubisco evolution would result in the separation of α-proteobacterial-type (e.g. *Rhodospirillum*) and β-proteobacterial-type (e.g. *Alcaligenes*) Rubiscos from the main cluster. One may then account for the *Alcaligenes*-type Rubisco in chromophytic and rhodophytic plastids as resulting from a single lateral gene transfer event into a common ancestor of the plastids in these algae. This mechanism is consistent with the 16S rRNA phylogeny, in which all plastids are sister groups within the cyanobacterial radiation, which is distinct from the proteobacterial lineages (Fig. 2.3).

In either gene transfer hypothesis, the recipient of the genetic exchange could have been either a free-living prokaryotic plastid ancestor (e.g. cyanobacterium) or a captive organelle. Intracellular genetic recombination may have occurred between ancestral plastids and mitochondria. Since mitochondria are likely to have evolved from symbiotic proteobacteria (Gray & Doolittle 1982), they may have provided a source of proteobacterial Rubisco genes to plastid ancestors. In any variation of the lateral transfer hypothesis, the presence of *Alcaligenes*-type Rubisco in all non-chlorophyll-*b* algae examined thus far suggests that gene transfer occurred in a common ancestor unique to chromophytic and rhodophytic plastids. Other examples of naturally occurring lateral gene transfer events have been reported (Doolittle *et al.* 1990).

Finally, the possibility exists that convergent evolution led to algal Rubisco genes with apparent affinity to distinct proteobacterial Rubiscos. That is, while all Rubiscos have a common evolutionary origin, the similarities observed do not reflect a close relationship but instead result from evolutionary convergence. However, this proposal seems unlikely given the high degree of sequence similarity observed within Rubiscos of either α-proteobacterial or β-proteobacterial classes. Other features supporting these groupings include shared gaps, insertions and extended amino and carboxyl termini (Boczar *et al.* 1989).

The differences observed in 16S rRNA and Rubisco phylogenies demonstrate that genes within a single genome can have different evolutionary histories. These differences appear to record a novel event involving the lateral transfer of genes during the evolution of plastids in non-chlorophytic algae. In addition to providing this insight, these observations also illustrate the fallibility of attempting to infer the phylogeny of a group of organisms using data from just a single class of molecular sequences.

Summary

To examine the evolutionary history of plastids from diverse algae, molecular sequence data were analyzed from genes encoding plastid 16S rRNA, and the large and small subunits of Rubisco. Separate phylogenies were derived using gene sequences obtained from representatives of chromophytic, chlorophytic and rhodophytic algae, terrestrial plants and certain prokaryotes. An unexpected result of this work is that different genes found on the same organellar genome can exhibit different evolutionary histories. Substantial differences were observed between 16S rRNA- and Rubisco-based phylogenies. Discordant phylogenies were observed for rhodophytic and chromophytic algae but not for chlorophytic algae or terrestrial plants. These differences appear to record a novel event early in the evolution of algal plastids. The interpretations of these results are discussed, including the possibility for the lateral transfer of genes into an ancestral chloroplast of certain algal groups. The importance of using multiple genes for phylogenetic analysis is emphasized.

Acknowledgments

This work was funded by grants from the National Sea Grant Biotechnology Program and National Science Foundation Systematic panel to R.A.C. and NIH Training Fellowships to T.P.D. (2T32GM07270) and L.K.H. (HD07183). We would like to thank Dr W. Hathway for editorial suggestions.

References

Anderson, K. & Caton, J. 1987. Sequence analysis of the *Alcaligenes eutrophus* chromosomally encoded ribulose bisphosphate carboxylase large and small subunit genes and their gene products. *J. Bacteriol.* **169**: 4547–58.

Assali, N.E., Mache, R. & Loiseaux-de Göer, S. 1990. Evidence for a composite phylogenetic origin of the plastid genome of the brown alga *Pylaiella littoralis* (L.) Kjellm. *Plant Mol. Biol.* **15**: 307–15.

Beakes, G.W. 1989. Oomycete fungi: their phylogeny and relationship to chromophyte algae. In: Green, J.C., Leadbeater, B.S.C. & Diver, W.L. (eds.) *The Chromophyte Algae: Problems and Perspectives.* Systematics Association Special Volume 38. Clarendon Press, Oxford. pp. 325–42.

Boczar, B.A., Delaney, T.P. & Cattolico, R.A. 1989. Gene for the ribulose-1,5-bisphosphate carboxylase small subunit protein of the marine chromophyte *Olisthodiscus luteus* is similar to that of a chemoautotrophic bacterium. *Proc. Natl. Acad. Sci. USA* **86**: 4996–9.

Brosius, J., Palmer, M.L., Poindexter, J.K. & Noller, H.F. 1978. Complete nucleotide sequence of a 16S ribosomal RNA gene from *Escherichia coli*. *Proc. Natl. Acad. Sci. USA* **75**: 4801–5.

Bryant, D.A., de Lorimier, R., Lambert, D.H., Dubbs, J.M., Stirewalt, V.L., Stevens, S.E., Jr, Porter, R.D., Tam, H. & Jay, E. 1985. Molecular cloning and nucleotide sequence of the alpha and beta subunits of allophycocyanin from the cyanelle genome of *Cyanophora paradoxa*. *Proc. Natl. Acad. Sci. USA* **82**: 3242–6.

Cavalier-Smith, T. 1989. The kingdom Chromista. In: Green, J.C., Leadbeater, B.S.C. & Diver, W.L. (eds). *The Chromophyte Algae: Problems and Perspectives*. Systematics Association Special Volume 38. Clarendon Press, Oxford. pp. 381–408.

Chapman, R.L. & Buchheim, M.A. 1991. Ribosomal RNA gene sequences: analysis and significance in the phylogeny and taxonomy of green algae. *Crit. Rev. Plant. Sci.* **10**: 343–68.

Christensen, T. 1962. Alger. In: Bocher, T.W., Lange, M. & Sorensen, T. (eds.) *Systematik Botanik*, vol. 2, no. 2. Munksgaard, Copenhagen. pp. 128–46.

Curtis, S.E. & Haselkorn, R. 1983. Isolation and sequence of the gene for the large subunit of ribulose-1,5-bisphosphate carboxylase from the cyanobacterium *Anabaena* 7120. *Proc. Natl. Acad. Sci. USA* **80**: 1835–9.

Delaney, T.P. & Cattolico, R.A. 1989. Chloroplast ribosomal DNA organization in the chromophytic alga *Olisthodiscus luteus*. *Curr. Genet.* **15**: 221–9.

Delaney, T.P. & Cattolico, R.A. 1991. Sequence and secondary structure of chloroplast 16S rRNA from the chromophyte alga *Olisthodiscus luteus*, as inferred from the gene sequence. *Nucleic Acids Res.* **19**: 6328.

Doolittle, W.F. & Feng, D.-F. 1991. A nearest neighbor procedure for relating progressively aligned amino acid sequences. *Methods Enzymol.* **183**: 659–609.

Doolittle, R.F., Feng, D.F., Anderson, K.L. & Alberro, M.R. 1990. A naturally occurring horizontal gene transfer from a eukaryote to a prokaryote. *J. Mol. Evol.* **31**: 383–8.

Douglas, S.E. & Durnford, D.G. 1989. The small subunit of ribulose-1,5-bisphosphate carboxylase is plastid encoded in the chlorophyll c-containing alga *Cryptomonas* φ. *Plant Mol. Biol.* **13**: 13–20.

Douglas, S.E. & Turner, S. 1991. Molecular evidence for the origin of plastids from a cyanobacterium-like ancestor. *J. Mol. Evol.* **33**: 267–73.

Douglas, S.E., Durnford, D.G. & Morden, C.W. 1990. Nucleotide sequence of the gene for the large subunit of ribulose-1,5-bisphosphate carboxylase/oxygenase from *Cryptomonas* φ: evidence supporting the polyphyletic origin of plastids *J. Phycol.* **26**: 500–8.

Dron, M., Rahire, M. & Rochaix, J.-D. 1982a. Sequence of the chloroplast 16S rRNA gene and its surrounding regions of *Chlamydomonas reinhardii*. *Nucleic Acids Res.* **10**: 7609–20.

Dron, M., Rahire, M. & Rochaix, J.-D. 1982b. Sequence of the chloroplast DNA region of *Chlamydomonas reinhardii* containing the gene of the large subunit of ribulose bisphosphate carboxylase and parts of its flanking genes. *J. Mol. Bio.*. **162**: 775–93.

Felsenstein, J. 1988. Phylogenies from molecular sequences: inference and reliability. *Annu. Rev. Genet.* **22**: 521–65.

Feng, D.-F. & Doolittle, R.F. 1991. Progressive alignment and phylogenetic tree construction of protein sequences. *Methods Enzymol.* **183**: 375–87.

Fitch, W.M. & Margoliash, E. 1967. Construction of phylogenetic trees. *Science* **155**: 279–84.

Gabrielson, P.W., Garbary, D.J. & Scagel, R.F. 1985. The nature of the ancestral red alga: inferences from a cladistic analysis. *BioSystems* **18**: 335–46.

Gibbs, S.P. 1978. The chloroplasts of *Euglena* may have evolved from symbiotic green algae. *Can. J. Bot.* **56**: 2883–9.

Gingrich, J.C. & Hallick, R.B. 1985. The *Euglena gracilis* chloroplast ribulose-1,5-bisphosphate carboxylase gene. I. Complete DNA sequence and analysis of the nine intervening sequences. *J. Biol. Chem.* **260**: 16156–61.

Giovannoni, S.J., Turner, S., Olsen, G.J., Barns, S., Lane, D.L. & Pace, N.R. 1988. Evolutionary relationships among cyanobacteria and green chloroplasts. *J. Bacteriol.* **170**: 3584–92.

Goldschmidt-Clermont, M. & Rahire, M. 1986. Sequence, evolution and differential expression of the two genes encoding variant small subunits of ribulose bisphosphate carboxylase/oxygenase in *Chlamydomonas reinhardtii*. *Mol. Biol.* **191**: 421–32.

Graf, L., Roux, E., Stutz, E. & Kössel, H. 1982. Nucleotide sequence of a *Euglena gracilis* chloroplast gene coding for the 16S rRNA: homologies to *E. coli* and *Zea mays* chloroplast 16S rRNA. *Nucleic Acids Res.* **10**: 6369–81.

Gray, M.W. & Doolittle, W.F. 1982. Has the endosymbiont hypothesis been proven? *Microbiol. Rev.* **46**: 1–42.

Gunderson, J.H. Elwood, H., Ingold, A., Kindle, K. & Sogin, M.L. 1987. Phylogenetic relationships between chlorophytes, chrysophytes, and oomycetes. *Proc. Natl. Acad. Sci. USA* **84**: 5823–7.

Gutell, R.R., Weiser, B., Woese, C.R. & Noller, H.F. 1985. Comparative analysis of 16-S-like ribosomal RNA. *Prog. Nucleic Acid Res. Mol. Biol.* **32**: 155–215.

Hardison, L.K., Boczar, B.A., Reynolds, A.E. & Cattolico, R.A. 1992. A description of the Rubisco large subunit gene and its transcript in *Olisthodiscus luteus*. *Plant Mol. Biol.* **18**: 595–9.

Hwang. S-R. & Tabita, F.R. 1989. Cloning and expression of the chloroplast-encoded *rbcL* and *rbcS* genes from the marine diatom *Cylindrotheca* sp. strain N1. *Plant Mol. Biol.* **13**: 69-79.

Hwang, S.-R. & Tabita, F.R. 1991. Cotranscription, deduced primary structure, and expression of the chloroplast-encoded *rbcL* and *rbcS* genes of the marine diatom *Cylindrotheca* sp. strain N1. *J. Biol. Chem.* **266**: 6271–9.

Jukes, T.H. & Cantor, C.R. 1969. Evolution of protein molecules. In: Munro, H.N. (ed.) *Mammalian Protein Metabolism*. Academic Press, New York. pp. 21–132.

McIntosh, L., Poulsen, C. & Bogorad, L. 1980. Chloroplast gene sequence for the large subunit of ribulose bisphosphate carboxylase of maize. *Nature (Lond.)* **288**: 556–66.

Markowicz, Y., Loiseaux-de Göer, S. & Mache, R. 1988. Presence of a 16S rRNA pseudogene in the bi-molecular plastid genome of the primitive brown alga *Pylaiella littoralis*: evolutionary implications. *Curr. Genet.* **14**: 499–608.

Markowicz, Y. & Loiseaux-de Göer, S. 1991. Plastid genomes of the Rhodophyta and Chromophyta constitute a distinct lineage which differs from that of the Chlorophyta and have a composite phylogenetic origin, perhaps like that of the Euglenophyta. *Curr. Genet.* **20**: 427–30.

Matsuoka, M., Kano-Murakami, Y., Tanaka, Y., Ozeki, Y. & Yamamoto, N.

1987. Nucleotide sequence of cDNA encoding the small subunit of ribulose-1,5-bisphosphate carboxylase from maize. *J. Biochem.* **102**: 673–6.

Mattox, K. & Stewart, K. 1984. Classification of the green algae: a concept based on comparative cytology. In: Irvine, D.E.G. & Johns, D.M. (eds.) *Systematics of the Green Algae.* Academic Press, London. 29–72.

Mazur, B.J. & Chui, C.-F. 1985. Sequence of a genomic DNA clone for the small subunit of ribulose bisphosphate carboxylase-oxygenase from tobacco *Nucleic Acids Res.* **13**: 2372–86.

Nargang, F.E., McIntosh, L. & Somerville, C. 1984. Nucleotide sequence of the ribulose bisphosphate carboxylase gene from *Rhodospirillum rubrum. Mol. Gen. Genet.* **193**: 220–4.

Nierzwicki-Bauer, S.A., Curtis, S.E. & Haselkorn, R. 1984. Cotranscription of genes encoding the small and large subunits of ribulose-1,5 bisphosphate carboxylase in the cyanobacterium *Anabaena* 7120. *Proc. Natl. Acad. Sci. USA* **81**: 5961–5.

Ohyama, K., Fukuzawa, H., Kohchi, T., Shirai, H., Sano, T., Sano, S., Umesono, K., Shiki, Y., Takeuchi, M., Chang, Z., Aota, S.-I., Inokuchi, H. & Ozeki, H. 1986. Complete nucleotide sequence of a liverwort, *Marchantia polymorpha* chloroplast DNA. *Plant Mol. Biol. Reporter* **4**: 148–75.

Olsen, G.J. 1987. Earliest phylogenetic branchings: comparing rRNA-based evolutionary trees inferred with various techniques. *Cold Spring Harbor Symp. Quant. Biol.* **52**: 825–37.

Olson, J.M. & Pierson, B.K. 1987. Evolution of reaction centers in photosynthetic prokaryotes. *Int. Rev. Cytol.* **108**: 209–48.

Palmer, J.D. 1985. Evolution of chloroplast and mitochondrial DNA in plants and algae. In: MacIntyre, R.J. (ed.) *Monographs in Evolutionary Biology: Molecular Evolutionary Genetics.* Plenum Publishing, New York. pp. 131–240.

Perasso, R., Baroin, A., Qu, L.H., Bachellerie, J.P. & Adoutte, A. 1989. Origin of the algae. *Nature (Lond.)* **339**: 142–4.

Preisig, H.R. 1989. The flagellar base ultrastructure and phylogeny of chromophytes. In: Green, J.C., Leadbeater, B.S.C. & Diver, W.L. (eds.) *The Chromophyte Algae: Problems and Perspectives.* Systematics Association Special Volume 38. Clarendon Press, Oxford, pp. 167–88.

Reichelt, B.Y. & Delaney, S.F. 1983. The nucleotide sequence for the large subunit of ribulose-1,5-bisphosphate carboxylase from a unicellular cyanobacterium, *Synechococcus* PCC 6301. *DNA* **2**: 121–9.

Reith, M. & Cattolico, R.A. 1986. Inverted repeat of *Olisthodiscus luteus* chloroplast DNA contains genes for both subunits of ribulose-1,5-bisphosphate carboxylase and the 32 000-dalton Qβ protein: phylogenetic implications. *Proc. Natl. Acad. USA* **83**: 8599–603.

Round, F.E. 1989. The chromophyte algae: problems and perspectives. A summarizing view. In: Green, J.C., Leadbeater, B.S.C. & Diver, W.L. (eds.) *The Chromophyte Algae: Problems and Perspectives.* Systematics Association Special Volume 38. Clarendon Press, Oxford. pp. 409–18.

Sailland, A., Amiri, I. & Freyssinet, G. 1986. Amino acid sequence of the ribulose-1,5-bisphosphate carboxylase/oxygenase small subunit from *Euglena. Plant Mol. Biol.* **7**: 213–18.

Schwarz, Z. & Kössel, H. 1980. The primary structure of 16S rDNA from *Zea mays* chloroplast is homologous to *E. coli* 16S rRNA. *Nature (Lond.)* **283**: 739–42.

Shivji, M.S. 1991. Organization of the chloroplast genome in the red alga *Porphyra yezoensis. Curr. Genet.* **19**: 49–54.

Shinozaki, K. & Sugiura, M. 1983*a*. The nucleotide sequence of the tobacco chloroplast gene for the large subunit of ribulose-1,5-bisphosphate carboxylase/oxygenase. *Gene* **20**: 91–102.

Shinozaki, K. & Sugiura, M. 1983*b*. The gene for the small subunit of ribulose-1,5-bisphosphate carboxylase/oxygenase is located close to the gene for the large subunit in cyanobacterium *Anacystis nidulans* 6301. *Nucleic Acids Res.* **11**: 6957–64.

Sogin, M.L., Elwood, H.J. & Gunderson, J.H. 1986. Evolutionary diversity of eukaryotic small-subunit rRNA genes. *Proc. Natl. Acad. Sci. USA* **83**: 1383–7.

Stackebrandt, E., Murray, R.G.E. & Trüper, H.G. 1988. *Proteobacteria* classis nov., a name for the phylogenetic taxon that includes the 'purple bacteria and their relatives'. *Int. J. Syst. Bacteriol.* **38**: 321–5.

Starnes, S.M., Lambert, D.H., Maxwell, E.S., Stevens, S.E., Jr, Porter, R.D. & Shively, J.M. 1985. Cotranscription of the large and small subunit genes of ribulose-1,5-bisphosphate carboxylase in *Cyanophora paradoxa. FEMS Microbiol. Lett.* **28**: 165–9.

Tohdoh, N. & Sugiura, M. 1982. The complete nucleotide sequence of a 16S ribosomal RNA gene from tobacco chloroplasts. *Gene* **17**: 213–18.

Tomioka, N. & Sugiura, M. 1983. The complete nucleotide sequence of a 16S ribosomal RNA gene from a blue-green alga, *Anacystis nidulans. Mol. Gen. Genet.* **191**: 46–50.

Turner, T., Burger-Wiersma, T., Giovannoni, S.J., Mur, L.R. & Pace, N.R. 1989. The relationship of a prochlorophyte *Prochlorothrix hollandica* to green chloroplasts. *Nature (Lond.)* **337**: 380–2.

Valentin, K. & Zetsche, K. 1989. The genes of both subunits of ribulose-1,5-bisphophate carboxylase constitute an operon on the plastome of a red alga. *Curr. Genet.* **16**: 203–9.

Valentin, K. & Zetsche, K. 1990*a*. Structure of the Rubisco operon from the unicellular red alga *Cyanidium caldarium*: evidence for a polyphyletic origin of the plastids. *Mol. Gen. Genet.* **222**: 425–30.

Valentin, K. & Zetsche, K. 1990*b*. Rubisco genes indicate a close phylogenetic relation between the plastids of Chromophyta and Rhodophyta. *Plant Mol. Biol.* **15**: 575–84.

Valentin, K., Cattolico, R.A. & Zetsche, K. 1992. Phylogenetic origin of the plastids In: Lewin, R.A. (ed.) *Origins of Plastids*. Chapman & Hall, New York. pp. 195–222.

Viale, A.M., Kobayashi, H. & Akazawa, T. 1989. Expressed genes for plant-type ribulose-1,5-bisphosphate carboxylase/oxygenase in the photosynthetic bacterium *Chromatium vinosum*, which possesses two complete sets of the genes. *J. Bacteriol.* **171**: 2391–400.

Witt, D. & Stackebrandt, E. 1988. Disproving the hypothesis of a common ancestry for the *Ochromonas danica* chrysoplast and *Heliobacterium chlorum. Arch. Microbiol.* **150**: 244–8.

Woese, C.R. 1987. Bacterial evolution. *Microbiol. Rev.* **51**: 221–71.

Yamada, T. 1988. Nucleotide sequence of the chloroplast 16S rRNA gene from the unicellular green alga *Chlorella ellipsoidea. Nucleic Acids Res.* **16**: 9865.

Yang, D., Oyaizu, Y., Oyaizu, H., Olsen, G.J. & Woese, C.R. 1985. Mitochondrial origins. *Proc. Natl. Acad. Sci. USA* **82**: 4443–7.

Yoshinaga, K., Ohta, T., Suzuki, Y. & Sugiura, M. 1988. *Chlorella* chloroplast
DNA sequence containing a gene for the large subunit of ribulose-1,5-
bisphosphate carboxylase/oxygenase and a part of a possible gene for the
beta subunit of RNA polymerase. *Plant Mol. Biol.* **10**: 245–50.

3
A modern concept of chrysophyte classification

HANS R. PREISIG

Previous concepts of classification

Until recently, most concepts of higher taxa in the Chrysophyceae have been based on some combination of vegetative morphological features and characteristics of the motile cells (such as flagellar number and position). For a survey on the various systematic treatments of the Chrysophyceae by earlier workers (Pascher 1914; Fritsch 1935; Bourrelly 1968; Fott 1971; Bold & Wynne 1978; Christensen 1980; Ettl 1980; and Kristiansen 1982) the reader is referred to Round (1986). In the present account (see Tables 3.1–3.3) special reference is given only to the more recent treatments by Starmach (1985) and Kristiansen (1986, 1990).

Within the class Chrysophyceae, Starmach (1985) distinguished three subclasses: the Heterochrysophycidae, the Acontochrysophycidae and the Craspedomonadophycidae. The Craspedomonadophycidae of Starmach included the choanoflagellates (families Monosigaceae and Salpingoecaceae) and the genus *Phalansterium* (family Phalansteriaceae) which are now known from electron microscopic studies to have no structural similarities with any group of algae (Hibberd 1986), and it is now generally accepted that they should be classified only as Protozoa. Starmach's scheme more or less followed that of Bourrelly (1968, 1981), in which three different lineages were recognized: the uniflagellate order Chromulinales in the subclass Heterochrysophycidae, the biflagellate order Ochromonadales in the subclass Heterochrysophycidae, and several aflagellate orders in the subclass Acontochrysophycidae.

Modern approaches recognize that uniflagellate members of Chrysophyceae usually have a second vestigial flagellum (see e.g. Hibberd 1976) and, in recent studies (e.g. Dop 1978, 1980*b*; Meyer 1986; O'Kelly & Wujek, this volume), zoospores have been found in several genera

Table 3.1. *Scheme of classification of Chrysophyceae by Starmach* (1985)

Class Chrysophyceae
  Subclass I. Heterochrysophycidae
    Order 1. Chromulinales
      Suborder 1. Chromulineae
        Family 1. Chromulinaceae
        Family 2. Chrysococcaceae
        Family 3. Bicosoecaceae
        Family 4. Pedinellaceae
        Family 5. Chrysocapsaceae
        Family 6. Hydruraceae
        Family 7. Chrysamoebaceae
        Family 8. Kybotionaceae
        Family 9. Myxochrysidaceae
      Suborder 2. Chrysosphaeriineae
        Family 1. Chrysosphaeraceae
        Family 2. Phaeobotrydaceae
      Suborder 3. Thallochysidineae
        Family 1. Thallochrysidaceae
        Family 2. Chrysomeridaceae

    Order 2. Ochromonadales
      Suborder 1. Ochromonadineae
        Family 1. Ochromonadaceae
        Family 2. Dinobryonaceae
        Family 3. Synuraceae
        Family 4. Ruttneraceae
        Family 5. Naegeliellaceae
      Suborder 2. Phaeothamnionineae
        Family 1. Phaeothamniaceae
      Suborder 3. Chrysapionineae
        Family 1. Chrysapionaceae

  Subclass II. Acontochrysophycidae
    Order 1. Chrysarachniales
      Family 1. Chrysarachniaceae
    Order 2. Stylococcales
      Family 1. Stylococcaceae
    Order 3. Chrysosaccales
      Family 1. Chrysosaccaceae
    Order 4. Phaeoplacales
      Family 1. Phaeoplacaceae
      Family 2. Sphaeridiotrichaceae

  Subclass III. Craspedomonadophycidae
    Order 1. Monosigales
      Family 1. Monosigaceae
      Family 2. Salpingoecaceae
      Family 3. Phalansteriaceae

Table 3.2. *Scheme of classification of*
Chrysophyceae and 'splinter groups'
(Pedinellophyceae and
Synurophyceae) by Kristiansen
(1986)

Class Chrysophyceae
  Order 1. Ochromonadales
    Family 1. Ochromonadaceae
    Family 2. Chromulinaceae
    Family 3. Dinobryaceae
    Family 4. Chrysococcaceae
    Family 5. Paraphysomonadaceae
  Order 2. Mallomonadales
    Family 1. Mallomonadaceae
  Order 3. Chrysamoebales
    Family 1. Rhizochrysidaceae
    Family 2. Chrysamoebaceae
    Family 3. Stylococcaceae
    Family 4. Kybotiaceae
    Family 5. Myxochrysidaceae
  Order 4. Chrysocapsales
    Family 1. Chrysocapsaceae
    Family 2. Chrysosaccaceae
    Family 3. Naegeliellaceae
    Family 4. Chrysochaetaceae
  Order 5. Hydrurales
    Family 1. Hydruraceae
  Order 6. Chrysosphaerales
    Family 1. Chrysosphaeraceae
    Family 2. Chrysapiaceae
    Family 3. Stichogloeaceae
    Family 4. Aurosphaeraceae
  Order 7. Phaeothamniales
    Family 1. Sphaeridiotrichaceae
    Family 2. Phaeoplacaceae
    Family 3. Phaeothamniaceae
  Order 8. Sarcinochrysidales
    Family 1. Sarcinochrysidaceae
    Family 2. Chrysomeridaceae
    Family 3. Phaeosacciaceae
    Family 4. Nematochrysidaceae
Class Pedinellophyceae
  Order 1. Pedinellales
    Family 1. Pedinellaceae
Class Dictyochophyceae
  Order 1. Dictyochales
    Family 1. Dictyochaceae

Table 3.3. *New scheme of classification of Chrysophyceae and*
'splinter groups' (Dictyochophyceae and Synurophyceae) (see
also Moestrup, this volume)

Class Chrysophyceae
  Pascher (1914, *nom. descript.*)
  Hibberd (1976, *nom. typificatum*)

  Order 1. Bicosoecales Grassé 1926 ('Bicoecinae', *nom. nud.*)
    Family 1. Cafeteriaceae Moestrup 1994 (this volume)
    Family 2. Bicosoecaceae F. Stein 1878 ('Bikoecida')

  Order 2. Chromulinales Pascher 1910
      (incl. Ochromonadales Pascher 1910)
    Family 1. Chromulinaceae Engler 1897
        (incl. Ochromonadaceae Lemmermann 1899)
    Family 2. Dinobryaceae Ehrenberg 1834 ('Dinobryina')
    Family 3. Paraphysomonadaceae Preisig & Hibberd 1983
    Family 4. Chrysolepidomonadaceae Peters & Andersen 1993
    Family 5. Chrysamoebaceae Poche 1913 ('Chrysamoebidae')
    Family 6. Myxochrysidaceae Pascher 1931
    Family 7. Chrysocapsaceae Pascher 1912
    Family 8. Stichogloeaceae Lemmermann 1899
        (incl. Chrysosphaeraceae Pascher 1914)
    Family 9. Phaeothamniaceae Hansgirg 1886 ('Phaeothamnieae')
    Family 10. Chrysothallaceae Huber-Pestalozzi 1941
        (incl. Phaeoplacaceae Bourrelly 1957)

  Order 3. Hibberdiales Andersen 1989
    Family 1. Hibberdiaceae Andersen 1989
    Family 2. Stylococcaceae Lemmermann 1899

  Order 4. Hydrurales Pascher 1931
    Family 1. Hydruraceae Rostafiński 1881 ('Hydrureae')

  Order 5. Sarcinochrysidales Gayral & Billard 1977
    Family 1. Ankylochrysidaceae O'Kelly & Billard (to be published)
    Family 2. Sarcinochrysidaceae Gayral & Billard 1977
    Family 3. Nematochrysopsidaceae Gayral & Billard 1977

  Order 6. Chrysomeridales O'Kelly & Billard (to be published)
    Family 1. Chrysomeridaceae Bourrelly 1957
    Family 2. Phaeosacciaceae Feldmann 1949 ('Phaeosaccionaceae')

Class Dictyochophyceae Silva 1980

  Order 1. Pedinellales Zimmermann et al. 1984
    Family 1. Pedinellaceae Pascher 1910
    Family 2. Cyrtophoraceae Pascher 1911

  Order 2. Rhizochromulinales O'Kelly & Wujek 1994 (this volume)
    Family 1. Rhizochromulinaceae O'Kelly & Wujek 1994 (this volume)

(*continued*)

Table 3.3. (*continued*)

Order 3. Dictyochales Haeckel 1894 ('Dictyocheae')
   Family 1. Dictyochaceae Lemmermann 1901

Class Synurophyceae Andersen 1987 (Cavalier-Smith 1986: 'Synurea')
  Order 1. Synurales Andersen 1987
   Family 1. Synuraceae Lemmermann 1899
   Family 2. Mallomonadaceae Diesing 1866 ('Mallomonadina')

---

previously thought to be aflagellate, so that the aflagellate lineage also appears to be artificial. This caused Kristiansen (1986, 1990) to disregard flagellar number as a major taxonomic criterion.

Pascher (1914, 1931) and other earlier workers on Chrysophyceae also emphasized the organization level of the vegetative thallus in classification. The dominant life stages (flagellate, rhizopodial, capsoid (= palmelloid), coccoid, filamentous) have normally been used at the ordinal level of algal classification. But we now know from other other algal groups such as the green algae that the level of organization is often difficult to correlate with other characteristics, and this seems also to be the case in the Chrysophyceae (see below).

Modern concept of classification

The present concept of classification (Table 3.3) includes new data from studies on the flagellar apparatus, mitosis, cytokinesis and other details of cell ultrastructure, as well as on photosynthetic pigments and other biochemical and physiological data (for more details see below; also compare: Kristiansen 1986, 1990; Andersen 1987, 1989b, 1991b; Preisig 1989). The new data, especially those from electron microscopy, motivated several authors to split the Chrysophyceae *sensu lato* (as circumscribed by e.g. Starmach 1985) into separate classes, such as the Bicosoecophyceae, Dictyochophyceae, Pedinellophyceae and Synurophyceae (cf. Silva 1980; Cavalier-Smith 1986; Andersen 1987; Kristiansen 1990). Moestrup (this volume) deals with these chrysophyte 'splinter groups' in detail. He accepts the Synurophyceae as a separate class, but argues that the *Bicosoeca* group does not deserve class status and should be classified only as a separate order (Bicosoecales) of the Chrysophyceae. The silico-flagellates and pedinellids, on the other hand, are placed by Moestrup as two orders (Dictyochales and Pedinellales) in the class Dictyochophyceae. This view is followed here.

The class Chrysophyceae as presently conceived comprises the following six orders: Bicosoecales, Chromulinales, Hibberdiales, Hydrurales, Sarcinochrysidales, Chrysomeridales (Table 3.3). Some taxa of uncertain position in the Chrysophyceae (e.g. the order Parmales) are also dealt with below.

Bicosoecales

For more details on this order see Moestrup (this volume).

Chromulinales

The Chromulinales is combined here with Ochromonadales. The two orders have been distinguished by previous workers mainly on the basis of differences in the number of flagella visible in the light microscope. It is now clear, however, that species of both *Chromulina* and *Ochromonas* possess two flagella. Both these genera are so similar in all respects, except for the length of the short flagellum, that in my opinion this minor difference does not justify separation of different families or even orders.

Because Chromulinaceae Engler 1897 has priority over Ochromonadaceae Lemmermann 1899, the combined family should bear the name Chromulinaceae. That being the case, I suggest Chromulinales Pascher 1910 rather than Ochromonadales Pascher 1910 should be selected for the ordinal name. For the family, some even earlier names would be available (Chrysomonadaceae F. Stein 1878, Dendromonadaceae F. Stein 1878 or Spumellaceae Kent 1881), but as pointed out by Silva (1980), Chromulinaceae has been much more widely used and therefore it seems appropriate to select this family name for conservation.

The Chromulinales are subdivided into ten families here: Chromulinaceae, Dinobryaceae, Paraphysomonadaceae, Chrysolepidomonadaceae, Chrysamoebaceae, Myxochrysidaceae, Chrysocapsaceae, Stichogloeaceae (= Chrysosphaeraceae), Phaeothamniaceae and Chrysothallaceae (= Phaeoplacaceae). The latter six families are mainly based on the level of organization and have previously often been included in separate orders (Rhizochrysidales Pascher 1925, Myxochrysidales Pascher 1931, Chrysocapsales Pascher 1912, Chrysosphaerales Pascher 1914, Phaeothamniales Bourrelly 1954 and Phaeoplacales Bourrelly 1954. They are now included in the Chromulinales primarily on the basis of similarities in motile cell structure (see below).

Chromulinaceae

The family Chromulinaceae comprises some 34 genera (Preisig 1995). The cells are usually naked, flagellate, solitary or colonial, and mostly pigmented (a few colorless genera also exist: see Preisig *et al.* 1991). The ultrastructure of only a few members of this family has been studied in any detail; this includes features of the flagellar apparatus (cf. *Ochromonas danica* Pringsheim: Bouck & Brown 1973; *Uroglena americana* Calkins: Owen *et al.* 1990*b*), and features of mitosis and cytokinesis (cf. *Ochromonas danica*: Slankis & Gibbs 1972, Tippit *et al.* 1980; *Uroglena/Uroglenopsis* sp.: Andersen 1989*b*). It is therefore not certain at present whether the observed characteristics are representative of the whole family.

In a few species (*Chromulina placentula* Belcher & Swale, *Syncrypta glomerifera* Clarke & Pennick) the cells are covered by simple organic scales (Belcher & Swale 1967; Clarke & Pennick 1975), but it must be stressed that these species are in need of further study to assess their taxonomic position beyond doubt. Species with elaborate tree-like organic scales have also been placed in the Chromulinaceae (as *Sphaleromantis* spp.) until recently (Manton & Harris 1966; Pienaar 1976; Starmach 1985), but re-examination by Andersen (1991*a*) and Peters & Andersen (1993) revealed that these organisms do not fit the original generic description of *Sphaleromantis*. These authors believe that the scaled organisms of Manton & Harris (1966) and Pienaar (1976) and the ones they studied are substantially different. For the latter they proposed a new genus, *Chrysolepidomonas*, which they placed in a separate family, Chrysolepidomonadaceae (see below).

Chrysonephele, a new genus recently described by Pipes *et al.* (1989), may also belong to the Chromulinaceae but differs in some unique features. The general cell ultrastructure is unquestionably chrysophycean. But unlike other members of this class, the long hairy flagellum has a lamellate swelling at its base, and this feature may indicate a possible evolutionary link with the Eustigmatophyceae. *Chrysonephele* also differs from members of the Chromulinaceae by its gelatinous, colonial habit consisting of thousands of statically embedded cells with vigorous flagellar activity but no actual mobility of either cells or colonies.

Dinobryaceae

In the Dinobryaceae the general cell structure is similar to that of members of Chromulinaceae, but the cells are surrounded by an envelope (lorica). Previously the 'uniflagellate' genera have often been placed in a separate family, Chrysococcaceae Lemmermann 1899 (e.g. Starmach 1985), but as discussed above, separation of families solely on the basis of flagellar number visible in the light microscope no longer appears to be justified.

Only a few species of this family have been studied in detail with regard to cell ultrastructure (cf. *Poterioochromonas malhamensis* Peterfi: Schnepf *et al.* 1977; *Dinobryon cylindricum* Imhof: Owen *et al.* 1990a; *Epipyxis pulchra* Hilliard & Asmund: Andersen & Wetherbee 1992). The flagellar apparatus is similar to that of other members of Chromulinales, but some differences in the structure and arrangement of flagellar roots and other components of the flagellar apparatus nevertheless occur (for a comparison, see Andersen 1989b, 1991b). It is certainly premature to comment on the taxonomic significance of these differences before more species have been studied on a comparative basis.

Paraphysomonadaceae

Silica-scaled genera such as *Chrysosphaerella*, *Paraphysomonas* and *Spiniferomonas* were previously classified with *Mallomonas* and *Synura* in the family Synuraceae, order Ochromonadales (e.g. by Starmach 1985), until Preisig & Hibberd (1983, 1986) showed that cell fine structure of the former three genera (and also of the new genus *Polylepidomonas*) is much more similar to that of *Ochromonas* and *Chromulina* than to *Mallomonas* and *Synura*. Andersen (1990) confirmed the close relationship of *Chrysosphaerella* with chromulinalean chrysophytes in a study on the flagellar apparatus. *Mallomonas* and *Synura* are now placed in a separate class, Synurophyceae (Cavalier-Smith 1986; Andersen 1987; Moestrup, this volume), whereas *Paraphysomonas* and related genera are accommodated as a separate family in the order Chromulinales. The most important feature by which this family is distinguished from other families of Chromulinales is the presence of siliceous scales on the cell surface.

Chrysolepidomonadaceae

Peters & Andersen (1993) established the family Chrysolepidomonadaceae for the new genus *Chrysolepidomonas*, so far including three species: *C.*

dendrolepidota Peters & Andersen (type species), *C. anglica* Peters & Andersen (=*Sphaleromantis tetragona* sensu Harris 1963; Manton & Harris 1966) and *C. marina* (Pienaar) Peters & Andersen (=*Sphaleromantis marina*, Pienaar 1976). The most characteristic feature of these algae is the presence of elaborate, tree- or basket-shaped, organic scales on the cell surface and flagella. It should be noted that simple organic scales may also occur in a few species classified in the Chromulinaceae (see above). The architecture of the flagellar apparatus of *Chrysolepidomonas* is similar to that of *Ochromonas*.

Chrysamoebaceae

Members of the family Chrysamoebaceae are naked and have a tendency to form lobed or branched cytoplasmic extensions. The vegetative cells are amoeboid during the greater part of the life history (rhizopodial organization). Flagellate stages occur in several genera (*Amphichrysis, Brehmiella, Chrysamoeba, Chrysostephanosphaera, Platychrysella, Rhizo-ochromonas*), whereas in genera such as *Chrysarachnion, Chrysastridium* (=*Chrysidiastrum*), *Leukapsis, Leukochrysis* and *Rhizochrysis*, flagellate stages have not been observed. The latter genera have often been placed in separate families (Rhizochrysidaceae Pascher ex Reichenow 1928 or Chrysarachniaceae Pascher 1931), or separate orders, but as discussed above it does not seem justified to define orders or families only by the absence of flagellate stages. It should also be noted that the 'aflagellate' genus *Rhizochrysis*, for instance, has been subsumed within the 'flagellate' genus *Chrysamoeba* by some authors (e.g. Starmach 1985). Similarly, separation of orders and families based solely on the flagellar number visible in the light microscope is not accepted in the present classification; therefore the 'biflagellate' rhizopodial family Brehmiellaceae Nicholls 1990 (including *Brehmiella* and *Rhizoochromonas*) is merged here with the 'uniflagellate' rhizopodial family Chrysamoebaceae.

On the basis of the characteristic amoeboid (rhizopodial) organization in Chrysamoebaceae, this family has often been placed together with other rhizopodial families (Stylococcaceae, Myxochrysidaceae) in a separate order, Rhizochrysidales Pascher 1925 (or Chrysamoebales).* O'Kelly & Wujek (this volume) showed that in the architecture of the flagellar

* If *Chrysamoeba* and *Rhizochrysis* are classified together in a separate order the correct name is Rhizochrysidales. Chrysamoebales has apparently not been validated, though it has often been used in the literature (e.g. Matvienko 1965; Starmach 1985; Kristiansen 1986, 1990).

apparatus and in other details of cell structure, *Chrysamoeba pyrenoidifera* Korshikov (Chrysamoebaceae) closely resembles chromulinalean chryso- phytes, whereas *Lagynion delicatulum* Skuja (Stylococcaceae) resembles hibberdialean chrysophytes (see below). Further studies must determine whether other members of these families are comparable to the species studied by O'Kelly & Wujek. Currently available data appear to provide no basis for the retention of the order Rhizochrysidales (or Chrysamoe- bales).

The marine amoebo-flagellate *Rhizochromulina marina* Hibberd & Chrétiennot-Dinet (Hibberd & Chrétiennot-Dinet 1979) somewhat re- sembles members of the chrysamoebaceae regarding the rhizopodial organization of the vegetative cells. But the zoospores are unique in being fusiform in shape and being genuinely uniflagellate with no emergent vestigial second flagellum. This monotypic genus has usually been treated as an 'aberrant' chrysophyte (Hibberd 1986), until O'Kelly & Wujek (this volume) showed in a thorough ultrastructural investigation that it is in fact more closely related to the pedinellids and silicoflagellates than to the chrysophytes *sensu stricto*. It is now classified as a separate order (Rhizochromulinales) and family (Rhizochromulinaceae) in the Dictyochophyceae (see Moestrup, this volume).

Myxochrysidaceae

The family Myxochrysidaceae comprises a single genus and species, *Myxochrysis paradoxa* Pascher, which is known only from the original description (Pascher 1916). In the normal condition this organism occurs as a multinucleate plasmodium surrounded by a thick calcified and iron-encrusted sheath. The protoplast may become cleft into numerous uni- or multinucleate portions which develop into thick-walled pseudocysts. These may germinate into uninucleate amoebae or into *Chromulina*-like zoospores, which after some time lose their single emergent flagellum and become amoebae. The amoebae grow and their nuclei divide and at all stages of their development fusions can take place leading to the large plasmodia that constitute the normal vegetative condition. Plasmodia may also be fragmented. Chloroplasts are usually present but may also be lacking. Stomatocysts (= statospores), the characteristic resting stage of many chrysophycean algae, are not known.

The taxonomic position is not clear. No other chrysophycean alga has such a complex life cycle. Plasmodia may also be formed by some other chrysophytes (*Chrysarachnion*, *Chrysastridium*). Both these genera are incompletely known, however, and are tentatively classified in the Chrys-amoebaceae (see above). Furthermore, in these cases there is not a uniform mass ('holoplasmodium') as in *Myxochrysis*, but the individual amoeboid cell bodies remain separate, only forming a common reticulopodial network ('meroplasmodium' sensu Grell *et al.* 1990, Grell 1991). Small plasmodia may also be formed rarely by *Rhizochromulina marina*, which is now known to be related to the silicoflagellates and pedinellids (see both Moestrup and O'Kelly & Wujek, this volume).

Chrysocapsaceae

The family Chrysocapsaceae contains chrysophycean algae which spend most of their life history as immotile naked cells surrounded by mucilage (capsoid organization). The vegetative cells often contain contractile vacuoles and eyespots may also be present. In most members of this family 'uniflagellate' or 'biflagellate' flagellate stages have been observed. The family Chrysosaccaceae Bourrelly 1957 for capsoid chrysophytes in which flagellate stages have not been observed is subsumed within Chryso-capsaceae Pascher 1912 here. At present the following genera are placed in this family: *Arthrogloea, Bourrellia, Chalkopyxis, Chrysocapsa, Chryso-capsella, Chrysochaete, Chrysopora, Chrysosaccus, Chrysotilos, Dermato-chrysis, Eirmodesmus, Entodesmis, Gloeochrysis, Heimiochrysis, Kremasto-chrysis* (including *Kremastochrysopsis*), *Naegeliella, Pascherella, Phaeaster, Phaeosphaera* (including *Chrysodictyon*) (compare Bourrelly 1981 and Starmach 1985).

Of these genera only one species of *Phaeaster* (*P. pascheri* Scherffel) has been examined in any detail, especially with respect to cell ultra-structure (Inouye *et al.* 1990). It should be noted, however, that this alga is unique in the whole of the Chrysophyceae as regards the very compressed cell body of the motile cells and the star-shaped chloroplast, and also shows several other unusual features (such as: unusual orientation of the short flagellum, basal bodies and roots; lack of both girdle lamella and ring nucleoid in the chloroplast). Almost certainly this genus has to be assigned to a family different from the other genera of Chrysocapsaceae when more genera have been studied in comparable detail.

In earlier classifications, the genus *Tetrasporopsis* has usually also been placed in the Chrysocapsaceae, but Entwisle & Andersen (1990) recently showed that the type species of this genus does in fact possess a cell wall and would be better placed in the family Stichogloeaceae (Chrysosphaer-aceae). *Tetrasporopsis*-like organisms without cell walls have also been

found (Entwisle & Andersen 1990) and these are placed in the new genus
Dermatochrysis in the Chrysocapsaceae. It is possible that several of the
genera currently still accommodated in the Chrysocapsaceae will also be
shown to have cell walls when ultrastructural studies are done.

The genera *Naegeliella* (Correns 1893) and *Chrysochaete* (Rosenberg
1941) have sometimes been included in separate families (Naegeliellaceae
Pascher 1925; Chrysochaetaceae Bourrelly 1957). In these genera the cells
are similarly organized as in other Chrysocapsaceae, but there are
characteristic, long gelatinous hairs extending beyond the mucilage.
Flagellate stages are also produced ('uniflagellate' in *Chrysochaete* and
'biflagellate' in *Naegeliella*). Both genera should certainly be grouped in
the same family; according to Geitler (1968) both should probably even
be combined in a single genus. Further studies must show whether
retention of a family separate from the Chrysocapsaceae is justified, and
whether perhaps the genus *Phaeoplaca* (see Chrysothallaceae below)
should also be assigned to the same family (see Dop 1978).

Algae of the family Hydruraceae, which are also primarily capsoid in
organization, were previously often placed together with Chrysocapsaceae
in a separate order, Chrysocapsales (e.g. by Kristiansen 1990), but now
the family Hydruraceae is accommodated in a different order, Hydrurales
(see below).

Stichogloeaceae

The family Stichogloeaceae comprises chrysophycean algae which spend
most of their life history in the coccoid phase, i.e. as cells surrounded by
a wall (coccoid organization). Cell walls and their formation during cell
division are important developmental features and are regarded to be of
considerable taxonomic significance. Therefore it appears to be justified
to place these organisms in a different family, at least until more details
are available on the ultrastructure, biochemistry and molecular genetics
of this group of chrysophytes.

Previously genera of coccoid Chrysophyceae were often split into different
orders (Chrysosphaerales Pascher 1914, Chrysapiales Bourrelly 1954,
Stichogloeales Bourrelly 1954, Tetrapiales Bourrelly 1981) and different
families (Stichogloeaceae Lemmermann 1899, Chrysosphaeraceae Pascher
1914, Chrysapiaceae Bourrelly 1957, Tetrapiaceae Bourrelly 1981), mainly
depending on whether 'uniflagellate', 'biflagellate' or 'aflagellate' reproductive stages were produced. In the present classification the number of
flagella visible in the light microscope is not considered to be such an

important character, and only one order and family is accepted. For reasons of priority, the correct name for the family is Stichogloeaceae.

The following genera are assigned to this family: *Chrysapion, Chrysosphaera, Epicystis, Koinopodium, Phaeobotrys, Phaeogloea, Phaeoschizochlamys, Podochrysis, Polypodochrysis, Selenophaea, Stichogloea, Tetrapion* and *Tetrasporopsis* (for the marine genera *Podochrysis* and *Polypodochrysis* see Magne 1975; the remainder occur in fresh water, see Bourrelly 1981 and Starmach 1985). It should be noted that most of these genera are only poorly known and further study must show how closely related they actually are. It should also be noted that one species previously assigned to *Chrysosphaera* (*C. magna* Belcher) has recently been shown to have some unique features and has been transferred to a new genus (*Hibberdia*) in a new family and order (see Hibberdiales below). Another species originally assigned to *Chrysosphaera* (*C. marina* Schussnig) has been suggested to be a non-motile stage of a prymnesiophycean alga (Parke 1961).

Sometimes the marine genera *Aurosphaera/Meringosphaera* and *Pelagococcus/Aureococcus* have been accommodated along with Stichogloeaceae (Chrysosphaeraceae) in the order Chrysosphaerales, but in this survey they are dealt with as Chrysophyceae of uncertain taxonomic position (see separate paragraph below).

Phaeothamniaceae

The family Phaeothamniaceae contains chrysophycean algae with a filamentous organization. The filaments may also pass over into a palmelloid condition, the cells rounding off and the membranes becoming gelatinous. In most genera, formation of *Chromulina*-like or *Ochromonas*-like zoospores has also been observed, but the flagellar apparatus has not been studied in any detail. Most genera have only rarely been recorded and their taxonomic position remains highly uncertain. Some filamentous chrysophycean algae previously placed in this family (or along with this family in a separate order, Phaeothamniales Bourrelly 1954) have been shown in ultrastructural studies to produce flagellate stages differing considerably from the *Chromulina*- and *Ochromonas*-type of organization and are now placed in different orders (see Sarcinochrysidales and Chrysomeridales below).

Pascher (1914) and other earlier workers called this family Chrysotrichaceae (and the order Chrysotrichales), but these names are illegitimate (cf. Silva 1980). The family Sphaeridiotrichaceae Bourrelly 1957 and the invalidly published 'Chrysocloniaceae' (Gayral & Haas 1969) are included in the Phaeothamniaceae here.

The following genera are now grouped in this family: *Chrysoclonium, Phaeothamnion, Sphaeridiothrix,* and *Tetrachrysis* (see Dop 1980*a, b*; Bourrelly 1981).

Apistonema, another genus often placed in the Phaeothamniaceae (see e.g. Starmach 1985), is probably not related to this group. The type species has not been studied in detail, but other species assigned to this genus are very similar to the benthic phases of certain prymnesio-phycean algae (Parke 1961; Leadbeater 1971). Yet other species of *Apistonema* have been suggested to be conspecific with the brown alga *Porterinema fluviatile* (Dop 1979). *Chrysonema, Nematochrysis* and *Thallo-chrysis* are also genera probably not related to Phaeothamniaceae, but instead apparently representing benthic phases of prymnesiophycean algae (Valkanov 1962; von Stosch 1967; Dop 1980*b*). *Stichochrysis,* a monotypic genus sometimes placed with *Sphaeridiothrix* in the same family (Bourrelly 1981), is now considered to be a diatom (Guillard & Hargraves 1993).

Chrysothallaceae

Members of the family Chrysothallaceae are parenchymatous in organ-ization. This family includes only one genus, *Phaeoplaca* Chodat 1926 (including *Chrysothallus* Meyer 1930). The family name Phaeoplacaceae Bourrelly 1957 has often been used in the literature, but Chrysothallaceae Huber-Pestalozzi 1941 has priority. Sometimes this family has been accommodated in a separate order, Phaeoplacales Bourrelly 1954.

The family Chrysothallaceae has an uncertain taxonomic position. The vegetative cells of *Phaeoplaca* show some features otherwise characteristic for the Chrysocapsaceae (Naegeliellaceae). Sometimes each cell is provided with a pseudocilium and contractile vacuoles may also be present (Dop 1978). On the other hand, there are thick fibrous cell walls around the vegetative cells unlike any member of the Chrysocapsaceae. Zoospores of the *Chromulina* type have also been observed, but the flagellar apparatus has not been studied in detail. *Phaeoplaca* clearly demonstrates that the organizational level is not a good character for defining orders or families in the Chrysophyceae. New studies applying modern methodo-logies of investigation are essential for achieving a more natural classification in this family.

Hibberdiales

Hibberdiaceae

The family Hibberdiaceae and order Hibberdiales have been established only recently (Andersen 1989*a*) for the new monotypic genus *Hibberdia* (type species: *H. magna* (Belcher) Andersen). This species has previously been accommodated in the 'coccoid' genus *Chrysosphaera* (as *C. magna* Belcher), but Andersen's investigations revealed conspicuous differences relative to other species of *Chrysosphaera* (Chrysosphaeraceae, see above) and also to other chrysophycean algae known at that time.

   Hibberdia magna has two stages in its life history. One is an immobile capsoid-like state and the second a free-swimming single-celled flagellate state. Vegetative cell division occurs in both the motile and immotile states, and therefore this alga cannot be considered to be a coccoid form. The immotile colony cannot properly be termed capsoid or palmelloid either, since flagella are present and beating inside the colonial gel, and this state can revert to a flagellate state at any time, releasing all cells as monads.

   The structure of the flagellar apparatus of *Hibberdia* is clearly different from that of *Ochromonas*, *Chromulina* and related genera. In *Hibberdia* the flagella are inserted at a much wider angle (similar to that of tribophytes), and the flagellar roots do not form a loop under the second flagellum. For more details on the flagellar apparatus, see Andersen (1989*a*, 1991*b*).

   Hibberdia is also unique in its pigment content (Alberte & Andersen 1986; Andersen 1989*a*). It contains antheraxanthin as a light-harvesting carotenoid, and because fucoxanthin also functions in this way, it has two different photosynthetic light-harvesting carotenoids. This is, to date, unique among the algae.

   Mitosis and cytokinesis of *Hibberdia* have not been analyzed in great detail. Preliminary observations suggest that the behavior of the nuclear envelope during mitosis is similar to that of *Hydrurus* (see below), but differs from that of other chrysophytes (such as *Ochromonas*, *Poterio-ochromonas* and *Uroglena/Uroglenopsis*; see Andersen 1989*b*), in that the nuclear envelope appears to be more or less intact during metaphase except for polar fenestrae, where the spindle microtubules penetrate. Unlike *Hydrurus*, however, the flagellar apparatus lies at the spindle poles and the spindle microtubules originate near the basal bodies. It has not been determined whether the spindle microtubules are nucleated by the rhizoplast.

Two species previously accommodated in other orders of Chrysophyceae are now also known to be clearly related to *Hibberdia*. R.A. Andersen (personal communication) revealed a hibberdialean flagellar apparatus architecture in *Chromophyton rosanoffii* Woronin (an alga previously thought to belong to the Chromulinales), and a similar general organization of the flagellar apparatus was also found in *Lagynion delicatulum* Skuja (Stylococcaceae) by O'Kelly & Wujek (this volume).

Stylococcaceae

Except for *Lagynion delicatulum*, no member of the amoeboid, loricate family Stylococcaceae has been studied in detail. It seems that this family (which also includes the Chrysopyxidaceae Lemmermann 1899, Chryso-thecaceae Huber-Pestalozzi 1941, Lagyniaceae Huber-Pestalozzi 1941, Chrysocrinaceae Krieger 1954, Derepyxidaceae Bourrelly 1957 and Kybotiaceae Bourrelly 1957; see Meyer 1986) should be classified together with the Hibberdiaceae in the Hibberdiales. But for a definite assignment it is necessary that more species (especially the type species of *Stylococcus*) be studied on a comparative basis.

It is likely that a number of genera and species now classified in other orders of Chrysophyceae will eventually be found to belong to the Hibberdiales too, as more details on cell organization are revealed.

Hydrurales

Hydruraceae

Three genera are included in the single family Hydruraceae of the order Hydrurales. *Hydrurus* and *Celloniella** are capsoid in organization, forming large gelatinous thalli of feathery, foliose or more irregular shape. *Phaeodermatium* forms parenchymatous crustose disks, but by gelatiniza-tion of cell walls, palmelloid stages, in which contractile vacuoles may occur, may also develop (Geitler 1927, 1971).

The zoospores produced by these genera provide the most distinctive features for this order and family (Hibberd 1977; Hoffman *et al.* 1986). They differ from other chrysophytes in being tetrahedral in shape. The long flagellum has so far not been clearly demonstrated to bear tubular flagellar hairs and the short flagellum is vestigial and lacks the two central

---

* According to R.A. Andersen (personal communication), *Celloniella* Pascher 1929 is congeneric with *Chrysonebula* Lund 1953.

microtubules. The flagellar roots do not form a loop under the second flagellum as in chromulinalean chrysophytes. The unique shape of the zoospores is associated with a complex skeletal system of microtubules extending from a broad flagellar root into the anterior processes of the cell. In contrast to most other Chrysophyceae, there is no photoreceptor system, and there is a greater number of contractile vacuoles and dictyosomes. The stomatocysts of *Hydrurus* and *Phaeodermatium* deviate from the normal chrysophycean type in being ellipsoid, with a delicate wing extending around half the periphery.

The processes of mitosis and cytokinesis in *Hydrurus* are also different from that of other Chrysophyceae (Vesk *et al.* 1984; Andersen 1989*b*). *Hydrurus* has a more or less intact nuclear envelope with polar fenestrae at metaphase, whereas in chromulinalean chrysophytes such as *Ochromonas*, the nucleus is completely dispersed at metaphase. As in *Ochromonas*, the basal bodies are indirectly associated with the spindle via a rhizoplast. In *Hydrurus* the basal bodies are above the nucleus and over the metaphase plate, whereas in *Ochromonas* they are above the nucleus but not over the metaphase plate.

Sarcinochrysidales

The order Sarcinochrysidales was formally described by Gayral & Billard (1977) to accommodate chrysophycean genera with phaeophycean affinities. In contrast to most genera of the other orders of Chrysophyceae which predominantly occur in freshwater habitats, these genera are exclusively marine (or brackish), and stomatocysts have never been observed. Further investigation demonstrated the existence of two natural groups within that order (O'Kelly & Floyd 1985; Gayral & Billard 1986; O'Kelly 1989; Billard *et al.* 1990), which will be split into two separate orders: Sarcinochrysidales *sensu stricto* and Chrysomeridales (C.J. O'Kelly & C. Billard, unpublished data).

The two orders are distinguished by the following characteristics (S = Sarcinochrysidales *sensu stricto*, C = Chrysomeridales):

(1) Eyespot in the motile cells: lacking (S) or present (C).
(2) Stalked phaeophycean-type pyrenoids: present (S) or lacking (C).
(3) Diatoxanthin and diadinoxanthin pigments: present (S) or lacking (C); if lacking they are normally replaced by one or more pigments of the zeaxanthin complex (asteraxanthin, violaxanthin, zeaxanthin) as in most other Chrysophyceae.

(4) Plastid DNA: with nucleoids scattered throughout the chloroplast (S), or with ring nucleoid lying just within the girdle lamella (C).
(5) Major sterol: C_{29}-sterol (S), or C_{30}-sterol (C).
(6) Helical structure in basal body transition region: lacking (S) or present (C).
(7) Basal plate elevated above the plane of the plasmalemma (S), or in the plane of the plasmalemma (C).
(8) Significant dissimilarities also exist in the flagellar root system.

The Sarcinochrysidales *sensu stricto* is subdivided into three families according to differences in vegetative organization:

Ankylochrysidaceae

Only one genus and species, *Ankylochrysis lutea* (van der Veer) Billard (= *Ankylonoton luteum* van der Veer), is known. This is a monad with a multilayered theca outside the plasmalemma (van der Veer 1970; C.J. O'Kelly & C. Billard, unpublished data).

Sarcinochrysidaceae

The family Sarcinochrysidaceae includes two genera, *Sarcinochrysis* and *Chrysoreinhardia* (C.J. O'Kelly & C. Billard, unpublished data), which both are primarily palmelloid in organization, forming microscopic or macroscopic cell aggregates. A filamentous variety of *Sarcinochrysis* is also known (Billard 1988). Zoospores are variable in shape.

Nematochrysopsidaceae

Nematochrysopsis, the only genus of the family Nematochrysopsidaceae, is filamentous in organization. Zoospores are formed singly per cell. Extracellular calcified pseudocysts, similar to those of *Chrysowaernella* and *Giraudyopsis* (Chrysomeridales), have been observed (Gayral & Lepailleur 1971).

Chrysomeridales

The many differences between the Chrysomeridales and Sarcinochrysidales (see above) seem to justify the separation of these orders. The architecture of the flagellar apparatus is particularly different between the two orders

and, in fact, there are more similarities in this respect between Chrysomeridales and brown algae than with sarcinochrysidalean chrysophytes
(O'Kelly 1989; C.J. O'Kelly & C. Billard, unpublished data). On the other
hand, classification of the Chrysomeridales in the Fucophyceae does not
appear to be appropriate on the basis of current knowledge, because they
lack so many of the fundamental characters by which the brown algae
can most readily be defined (i.e. unilocular sporangia, alginates, physodes,
plasmodesmata, etc.).

Chrysomeridaceae

At present the family Chrysomeridaceae includes one genus with a
palmelloid organization, *Chrysoderma*, and four filamentous genera,
Chrysomeris, *Chrysonephos*, *Chrysowaernella* and *Rhamnochrysis* (Gayral
& Billard 1986; C.J. O'Kelly & C. Billard, unpublished data). The basal
attachment of the filaments is unicellular and chloroplast pyrenoids are
lacking or inconspicuous. Extracellular calcified pseudocysts, similar to
those of *Nematochrysopsis* (Sarcinochrysidales) and *Giraudyopsis* (see
below), are formed by *Chrysowaernella* (Gayral & Lepailleur 1971). The
flagellar apparatus of these genera has not been examined in detail.
Further studies must ascertain whether they can all be retained in the
same family.

Phaeosacciaceae

Two parenchymatous and probably closely related genera, *Antarctosaccion*
and *Phaeosaccion*, were assigned to the family Phaeosacciaceae by Gayral
& Billard (1986). The monotypic genus *Giraudyopsis* also appears to
belong here (C.J. O'Kelly & C. Billard, unpublished data). Young stages
of *Giraudyopsis* and *Phaeosaccion* are very much alike and there is also
close resemblance in the two genera in the way fully developed algae are
attached to the substrate (forming a multicellular discoidal base). In
Giraudyopsis erect filaments radiate from this base, whereas in *Phaeosaccion*
tubular to saccate, monostromatic thalli are formed. The chloroplast
pyrenoids of *Giraudyopsis* and *Phaeosaccion* are also of the same type
(membrane limited, embedded). Ultrastructural cell morphology is known
from both these genera (Chen *et al.* 1974; Billard 1984), but the absolute
configuration of the flagellar apparatus has been studied only in zoospores
of *Giraudyopsis* (O'Kelly & Floyd 1984). For a description of *Phaeosaccion*
from the biochemical viewpoint, see Craigie *et al.* (1971).

Organisms of uncertain taxonomic position in the Chrysophyceae

Parmales

The order Parmales was described by Booth & Marchant (1987) for distinctive siliceous organisms they found in abundance in the plankton of polar and sub-polar seas. The siliceous cell wall is constructed of five to eight distinctively shaped plates, and inside the cell there is a large chloroplast with three thylakoid lamellae, a girdle lamella and chloroplast endoplasmic reticulum. Chloroplast morphology resembles that of the Chrysophyceae but also that of other groups of chromophyte algae (e.g. diatoms, eustigmatophytes or tribophytes). At present, nothing is known about the life cycles or pigmentation of the Parmales, and more information is clearly required to confirm the possible affinity with chrysophycean algae (see also Mann & Marchant 1989).

Booth & Marchant (1987, 1988) subdivided the order into two families. In Pentalaminaceae, comprising the monotypic genus *Pentalamina*, the cell covering is composed of five plates. In the two genera assigned to the Triparmaceae, *Tetraparma* (1 species) and *Triparma* (5 species), the cell covering is composed of eight plates.

Aurosphaera/Meringosphaera

In the marine genus *Aurosphaera*, consisting of three rarely recorded species, the solitary coccoid cells contain two or three yellow chloroplasts and the cell surface is covered by numerous siliceous bristles. Reproduction is by autospores and by *Chromulina*-like zoospores. The internal cell structure has not been studied by electron microscopy. Schiller (1916, 1925) assigned the genus to a separate family, Aurosphaeraceae, which he considered to be related to the silicoflagellates. Alternatively, this family and genus have been classified in the chrysophycean order Chrysosphaerales (Bourrelly 1957; Leadbeater 1974). Further studies are necessary to ascertain the taxonomic position.

The genus *Meringosphaera*, containing several widely distributed species of the marine and brackish plankton, resembles *Aurosphaera* in its coccoid cell organization and the presence of siliceous bristles on the cell surface. However, the chloroplasts of *Meringosphaera* are usually greenish in color and in at least one species (*M. aculeata* Pascher) there is a cell wall composed of two halves (Pascher 1932). Reproduction appears to be exclusively by autosporulation. *Meringosphaera* has often been assigned to the Xanthophyceae (e.g. Ettl 1978; Sieburth 1979; Silva 1979), but a relationship with chrysophycean algae has also been suggested (Norris 1971; Leadbeater 1974; Chrétiennot-Dinet 1990). For a correct taxonomic placement of this genus, detailed biochemical, ultrastructural and life history studies must be done. These should include comparative studies on the possibly related genera *Actinellipsoidion*, *Raphidosphaera* and *Skiadosphaera*.

Pelagococcus/Aureococcus

The minute green-gold coccoid cells of the monospecific marine genus *Pelagococcus* have a simple cell structure, surrounded by a distinct three-layered wall (Lewin *et al.* 1977; Vesk & Jeffrey 1987). Motile cells and stomatocysts are unknown. Organelle ultrastructure (e.g. Golgi body, presence of a girdle lamella in the chloroplast) is in conformity with chrysophycean organization, but the pigment composition (with chlorophyll c_3, two fucoxanthin derivatives and diadinoxanthin as the major secondary carotenoids) suggests affinity with some prymnesiophytes (typical chrysophytes contain fucoxanthin and violaxanthin as their major carotenoids). Mitosis is atypical of chrysophytes, prymnesiophytes and all other algal groups so far examined. Unusual features include: the *de novo* appearance of centrioles prior to mitosis, and the formation of a small, extranuclear spindle. Identification of *Pelagococcus* has also been shown to be possible by the application of species-specific immunological antibodies (Campbell *et al.* 1989).

The marine, monospecific genus *Aureococcus* (well known for its production of recurrent brown tides) has many structural similarities with *Pelagococcus* (Sieburth *et al.* 1988). Some authors considered the two to be congeneric or perhaps even conspecific (e.g. Hallegraeff 1991). The main difference is that cells of *Aureococcus* do not possess a cell wall, but are coated with extracellular polysaccharide material. Pigment composition also bears a close resemblance (Bidigare 1989), and fatty acid composition in *Aureococcus* is very comparable to that in a broad spectrum of algal species, including other chrysophytes (Bricelj *et al.* 1989).

Conclusions

The exclusively marine orders Sarcinochrysidales and Chrysomeridales, as well as the exclusively marine taxa of uncertain position (Parmales, *Aurosphaera*, *Pelagococcus/Aureococcus*) appear to deviate clearly from the 'typical' chrysophycean pattern of organization. It is possible that these marine organisms formed separate lines of morphological development from those of freshwater forms.

Both the freshwater orders Chromulinales, Hibberdiales and Hydrurales, and the marine orders Sarcinochrysidales and Chrysomeridales appear to be well defined on the basis of ultrastructural characteristics of the motile cells, especially with regard to the architecture of the flagellar apparatus. The latter is considered to be highly conserved in an evolutionary sense and therefore provides a good criterion for classification purposes. Information on mitosis and cytokinesis, as well as on physiology and biochemistry (photosynthetic pigments, fatty acids, sterols, etc.), also seems to be important in characterizing the different taxonomic groups.

Much of the available information is restricted to a few common or easily cultivated species. More attention needs to be given to other chrysophycean species, so that our understanding of classification is not too narrowly based. It is also especially important that members of the most poorly known taxa of higher rank (such as the chromulinalean families Myxochrysidaceae, Chrysocapsaceae, Stichogloeaceae, Phaeothamniaceae) are studied in more detail. It may well be that organisms placed in these groups have in fact a more natural taxonomic position in other families and orders.

In future, it is probable that information gained from new methods of investigation (such as macromolecular sequencing) will also be used increasingly for classification purposes within this algal class.

Summary

Criteria for subdivision of the Chrysophyceae involving both gross and ultrastructural detail are outlined. The structure of the flagellar apparatus is considered to be of major importance, but other features are also taken into account. The author is currently working on a new edition of the volume on Chrysophyceae for the monography series *Süsswasserflora von Mitteleuropa* and it is shown that this present concept of classification differs fundamentally from that of the previous edition by Starmach (1985).

Acknowledgments

Financial support from the Swiss National Science Foundation (grant no. 31-9129.87) is acknowledged.

References

Alberte, R.S. & Andersen, R.A. 1986. Antheraxanthin, a light-harvesting carotenoid found in a chromophyte alga. *Plant Physiol.* **80**: 583–7.

Andersen, R.A. 1987. Synurophyceae *classis* nov., a new class of algae. *Am. J. Bot.* **74**: 337–53.

Andersen, R.A. 1989*a*. Absolute orientation of the flagellar apparatus of *Hibberdia magna* comb. nov. (Chrysophyceae). *Nord. J. Bot.* **8**: 653–69.

Andersen, R.A. 1989*b*. The Synurophyceae and their relationship to other golden algae. *Nova Hedwigia Beih.* **95**: 1–26.

Andersen, R.A. 1990. The three-dimensional structure of the flagellar apparatus of *Chrysosphaerella brevispina* (Chrysophyceae) as viewed by high-voltage electron microscopy stereo-pairs. *Phycologia* **29**: 86–97.

Andersen, R.A. 1991*a*. Taxonomy and ultrastructure of a new genus and family of Chrysophyceae whose cells are covered with organic scales. *J. Phycol.* **27**(Suppl.): 5.

Andersen, R.A. 1991*b*. The cytoskeleton of chromophyte algae. *Protoplasma* **164**: 143–59.

Andersen, R.A. & Wetherbee, R. 1992. Microtubules of the flagellar apparatus are active during prey capture in the chrysophycean alga *Epipyxis pulchra*. *Protoplasma* **166**: 8–20.

Belcher, J.H. & Swale, E.M.F. (1967). *Chromulina placentula* sp. nov. (Chrysophyceae), a freshwater nanoplankton flagellate. *Br. Phycol. Bull.* **3**: 257–67.

Bidigare, R.R. 1989. Photosynthetic pigment composition of the brown tide alga: unique chlorophyll and carotenoid derivatives. In: Cosper, E.M., Bricelj, V.M. & Carpenter, E.J. (eds.) *Novel Phytoplankton Blooms.* Coastal and Estuarine Studies 35. Springer, Berlin. pp. 57–75.

Billard, C. 1984. Recherches sur les Chrysophyceae marines de l'ordre Sarcinochrysidales: biologie, systématique, phylogénie. Doctoral thesis, University of Caen. 233 pp.

Billard, C. 1988. Les Sarcinochrysidales des côtes françaises. *Bull. Soc. Linn. Normandie* **110/111**: 61–9.

Billard, C., Dauguet, J.-C., Maume, D. & Bert, M. 1990. Sterols and chemotaxonomy of marine Chrysophyceae. *Bot. Mar.* **33**: 225–8.

Bold, H.C. & Wynne, M.J. 1978. *Introduction to the Algae.* Prentice-Hall, Englewood Cliffs, N.J. 706pp.

Booth, B.C. & Marchant, H.J. 1987. Parmales, a new order of marine chrysophytes, with descriptions of three new genera and seven new species. *J. Phycol.* **23**: 245–60.

Booth, B.C. & Marchant, H.J. 1988. Triparmaceae, a substitute name for a family in the order Parmales (Chrysophyceae). *J. Phycol.* **24**: 124.

Bouck, G.B. & Brown, D.L. 1973. Microtubule biogenesis and cell shape in *Ochromonas*. I. The distribution of cytoplasmic and mitotic microtubules. *J. Cell Biol.* **56**: 340–59.

Bourrelly, P. 1954. Phylogénie et systématique des Chrysophycées. In: *Rapports et Communications de l'Huitième Congrès International de Botanique* [*Paris*], Sect. 17. pp. 117–18.

Bourrelly, P. 1957. Recherches sur les Chrysophycées. *Rev. Algol., Mém. Hors-Sér.* **1**: 1–412.

Bourrelly, P. 1968. *Les alges d'eau douce*, vol. 2. Boubée, Paris. 438pp.

Bourrelly, P. 1981. *Les alges d'eau douce*, vol. 2 (revised and enlarged edn). Boubée, Paris. 517pp.

Bricelj, V.M., Fisher, N.S., Guckert, J.B. & Chu, F.-L.E. 1989. Lipid composition and nutritional value of the brown tide alga *Aureococcus anophagefferens*. In: Cosper, E.M., Bricelj, V.M. & Carpenter, E.J. (eds,) *Novel Phytoplankton Blooms. Coastal and Estuarine Studies* 35. Springer, Berlin. pp. 85–100.

Campbell, L., Shapiro L.P., Haugen, E.M. & Morris, L. 1989. Immunochemical approaches to the identification of the ultraplankton: assets and limitations. In: Cosper, E.M., Bricelj, V.M. & Carpenter, E.J. (eds.) *Novel Phytoplankton Blooms. Coastal and Estuarine Studies* 35. Springer, Berlin. pp. 39–56.

Cavalier-Smith, T. 1986. The kingdom Chromista: origin and systematics. *Prog. Phycol. Res.* **4**: 309–47.

Chen, L.C.-M., McLachlan, J. & Craigie, J.S. 1974. The fine structure of the marine chrysophycean alga *Phaeosaccion collinsii*. *Can. J. Bot.* **52**: 1621–4.

Chodat, R. 1926. Algues de la région de Grand Saint-Bernard: III. *Bull. Soc. Bot. Genève* **17**: 202–17.

Chrétiennot-Dinet, M.-J. 1990. Chlorarachniophycées, Chlorophycées, Chrysophycées. Cryptophycées, Euglénophycées, Eustigmatophycées, Prasinophycées, Prymnesiophycées, Rhodophycées et Tribophycées. In: Sournia, A. (ed.) *Atlas du phytoplancton marin*, vol. 3. Éditions de Centre National de la Recherche Scientifique, Paris. pp. 1–261.

Christensen, T. 1980. *Algae: A Taxonomic Survey*. Aio Tryk, Odense, 228pp.

Clarke, K.J. & Pennick, N.C. 1975. *Syncrypta glomerifera* sp. nov., a marine member of the Chrysophyceae bearing a new form of scale. *Br. Phycol. J.* **10**: 363–70.

Correns, C. 1893. Über eine neue braune Süsswasseralge, *Naegeliella flagellifera*, nov. gen. et spec. *Ber. Dtsch. Bot. Ges.* **10**: 629–36.

Craigie, J.S., Leigh, C., Chen, L.C.-M. & McLachlan, J. 1971. Pigments, polysaccharides, and photosynthetic products of *Phaeosaccion collinsii*. *Can. J. Bot.* **49**: 1067–74.

Dop, A.J. 1978. Systematics and morphology of *Chrysochaete brittanica* (Godward) Rosenberg and *Phaeoplaca thallosa* Chodat (Chrysophyceae). *Acta Bot. Neerl.* **27**: 35–60.

Dop, A.J. 1979. *Porterinema fluviatile* (Porter) Waern (Phaeophyceae) in the Netherlands. *Acta. Bot. Neerl.* **28**: 449–58.

Dop, A.J. 1980a. The genera *Phaeothamnion* Lagerheim, *Tetrachrysis* gen. nov. and *Sphaeridiothrix* Pascher et Vlk (Chrysophyceae). *Acta Bot. Neerl.* **29**: 65–86.

Dop, A.J. 1980b. Benthic Chrysophyceae from the Netherlands. PhD thesis, University of Amsterdam. 141pp.

Ehrenberg, C.G. 1834. Dritter Beitrag zur Erkenntnis grosser Organisation in der Richtung des kleinsten Raumes. *Abh. Königl. Akad. Wiss. Berlin, Phys. Kl.* [1833/4]: 145–336.

Engler, A. 1897. Bemerkung, betreffend der in Abteilung I. 2 noch nicht berücksichtigten Chlorophyceae und Phaeophyceae. In: Engler, A. & Prantl, K. (eds.) *Die natürlichen Pflanzenfamilien*, part 1, section 2. Engelmann, Leipzig. p. 570.

Entwisle, T.J. & Andersen, R.A. 1990. A re-examination of *Tetrasporopsis* (Chrysophyceae) and the description of *Dermatochrysis* gen. nov.

(Chrysophyceae): a monostromatic alga lacking cell walls. *Phycologia* **29**: 263–74.

Ettl, H. 1978. Xanthophyceae: I. In: Ettl, H., Gerloff, J. & Heynig, H. (eds.) *Süsswasserflora von Mitteleuropa*, vol. 3. Fischer, Stuttgart. 530pp.

Ettl, H. 1980. *Grundriss der allgemeinen Algologie*. Fischer, Stuttgart. 549pp.

Feldmann, J. 1949. L'ordre des Scytosiphonales. *Mém. Hors-Sér. Soc. Hist. Nat. Afr. Nord* **2**: 103–15.

Fott, B. 1971. *Algenkunde*, 2nd edn. Fischer, Jena. 581pp.

Fritsch, F.E. 1935. *The Structure and Reproduction of the Algae*, vol. 1. Cambridge University Press, Cambridge. 791pp.

Gayral, P. & Billard, C. 1977. Synopsis du nouvel ordre des Sarcinochrysidales (Chrysophyceae). *Taxon* **26**: 241–5.

Gayral, P. & Billard, C. 1986. A survey of the marine Chrysophyceae with special reference to the Sarcinochrysidales. In: Kristiansen, J. & Andersen, R.A. (eds.) *Chrysophytes: Aspects and Problems*. Cambridge University Press, Cambridge. pp. 37–48.

Gayral, P. & Haas, C. 1969. Étude comparée des genres *Chrysomeris* Carter et *Giraudyopsis* P. Dang., position systématique des Chrysomeridaceae (Chrysophyceae). *Rev. Gen. Bot.* **76**: 659–66.

Gayral, P. & Lepailleur, H. 1971. Étude de deux Chrysophycées filamenteuses: *Nematochrysopsis roscoffensis* Chadefaud, *Nematochrysis hieroglyphica* Waern. *Rev. Gen. Bot.* **78**: 61–74.

Geitler, L. 1927. Die Schwärmer und Kieselcysten von *Phaeodermatium rivulare*. *Arch. Protistenk.*, **58**: 272–80.

Geitler, L. 1968. Der Zellbau der Chrysophycee *Chrysochaete*. *Österr. Bot. Z.* **115**: 134–43.

Geitler, L. 1971. Weitere Untersuchungen über ein Massenvorkommen von *Chrysocapsella granifera* im Lunzer Untersee mit Zonenbildung. *Arch. Hydrobiol.* **68**: 516–18.

Grell, K.G. 1991. *Leucodictyon marinum* n. gen., n. sp., a plasmodial protist with zoospore formation from the Japanese coast. *Arch. Protistenk.* **140**: 1–21.

Grell, K.G., Heini, A. & Schüller, S. 1990. The ultrastructure of *Reticulosphaera socialis* Grell (Heterokontophyta). *Eur. J. Protistol.* **26**: 37–54.

Guillard, R.R.L. & Hargraves, P.E. 1993. *Stichochrysis immobilis* is a diatom, not a chrysophyte. *Phycologia* **32**: 234–6.

Haeckel, E. 1894. *Systematische Phylogenie der Protisten und Pflanzen*. Berlin. 400pp.

Hallegraeff, G.M. 1991. *Aquaculturists' Guide to Harmful Australian Microalgae*. CSIRO Division of Fisheries, Hobart, Tasmania, Australia. 112pp.

Hansgirg, A. 1886. Prodromus der Algenflora von Böhmen. Erster Theil enthaltend die Rhodophyceen, Phaeophyceen und einen Theil der Chlorophyceen. *Arch. Naturwiss. Landesdurchf. Böhmen* **5**: 1–96.

Harris, K. 1963. Observations on *Sphaleromantis tetragona*. *J. Gen. Microbiol.* **33**: 345–8.

Hibberd, D.J. 1976. The ultrastructure and taxonomy of the Chrysophyceae and Prymnesiophyceae (Haptophyceae): a survey with some new observations on the ultrastructure of the Chrysophyceae. *Bot. J. Linn. Soc.* **72**: 55–80.

Hibberd, D.J. 1977. The cytology and ultrastructure of *Chrysonebula holmesii*

Lund (Chrysophyceae), with special reference to the flagellar apparatus. *Br. Phycol. J.* **12**: 369–83.

Hibberd, D.J. 1986. Ultrastructure of the Chrysophyceae: phylogenetic implications and taxonomy. In: Kristiansen, J. & Andersen, R.A. (eds.) *Chrysophytes: Aspects and Problems.* Cambridge University Press, Cambridge. pp. 23–36.

Hibberd, D.J. & Chrétiennot-Dinet, M.-J. 1979. The ultrastructure and taxonomy of *Rhizochromulina marina* gen. et sp. nov., an amoeboid marine chrysophyte. *J. Mar. Biol. Assoc. U.K.* **59**: 179–93.

Hoffman, L.R., Vesk. M. & Pickett-Heaps, J.D. 1986. The cytology and ultrastructure of zoospores of *Hydrurus foetidus* (Chrysophyceae). *Nord. J. Bot.* **6**: 105–22.

Huber-Pestalozzi, G. 1941. Das Phytoplankton des Süsswassers. Systematik und Biologie, part 2(1): Chrysophyceen, farblose Flagellaten, Heterokonten. In: Thienemann, A. (ed.) *Die Binnengewässer*, vol. 16, part 2(1). Schweizerbart, Stuttgart. pp. 1–365.

Inouye, I., Zhang, X., Enomoto, M. & Chihara, M. 1990. Cell structure of an unusual chrysophyte *Phaeaster pascheri* with particular emphasis on the flagellar apparatus architecture. *Jpn. J. Phycol.* **38**: 11–24.

Kent, W.S. 1881. *A Manual of the Infusoria.* Bogue, London. 913pp.

Krieger, W. 1954. Part VII: Chrysophyta. In: Melchior, H. & Werdermann, E. (eds.) *A. Engler's Syllabus der Pflanzenfamilien*, 12th edn, vol. 1. Borntraeger, Berlin. pp. 73–85.

Kristiansen, J. 1982. Chromophycota – Chrysophyceae. In: Parker, S.P. (ed.) *Synopsis and Classification of Living Organisms.* McGraw-Hill, New York. pp. 81–6.

Kristiansen, J. 1986. The ultrastructural bases of chrysophyte systematics and phylogeny. *CRC Crit. Rev. Plant Sci.* **4**: 149–211.

Kristiansen, J. 1990. Phylm Chrysophyta. In: Margulis, L., Corliss, J.O., Melkonian, M. & Chapman, D.J. (eds.) *Handbook of Protoctista.* Jones & Bartlett, Boston. pp. 438–53.

Leadbeater, B.S.C. 1971. Preliminary observations on differences of scale morphology at various stages in the life cycle of '*Apistonema–Syracosphaera*' sensu von Stosch. *Br. Phycol. J.* **5**: 57–69.

Leadbeater, B.S.C. 1974. Ultrastructural observations on nanoplankton collected from the coast of Jugoslavia and the Bay of Algiers. *J. Mar. Biol. Assoc. U.K.* **54**: 179–96.

Lemmermann, E. 1899. Das Phytoplankton sächsischer Teiche. *Forschungsber. Biol. Stat. Plön* **7**: 96–135.

Lemmermann, E. 1901. Beiträge zur Kenntnis der Planktonalgen. XII. Notizen über einige Schwebalgen. XIII. Das Phytoplankton des Ryck und des Greifswalder Boddens. *Ber. Dtsch. Bot. Ges.* **19**: 85–95.

Lewin, J., Norris, R.E., Jeffrey, S.W. & Pearson, B.A. 1977. An aberrant chrysophycean alga, *Pelagococcus subviridis* gen. nov. et sp. nov., from the North Pacific Ocean. *J. Phycol.* **13**: 259–66.

Lund, J.W.G. 1953. New or rare British Chrysophyceae. II. *Hyalobryon polymorphum* n.sp. and *Chrysonebula holmesii* n.gen., n.sp. *New Phytol.* **52**: 114–23.

Magne, F. 1975. Contribution à la connaissance des Stichogloeacées (Chrysophycées, Stichogloeales). *Cah. Biol. Mar.* **16**: 531–9.

Mann, D.G. & Marchant, H.J. 1989. The origins of the diatom and its life cycle. In: Green, J.C., Leadbeater, B.S.C. & Diver, W.L. (eds.) *The*

Chromophyte Algae: Problems and Perspectives. Systematics Association
Special Volume 38. Clarendon Press, Oxford. pp. 307–23.

Manton, I. & Harris, K. 1966. Observations on the microanatomy of the brown
flagellate *Sphaleromantis tetragona* Skuja with special reference to the
flagellar apparatus and scales. *J. Linn. Soc. (Bot.)* **59**: 397–403.

Matvienko, O. [A.] M. 1965. Zolotisti vodorosti: Chrysophyta. *Vyznachnyk
prisnovodnykh vodorostej Ukrajns'koj R.S.R.*, vol. 3(1). Akademia Nauk
Ukrajns'koj R.S.R., Naukova Dumka, Kiev. 367pp.

Meyer, K.I. 1930. Einige neue Algenformen des Baikalsees. *Arch. Protistenk.* **72**:
158–75.

Meyer, R.L. 1986. A proposed phylogenetic sequence for the loricate
rhizopodial Chrysophyceae. In: Kristiansen, J. & Andersen, R.A. (eds.)
Chrysophytes: Aspects and Problems. Cambridge University Press,
Cambridge. pp. 75–85.

Nicholls, K.H. 1990. Life history and taxonomy of *Rhizoochromonas
endoloricata* gen. et sp. nov., a new freshwater chrysophyte inhabiting
Dinobryon loricae. *J. Phycol.* 26: 558–63.

Norris, R.E. 1971. Extant siliceous microalgae from the Indian Ocean. In:
Farinacci, A. (ed.) *Proceedings of the Second Planktonic Conference, Rome
1970*, vol. 2. Edizioni Tecnoscienza, Rome. pp. 911–19.

O'Kelly, C.J. 1989. The evolutionary origin of the brown algae: information
from studies of motile cell ultrastructure. In: Green, J.C., Leadbeater,
B.S.C. & Diver, W.L. (eds.) *The Chromophyte Algae: Problems and
Perspectives.* Systematics Association Special Volume 38. Clarendon Press,
Oxford. pp. 255–78.

O'Kelly, C.J. & Billard, C. Phylogeny of golden algae and taxonomic positions
of the orders Sarcinochrysidales and Chrysomeridales ord. nov., as
provided by ultrastructural studies on *Sarcinochrysis marina* zoospores. In
preparation.

O'Kelly, C.J. & Floyd, G.L. 1985. Absolute configuration analysis of the
flagellar apparatus in *Giraudyopsis stellifer* (Chrysophyceae,
Sarcinochrysidales) zoospores and its significance in the evolution of the
Phaeophyceae. *Phycologia* 24: 263–74.

Owen, H.A., Mattox, K.R. & Stewart, K.D. 1990*a*. Fine structure of the
flagellar apparatus of *Dinobryon cylindricum* (Chrysophyceae). *J. Phycol.*
26: 131–41.

Owen, H.A., Stewart, K.D. & Mattox, K.R. 1990*b*. Fine structure of the
flagellar apparatus of *Uroglena americana* (Chrysophyceae). *J. Phycol.* 26:
142–9.

Parke, M. 1961. Some remarks concerning the class Chrysophyceae. *Br. Phycol.
Bull.* **2**: 47–55.

Pascher, A. 1910. Der Grossteich bei Hirschberg in Nord-Böhmen. *Monogr.
Abhandl. Int. Rev. Ges. Hydrobiol. Hydrogr.* **1**: 1–66.

Pascher, A. 1911. *Cyrtophora*, eine neue tentakeltragende Chrysomonade aus
Franzensbad und ihre Verwandten. *Ber. Dtsch. Bot. Ges.* **29**: 112–25.

Pascher, A. 1912. Über Rhizopoden- und Palmellastadien bei Flagellaten
(Chrysomonaden), nebst einer Übersicht über die braunen Flagellaten.
Arch. Protistenk. **25**: 153–200.

Pascher, A. 1914. Über Flagellaten und Algen. *Ber. Dtsch. Bot. Ges.* **32**: 136–60.

Pascher, A. 1916. Fusionsplasmodien bei Flagellaten und ihre Bedeutung für
die Ableitung der Rhizopoden von den Flagellaten. *Arch. Protistenk.* **37**:
31–64.

Pascher, A. 1925. Die braune Algenreihe der Chrysophyceen. *Arch. Protistenk.* **52**: 489–564.

Pascher, A. 1929. Über die Beziehungen zwischen Lagerform und Standortsverhältnissen bei einer Gallertalge (Chrysocapsale). *Arch. Protistenk.* **68**: 637–68.

Pascher, A. 1931. Systematische Übersicht über die mit Flagellaten in Zusammenhang stehenden Algenreihen und Versuch einer Einreihung dieser Algenstämme in die Stämme des Pflanzenreiches. *Beih. Bot. Centralb.* **48**: 317–32.

Pascher, A. 1932. Zur Kenntnis mariner Planktonten. I. *Meringosphaera* und ihre Verwandten. *Arch. Protistenk.* **77**: 195–218.

Peters, M.C. & Andersen, R.A. (1993). The fine structure and scale formation of *Chrysolepidomonas dendrolepidota* gen. et sp. nov. (Chrysolepido-monadaceae fam. nov., Chrysophyceae). *J. Phycol.* **29**: 469–75.

Pienaar, R.N. 1976. The microanatomy of *Sphaleromantis marina* sp. nov. (Chrysophyceae). *Br. Phycol. J.* **11**: 83–92.

Pipes, L., Tyler, P.A. & Leedale, G.F. 1989. *Chrysonephele palustris* gen. et sp. nov. (Chrysophyceae), a new colonial chrysophyte from Tasmania. *Nova Hedwigia Beih.* **95**: 81–97.

Poche, F. 1913. Das System der Protozoa. *Arch. Protistenk.* **30**: 125–321.

Preisig, H.R. 1989. The flagellar base ultrastructure and phylogeny of chromophytes. In: Green, J.C., Leadbeater, B.S.C. & Diver, W.L. (eds.) *The Chromophyte Algae: Problems and Perspectives.* Systematics Association Special Volume 38. Clarendon Press, Oxford. pp. 167–87.

Preisig, H.R. 1995. Chromulinaceae. In: Parker, B.C. (ed). *Encyclopedia of algal genera.* Dioscorides Press, Portland, Oregon (in press).

Preisig, H.R. & Hibberd, D.J. 1983. Ultrastructure and taxonomy of *Paraphysomonas* (Chrysophyceae) and related genera (3). *Nord. J. Bot.* **3**: 695–723.

Preisig, H.R. & Hibberd, D.J. 1986. Classification of four genera of Chrysophyceae bearing silica scales in a family separate from *Mallomonas* and *Synura.* In: Kristiansen, J. & Andersen, R.A. (eds.) *Chrysophytes: Aspects and Problems.* Cambridge University Press, Cambridge. pp. 71–4.

Preisig, H.R., Vørs, N. & Hällfors, G. 1991. Diversity of heterotrophic heterokont flagellates. In: Patterson, D.J. & Larsen, J. (eds.) *The Biology of Free-Living Heterotrophic Flagellates.* Systematics Association Special Volume 43. Clarendon Press, Oxford. pp. 361–99.

Reichenow, E. 1928. [*F. Doflein's*] *Lehrbuch der Protozoenkunde,* 5th edn, part 2. Fischer, Jena. pp. 437–862.

Rosenberg, M. 1941. *Chrysochaete,* a new genus of the Chrysophyceae, allied to *Naegeliella. New Phytol.* **40**: 304–15.

Rostafiński, J. 1881. Tymczasowa wiadomość o czerwonym i żółtym śniegu tudzież o nowo-odkrytéj grupie wodorostów brunatnych znalezionych w Tatrach. *Rozpr. Spraw. Posiedzeń Wydz. Mat.-Przyr. Akad. Umiejetn.* **8**: 8–13.

Round, F.E. 1986. The Chrysophyta: a reassessment. In: Kristiansen, J. & Andersen, R.A. (eds.) *Chrysophytes: Aspects and Problems.* Cambridge University Press, Cambridge. pp. 3–22.

Schiller, J. 1916. Eine neue kieselschalige Protophyten-Gattung aus der Adria. *Arch. Protistenk.* **36**: 303–10.

Schiller, J. 1925. Die planktontischen Vegetationen des adriatischen Meeres. B.

74 *Chrysophyte algae*

Chrysomonadina, Heterokontae, Cryptomonadina, Eugleninae, Volvocales. 1. Systematischer Teil. *Arch. Protistenk.* **53**: 59–123.

Schnepf, E., Deichgräber, G., Röderer, G. & Herth, W. 1977. The flagellar root apparatus, the microtubular system and associated organelles in the chrysophycean flagellate *Poterioochromonas malhamensis* Peterfi (syn. *Poteriochromonas stipitata* Scherffel and *Ochromonas malhamensis* Pringsheim). *Protoplasma* **92**: 87–107.

Sieburth, J.McN. 1979. *Sea Microbes.* Oxford University Press, Oxford. 491pp.

Sieburth, J.McN. & Johnson, P.W. 1989. Picoplankton ultrastructure: a decade of preparation for the brown tide alga, *Aureococcus anophagefferens.* In: Cosper, E.M., Bricelj, V.M. & Carpenter, E.J. (eds.) *Novel Phytoplankton Blooms.* Coastal and Estuarine Studies 35. Springer, Berlin. pp. 1–21.

Sieburth, J.McN., Johnson, P.W. & Hargraves, P.E. 1988. Ultrastructure and ecology of *Aureococcus anophagefferens* gen. et sp. nov. (Chrysophyceae): the dominant picoplankter during a bloom in Narragansett Bay, Rhode Island, summer 1985. *J. Phycol.* **24**: 416–25.

Silva, P.C. 1979. Review of the taxonomic history and nomenclature of the yellow-green algae. *Arch. Protistenk.* **121**: 20–63.

Silva, P.C. 1980. Names of classes and families of living algae. *Regnum Vegetabile* **103**: 1–156.

Starmach, K. 1985. Chrysophyceae und Haptophyceae. In: Ettl, H., Gerloff, J., Heynig, H. & Mollenhauer, D. (eds.) *Süsswasserflora von Mitteleuropa,* Vol. 1. Fischer, Stuttgart. pp. 1–515.

Stein, F. von 1878. *Der Organismus der Infusionsthiere,* Part 3(1). Engelmann, Leipzig. 154pp.

Stosch, H.A. von 1967. Haptophyceae. In: Ruhland, W. (ed.) *Sexuality, Reproduction, Alternation of Generations.* Encyclopedia of Plant Physiology 18. Springer, Berlin. pp. 646–56.

Tippit, D.H., Pillus, L. & Pickett-Heaps, J.D. 1980. Organization of spindle microtubules in *Ochromonas danica. J. Cell Biol.* **87**: 531–45.

Valkanov, A. 1962. Über die Entwicklung von *Hymenomonas coccolithophora* Conrad. *Rev. Algol., N.S.* **6**: 218–26.

Veer, J. van der 1970. *Ankylochrysis luteum* (Chrysophyta), a new species from the Tamar Estuary, Cornwall. *Acta Bot. Neerl.* **19**: 616–36.

Vesk, M. & Jeffrey, S.W. 1987. Ultrastructure and pigments of two strains of the picoplanktonic alga *Pelagococcus subviridis* (Chrysophyceae). *J. Phycol.* **23**: 322–36.

Vesk, M., Hoffman, L.R. & Pickett-Heaps, J.D. 1984. Mitosis and cell division in *Hydrurus foetidus* (Chrysophyceae). *J. Phycol.* **20**: 461–70.

West, G.S. & Fritsch, F.E. 1927. *A Treatise on the British Freshwater Algae* (new and revised edition by F.E. Fritsch). Cambridge University Press, Cambridge. 534pp.

Zimmermann, B., Moestrup, Ø. & Hällfors, G. 1984. Chrysophyte or heliozoon: ultrastructural studies on a cultured species of *Pseudopedinella* (Pedinellales ord. nov.), with comments on species taxonomy. *Protistologica* **20**: 591–612.

4

Current status of chrysophyte 'splinter groups': synurophytes, pedinellids, silicoflagellates

ØJVIND MOESTRUP

Introduction

Classification of the chrysophytes *sensu lato* is presently in a state of flux which increasingly recalls the problems encountered in classification of the green algae. Christensen (1980) included all chrysophytes in a single class Chrysophyceae, with seven orders. A much more restricted concept was suggested by Hibberd (1976) who proposed to include in the Chrysophyceae only species with the *Ochromonas* type of organization. Hibberd (1986) excluded silicoflagellates, pedinellids and bicosoecids, and his ideas were accepted by Kristiansen (1986, 1990). Subsequently, Andersen (1987), on the basis of pigment and ultrastructural data, suggested separating the *Synura* group into a class of its own, the Synurophyceae.

Below I first summarize some of the features of the pedinellids, synurophytes and silicoflagellates, as presently known. A classification into three classes is then suggested. It is argued that photoautotrophs such as *Ochromonas* and its allies in the Ochromonadales are not necessarily the most primitive chrysophytes. It is equally or even more parsimonious, considering the endosymbiosis theory for the origin of chloroplasts, that heterotrophic, apoplastidic species such as members of the *Bicosoeca* group or, less likely, apoplastidic pedinellids such as *Actinomonas* and *Pteridomonas*, may represent the most primitive chrysophytes.

The idea of the *Bicosoeca* group as a separate class is no longer justified. Members of this group are closely related to the Ochromonadales, differing mainly in the presence of non-silicified resting stages instead of silicified stomatocysts. This situation has a parallel in the Haptophyceae (see below). The flagellar apparatus of the Bicosoecales is strikingly similar to that of *Ochromonas* and its allies, and the most primitive chrysomonads may be non-loricate bicosoecids such as *Cafeteria* and *Pseudobodo*.

The individual 'splinter groups'

Synurophytes

Number of genera/species The Synurophyceae was proposed as a class for *Synura*, *Mallomonas* and their allies by Cavalier-Smith (1986), based on ideas of Andersen (1987). The class comprises at least four genera (Table 4.1) with *c*. 116 species. Synurophytes are unicellular or colony-forming flagellates which occur almost exclusively in freshwater. They often form blooms (see Nicholls, this volume).

Table 4.1. *Synurophytes*

Synura Ehrenberg 1833	14 spp.
Mallomonas Perty 1852	*c*. 100 spp.
Chrysodidymus Prowse 1962	1 sp.
Tessellaria Playfair 1918	1 sp.

Chloroplasts Usually two per cell; sometimes one large bilobed chloroplast.

Girdle lamella Present.

Pyrenoids Rarely reported, but in *Synura sphagnicola* a single large pyrenoid bulges from the inner face of each chloroplast (Hibberd 1978). Two similarly located pyrenoids are present in the bilobed chloroplast of *Mallomonas tonsurata* (Zhang *et al.* 1990).

Diagnostic pigments Chlorophylls *a*, c_1, and fucoxanthin.

Flagella Two flagella inserted anteriorly. The long flagellum carries tripartite hairs in two opposed rows. The flagella are inserted in an almost parallel fashion to each other. An angle of *c*. 15° is illustrated between the basal bodies of *Synura uvella* (Andersen 1987), *c*. 10° in *Synura petersenii* (Manton 1955), and from a few degrees to *c*. 20° in the published micrographs on *Tessellaria* (Pipes *et al.* 1991); they are practically parallel in *Mallomonas tonsurata* (Zhang *et al.* 1990). Manton (1955) found a loose-fitting sheath around the hairy flagellum of *Synura petersenii*, but this was not confirmed by Schnepf & Deichgräber (1969). More basally located swellings (sometimes wing-like) have been illustrated in both flagella of *Tessellaria* (Pipes *et al.* 1991) and *Synura uvella* (Andersen 1987).

Flagellar scales One or both flagella sometimes covered with organic annular or rod-like scales manufactured in the cisternae of the Golgi apparatus (e.g. Hibberd 1973; Moestrup 1982). The scales may also be present on the cell body (Moestrup 1982).

Transitional helix A notably long transitional helix, consisting typically of six to nine gyres, occurs in *Synura* and *Tessellaria*. No transitional helix was detected in *Mallomonas tonsurata* (Zhang *et al.* 1990).

Flagellar connecting fibres Two or three connecting fibres appear to be present (e.g. Andersen 1985; Zhang *et al.* 1990; Pipes *et al.* 1991).

Flagellar roots A crossbanded root (rhizoplast) extends from the flagellar bases to the nucleus, passing along the nuclear surface for some distance and probably terminating here. A microtubular root extends from the rhizoplast and forms a clockwise loop around the flagella (the r_1 root). Numerous cytoskeletal microtubules arise from this root. A second microtubular root is also present, probably the homologue of the r_3 root of other heterokonts. *Tessellaria* appears to lack the latter root (Pipes *et al.* 1991).

Photoreceptor Eyespot absent. The base of the smooth flagellum may carry a large swelling with electron-opaque contents (*Mallomonas, Synura*; absent in *Tessellaria*). It is not associated with the cell surface.

Contractile vacuoles In the cell posterior.

Nucleus/chloroplast connections Little confluence between the nuclear and chloroplast envelopes (see drawing in Mignot & Brugerolle 1982).

Body scales Silica scales and bristles, formed endogenously in cisternae located on the surface of a chloroplast (*Mallomonas, Synura*) or distributed throughout the cell (*Tessellaria*).

Tentacles Absent.

Mitosis A few micrographs were provided by Andersen (1989), who found a partially intact nuclear envelope at metaphase. The replicated flagella were located over the mitotic poles, associated with the poles via the replicated rhizoplasts.

Sexual reproduction Isogamy (hologamy). Gametes fuse at their posterior ends (reference in Andersen 1987).

Cysts Resting stages are stomatocysts, i.e. endogenously formed siliceous cysts with a small pore and an organic plug.

Nutrition Photoautotrophs. Mixotrophy and osmotrophy are unknown.

Classification Until recently, synurophytes were classified in the family Mallomonadaceae (= Synuraceae) of the Ochromonadales (class Chrysophyceae). Andersen, however, impressed by the differences in ultrastructure between the Mallomonadaceae and chrysophytes such as *Ochromonas*, suggested that they form a separate class Synurophyceae, with the single order Synurales (Andersen 1987).

Main papers Andersen (1987). The genus *Mallomonas* was recently treated in detail by Asmund & Kristiansen (1986: a taxonomic survey) and Siver (1991): morphology, taxonomy and ecology).

Pedinellids

Number of genera/species The pedinellids constitute a well-defined group of flagellates, presently comprising nine genera and *c.* 26 species (Table 4.2). There is considerable uncertainty regarding the number of species, due to a lack of reliable morphological characters. All pedinellids are unicellular with a single anterior flagellum, but a second flagellum is present internally as a very short basal body. Pedinellids occur in both marine and freshwater environments, sometimes as blooms (Thomsen 1988: 75×10^6 cells per litre of *Pseudopedinella tricostata* under the ice in Denmark 1985).

Chloroplasts Pedinellids differ from all other chrysophytes in chloroplast conditions. They constitute the only group of heterokont flagellates in

Table 4.2. *Pedinellids*

Actinomonas Kent 1880
A. mirabilis Kent 1880
A. vernalis Stokes 1885
Apedinella Throndsen 1971
A. radians (Lohmann) Campbell 1973
Ciliophrys Cienkowski 1876
C. infusionum Cienkowski 1876
C. australis Schewiakoff 1893
Cyrtophora Pascher 1911
C. pedicellata Pascher 1911
Palatinella Lauterborn 1906
P. cyrtophora Lauterborn 1906
P. minor (Wisłouch) Matvienko
Parapedinella Pedersen & Thomsen 1986
P. reticulata Pedersen & Thomsen 1986
Pedinella Wyssotzki 1887
P. hexacostata Wyssotzki 1887
Pseudopedinella Carter 1937
P. pyriformis Carter 1937
P. elastica Skuja 1948
P. tricostata (Rouchijajnen) Thomsen 1987
P. erkensis Skuja 1948
P. ambigua Bourrelly 1957
Pteridomonas Penard 1890
P. pulex Penard 1890
P. danica Patterson & Fenchel 1985
P. salina Bourrelly 1953
[*Stylomonas* Korshikov 1926]

which some members (*Pseudopedinella, Pedinella, Apedinella*) possess chloroplasts (three or six) while others (*Actinomonas, Ciliophrys, Parapedinella, Pteridomonas*) are entirely without chloroplasts. The incompletely studied genera *Cyrtophora* and *Palatinella* were described as having a single lobed chloroplast.

Girdle lamella Present.

Pyrenoids Present in *Apedinella* and *Pseudopedinella*, one per chloroplast.

Diagnostic pigments Chlorophylls a, c_1, c_2, fucoxanthin (Andersen 1989).

Flagella   Long flagellum extended into a wing supported by an internal rod. The wing is well developed in *Actinomonas, Apedinella, Parapedinella, Pedinella,* and *Pseudopedinella.* It is lacking in *Pteridomonas,* which has a very reduced paraxonemal rod. The long flagellum in all genera is covered by tripartite flagellar hairs in two opposed rows. The two flagella are inserted at a slight angle to each other: *c.* 17° in *Pseudopedinella tricostata* (Thomsen 1988: fig. 19), *c.* 20° in *Pteridomonas danica* (Patterson & Fenchel 1985: fig. 9).

Flagellar scales   Small annular, probably unmineralized scales are present on the flagellar surface of *Actinomonas mirabilis, Parapedinella reticulata* and *Pteridomonas danica* (Larsen 1985; Pedersen *et al.* 1986; N. Vørs, personal communication).

Transitional helix   Absent. A somewhat similar structure, probably two rings, is located underneath the transverse septum in *Apedinella radians, Pteridomonas danica* and *Ciliophrys infusionum.*

Flagellar connecting fibers   Not known.

Flagellar roots   Microtubular roots unknown. Fine fibers emanating from the basal bodies to the nuclear surface may represent a rhizoplast homologue (Hibberd 1979).

Photoreceptor   Eyespot absent.

Contractile vacuoles   Several contractile vacuoles were seen by Swale (1969) in *Pedinella hexacostata,* usually in the anterior part of the cell but also further back.

Nucleus/chloroplast connections   The outer nuclear membrane is continuous with the outer chloroplast membrane in or near the pyrenoid region (*Apedinella, Pseudopedinella*). No connections were mentioned or illustrated in the pyrenoid-lacking species *Pedinella hexacostata* (Swale 1969).

Body scales   The cell body of *Parapedinella* and *Apedinella* is covered with non-mineralized scales formed in the cisternae of the Golgi apparatus.

Tentacles Pedinellids differ from all other heterokont algae by the presence of tentacles supported internally by triads of microtubules. The microtubules emanate from the nuclear surface, the proximal terminus embedded in dense material.

Mitosis Not studied in detail. In *Apedinella* (Moestrup, unpublished data) the metaphase nucleus has large gaps in the nuclear envelope. Each mitotic pole contains a large Golgi body, but a rhizoplast (as in *Ochromonas*) was not seen. Further data on mitosis are lacking.

Sexual reproduction Unknown.

Cysts Resting stages (cysts) are found in several genera. They appear to be unmineralized. Thomsen (1988) found no trace of silica in the cysts of *Pseudopedinella tricostata*, examined by X-ray analytical electron microscopy. There is a single record of stomatocysts of chrysophyte type in *Palatinella*, but the phylogenetic affinities of this genus are obscure.

Nutrition The chloroplast-containing genera *Apedinella* and *Pseudopedinella* are autotrophic. *Pedinella* is mixotrophic, while genera lacking a chloroplast, (*Actinomonas, Ciliophrys, Parapedinella* and *Pteridomonas*), are phagotrophic. Two additional genera, *Cyrtophora* and *Palatinella*, contain chloroplasts, but their affinities to the pedinellids are obscure due to lack of ultrastructural studies.

Classification The pedinellids may be classified in the class Chrysophyceae as a family of the order Ochromonadales (Christensen 1980), or as a separate order Pedinellales (Zimmermann *et al.* 1984). The phagotrophic pedinellids were included by Lee *et al.* (1985) in the protozoan class Heliozoea (forming the order Ciliophryida), almost certainly a polyphyletic class (Patterson & Fenchel 1985). In other classifications pedinellids are treated as a separate class, for which the correct botanical name is Pedinellophyceae Cavalier-Smith 1986 (= Pedinellea Cavalier-Smith 1986: incorrect ending according to the Botanical Code of Nomenclature). Pedinellophyceae Kristiansen 1986 is invalid (no Latin diagnosis).

Main papers Pseudopedinella: Zimmermann *et al.* (1984), Thomsen (1988); *Actinomonas*: Larsen (1985); *Pteridomonas*: Patterson & Fenchel (1985); *Apedinella*: Koutoulis *et al.* (1988); *Ciliophrys*: Davidson (1982); *Pedinella*: Swale (1969); phylogeny of the Pedinellales: Smith & Patterson (1986).

Silicoflagellates

Number of species The silicoflagellates constitute a small group of marine flagellates which occur in all seas. They may occasionally attain bloom proportions; 303 000 cells per litre were recorded by Tangen (1974) from coastal waters off Norway, and 2.6 million cells per litre in the Kieler Bight (Neuer *in* Moestrup & Thomsen 1990). Silicoflagellates are unicellular flagellates characterized particularly by the presence of an external silicified skeleton. The morphology of the skeleton is used as a specific character. The few extant species, three according to Deflandre (1952) and Moestrup & Thomsen (1990) (Table 4.3), were referred by these authors to a single genus, *Dictyocha*. Several fossil genera are known.

Table 4.3. *Silicoflagellates*

Dictyocha Ehrenberg 1837
D. fibula Ehrenberg 1837
D. speculum Ehrenberg 1837
D. octonaria Ehrenberg 1844
[*D. crux* Ehrenberg 1840?]

Chloroplasts Each cell contains numerous chloroplasts.

Girdle lamella Present.

Pyrenoids Each chloroplast has a pyrenoid.

Diagnostic pigments Chlorophylls *a* and *c*, carotenes, diadinoxanthin, fucoxanthin, diatoxanthin, lutein (van Valkenburg 1970).

Flagella Two unicellular stages are known in the life cycle of both *Dictyocha fibula* and *D. speculum*. Both stages are biflagellate (confirmed in *D. speculum* only), with a long hairy anterior flagellum and a very short second flagellum. The short flagellum of the skeleton-bearing stage is reduced to an internal basal body. The long flagellum is extended into a wing supported by a dense rod. The angle between the basal bodies is 30–35° in the skeleton-bearing stage, somewhat less in the naked stage.

Flagellar scales Unknown.

Transitional helix Absent.

Flagellar connecting fibers, flagellar roots Unknown. Very short microtubular roots are perhaps present. Thin fibrillar structures may represent a rhizoplast.

Photoreceptor Eyespot absent.

Contractile vacuoles Absent.

Nucleus/chloroplast connections No direct membrane connection.

Body scales/skeleton Scales absent, but external silicified skeleton present. Ultrastructural studies on the origin of the skeleton have yet to be performed.

Tentacles The skeleton-bearing stage has numerous, sometimes branching tentacles emanating from all parts of the cell surface. The tentacles are supported internally by bundles of microtubules which attach onto the nuclear envelope. There is no fixed number of microtubules per tentacle.

Mitosis Ultrastructural studies are lacking.

Cysts/life cycle Double skeletons have been interpreted as zygotes but cell fusion has not been seen. The double skeletons almost certainly represent dividing cells. Cysts are unknown. The life cycle comprises at least three morphological stages: a uninucleate skeleton-bearing stage, a uninucleate naked stage, a naked multinucleate stage up to 500 μm in diameter, and a multinucleate amoeboid stage (Moestrup & Thomsen 1990). Attempts to complete the life cycle in the laboratory have failed (Henriksen *et al.* 1993).

Nutrition Photoautotroph. Mixotrophy is unknown.

Main papers van Valkenburg & Norris (1970), Moestrup & Thomsen (1990), Henriksen *et al.* (1993).

Classification of the 'splinter' groups

Many of the presently used classification systems of chrysophytes *sensu lato* treat the order Ochromonadales and the family Ochromonadaceae as the most primitive chrysophytes. This is based on the assumption that autotrophic/mixotrophic genera such as *Ochromonas* and *Chromulina* represent the most primitive chrysophytes, an idea which goes back to the time of Pascher (1910, 1914). It does not reflect present ideas on the origin of chloroplasts, however. According to the endosymbiont theory that has been advocated particularly since about 1970, chloroplasts represent an autotrophic prokaryotic or eukaryotic endosymbiont phago-cytosed by a heterotrophic (phagotrophic) flagellate and subsequently transformed into a chloroplast. If this theory applies also to chrysophytes, the ancestors of this group should probably be sought among phagotrophic heterokont flagellates, which lack all trace of a chloroplast.

Very few phagotrophic heterokonts are known, and only two groups, the bicosoecids and the pedinellids, include species which lack chloroplasts entirely. Species of the genera *Paraphysomonas* and *Spumella* (= *Monas*) are also phagotrophic, but all species examined so far possess a 'leuko-plast', which appears to represent a reduced chloroplast (Moestrup & Andersen 1991). The pedinellids and bicosoecids will be discussed separately.

Pedinellids are unique among the Chrysophyceae *sensu lato* in com-prising both autotrophic and mixotrophic forms as well as genera in which the few known species lack all traces of a chloroplast. It appears unlikely, however, that the pedinellids represent the ancestral stock from which other chrysophytes evolved. This is because the flagellar apparatus of pedinellids differs markedly from that found in other heterokont groups (brown algae, xanthophytes, oomycetes, hyphochytriomycetes, etc). The four microtubular flagellar roots typically present in these groups appear to be missing in pedinellids, and a rhizoplast is also absent, or present in a very reduced form. Also, ultrastructural data have failed to indicate whether heterotrophic or autotrophic forms represent the most primitive pedinellids (Smith & Patterson 1986). The pedinellids resemble silico-flagellates in having microtubule-supported tentacles, in which the micro-tubules are nucleated on the nuclear envelope, and the same feature has now been discovered in the monotypic aberrant chrysophyte genus *Rhizochromulina* (O'Kelly & Wujek, this volume). Additional similarities between the three groups (pedinellids, silicoflagellates, *Rhizochromulina*) include three features not found in other groups of heterokonts: (1) the

Table 4.4. *Bicosoecids*

Cafeteria Fenchel & Patterson 1988
   C. roenbergensis Fenchel & Patterson 1988
   C. ligulifera Larsen & Patterson 1990
   C. marsupialis Larsen & Patterson 1990
   C. minuta (Ruinen) Larsen & Patterson 1990

Pseudobodo Griessmann 1913
   P. tremulans Griessmann 1913
   P. minimus Ruinen 1938

Bicosoeca James-Clark 1866
   B. gracilipes James-Clark 1867
   B. mignotii Moestrup, Thomsen & Hibberd 1992
   37 additional species as listed by Preisig *et al.* (1991)

---

flagellar transition region contains a ringlike structure (perhaps a ring of opaque bodies) outside the axoneme in the pedinellid *Actinomonas* (Larsen 1985), the silicoflagellate *Dictyocha* (Moestrup & Thomsen 1990) and in *Rhizochromulina* (O'Kelly & Wujek, this volume); (2) the same region contains a double ring below the transitional plate in three pedinellids and in *Rhizochromulina*; (3) a flagellar wing supported by a dense rod is present in pedinellids and silicoflagellates. Present evidence thus indicates that the three groups form a natural cluster. In the classification suggested in Table 4.5, the three groups are classified as orders within the class Dictyochophyceae. The order Rhizochromulinales is somewhat intermediate between the Pedinellales and the Dictyochales, lacking, as do the pedinellids, a silicified skeleton but possessing microtubular tentacles similar to those of the silicoflagellates. The class Dictyochophyceae (as circumscribed here) occupies a somewhat isolated position among the heterokont protists.

The bicosoecids, a group presently comprising 45 described species (Table 4.4), represent the other group of heterokont flagellates in which a chloroplast is entirely absent, and members of this group are more likely to represent or to be related to the ancestors of autotrophic chrysophytes such as *Ochromonas*. Careful examination of published micrographs has shown considerable similarity between the flagellar apparatus of bicosoecids and chrysophytes such as *Epipyxis*, *Ochromonas* and *Chrysosphaerella*. The flagellar roots of the latter organisms have been termed r_1, r_2, r_3 and r_4 (Andersen 1987). The r_1 and r_2 roots are associated with the hairy (anterior) flagellum, while roots r_3 and r_4 emanate from the base of the smooth, often posterior flagellum. The r_1 root serves as a microtubule organizing center, and the equivalent root is visible also in published micrographs of bicosoecids: *Bicosoeca maris* (Moestrup & Thomsen 1976 'root 1') and *Cafeteria* (Fenchel & Patterson 1988: fig. 8, 'microtubule ribbon 4') in both cases serving as microtubule organizing centers (for 'microtubule ribbon 5' in *Cafeteria*). The equivalents of r_2 were also illustrated in *B. maris* ('root 3' in Moestrup & Thomsen 1976) and, although not commented upon, may also be seen in Fenchel & Patterson's (1988) fig. 8 of *Cafeteria*. Root 3 of the chrysophyte has recently been shown to function in the most exceptional way. In mixo-trophic species such as *Epipyxis*, and probably in all mixotrophic species, it functions in food uptake, the root microtubules sliding along each other to form a loop through which the prey is engulfed (Andersen & Wetherbee 1992). In bicosoecids there is little doubt that the large L-shaped microtubular root (8 + 3 microtubules in *B. maris*) is the homologue of the r_3 root in *Epipyxis*. One end of the root is associated with the hairy flagellum basal body (Mignot 1974; Moestrup & Thomsen 1976), but the opposite side of the root comes very close to the smooth flagellum basal body. Distally, the root forms a loop which supports the peristome through which prey is ingested. Interestingly, in *Chrysosphaerella* one of the microtubules of this root (the 'f' microtubule) is associated with a thin curved plate (Andersen 1990; figs. 12–16) and the apparently homologous microtubule of *B. maris* (microtubule 1) also carries a thin curved plate. The r_4 root of *Epipyxis* and other autotrophic chrysophytes has its homologue also in the bicosoecids ('r_2' in *B. maris*: Moestrup & Thomsen 1976; 'band 3' in *Cafeteria*: Fenchel & Patterson 1988).

   The similarities between the flagellar apparatus of bicosoecids and the phototrophic chrysophytes are not reflected in present classifications. Bicosoecids are often considered to represent a separate class of somewhat uncertain affinity to the phototrophic forms (Hibberd 1986). As shown above, however, a careful comparison of the flagellar apparatus indicates a close relationship between the two groups. The main difference is the presence, in many of the phototrophic chrysophytes, of a silicified cyst, the stomatocyst. Cysts of bicosoecids have not been studied in detail, but are believed to be organic (Klug 1936; Fott 1946; Willén 1963). The phylogenetic importance of this feature is somewhat questionable, however. Very similar silicified cysts have been described in two genera of the Haptophyceae, *Isochrysis* and *Dicrateria*, by Parke (1949), and these are almost certainly homologous with chrysophycean stomatocysts. *Isochrysis* and *Dicrateria* are often considered to be primitive members of the

Table 4.5. *Proposed classification of chrysophytes*
sensu lato

Class 1. Chrysophyceae Pascher 1914
  Order 1. Bicosoecales Grassé 1926
    Cafeteriaceae fam. nov.
    Bicosoecaceae Stein 1878
  Order 2. Chromulinales Pascher 1910
  :

Class 2. Synurophyceae Cavalier-Smith 1986
  Order 1. Synurales Andersen 1987

Class 3. Dictyochophyceae Silva 1982
  Order 1. Pedinellales Zimmermann *et al.* 1984
  Order 2. Rhizochromulinales O'Kelly & Wujek, this volume
  Order 3. Dictyochales Haeckel 1894

Haptophyceae, the two known species of *Dicrateria* lacking all indications of a haptonema. More advanced genera such as *Prymnesium* form cysts in which the wall is composed of layers of scales with electron-dense siliceous material deposited on the outer scales (Pienaar 1980, 1981). It appears likely that the common ancestors of the haptophytes and chrysophytes possessed a stomatocyst which was subsequently lost independently in several lines, including the bicosoecids.

There is, in my opinion, no longer any justification for considering the *Bicosoeca* group a separate class. The bicosoecids and the ochromonadalean chrysophytes are very similar to each other; the *Bicosoeca* group may represent the ancestral stock (or, because of the non-silicified cysts, a relative of the ancestral stock) which by phagocytosis took up a eukaryotic prey (as in present-day *Bicosoeca*: fig. 55 in Moestrup & Thomsen 1976) that was subsequently transformed into a chloroplast.

The classification in Table 4.5 proposes the synurophytes as a separate class, although it is clear that the Chrysophyceae and the Synurophyceae share many features. It may be argued that the synurophytes should be considered an order within the Chrysophyceae. I currently prefer a three-class system: (1) the Chrysophyceae with the ancestral type of heterokont flagellar apparatus including four microtubular flagellar roots; (2) the Synurophyceae with a somewhat modified system including only two microtubular roots; and (3) the Dictyochophyceae without distinct roots. The Synurophyceae is apparently a derived group, the parallel or nearly parallel insertion of the flagella and the lack of two microtubular roots having been derived from the type of flagellar apparatus which characterizes most other heterokonts.

In the classification in Table 4.5 I have included the *Bicosoeca* group as an order with two families: Cafeteriaceae fam. nov., whose few known members lack a lorica (the genera *Cafeteria* and *Pseudobodo*), and Bicosoecaceae with the loricate genus *Bicosoeca*. The establishment of the new family Cafeteriaceae is a parallel to the classification of autotrophic chrysophycean forms (Chromulinaceae (naked) versus Dinobryaceae (loricated)) (see Preisig, this volume).

Appendix

Cafeteriaceae fam. nov.

Biflagellate apochlorotic chrysophytes in which the posterior smooth flagellum may be used for temporary attachment. Cells with more or less prominent lip, used in food uptake. Lorica absent – a naked parallel to the Bicosoecaceae. Type genus: *Cafeteria* Fenchel & Patterson.

Chrysophyceae biflagellatae, apochloroticae saepe flagello posteriore glabro ad tempus affixae, alimenta labello ingestorio majore vel minore recipientes. Bicosoecaceis affines sed loricis carentes. Genus typificum: *Cafeteria* Fenchel & Patterson.

Summary

The algal flagellate groups commonly referred to as synurophytes, pedinellids and silicoflagellates are briefly reviewed. A three-class system for classification of chrysophytes *sensu lato* is proposed comprising Chrysophyceae, Synurophyceae and Dictyochophyceae. The bicosoecids, sometimes considered to represent a class of their own, are shown to be very similar to autotrophic/mixotrophic members of the Chrysophyceae, and it is argued that the most primitive members of the Chrysophyceae – and perhaps of the chrysophytes *sensu lato* – may be chloroplast-lacking, non-loricate members of the Bicosoecales.

References

Andersen, R.A. 1985. The flagellar apparatus of the golden alga *Synura uvella*: four absolute orientations. *Protoplasma* **128**: 94–106.
Andersen, R.A. 1987. Synurophyceae classis nov., a new class of algae. *Am. J. Bot.* **74**: 337–53.

Andersen, R.A. 1989. The Synurophyceae and their relationship to other golden algae. *Nova Hedwigia Bei.* **95**: 1–26.
Andersen, R.A. 1990. The three-dimensional structure of the flagellar apparatus of *Chrysosphaerella brevispina* (Chrysophyceae) as viewed by high voltage electron microscopy stereo pairs. *Phycologia* **29**: 86–97.
Andersen, R.A. & Wetherbee, R. 1992. Microtubules of the flagellar apparatus are active during prey capture in the chrysophycean alga *Epipyxis pulchra.* *Protoplasma* **166**: 8–20.
Asmund, B. & Kristiansen, J. 1986. The genus *Mallomonas* (Chrysophyceae). *Opera Bot.* **85**: 1–128.
Cavalier-Smith, T. 1986. The kingdom Chromista: origin and systematics. *Prog. Phycol. Res.* **4**: 309–47.
Christensen, T. 1980. *Algae, A Taxonomic Survey*, vol. 1. AiO Tryk, Odense, Denmark. 216pp.
Davidson, L.A. 1982. Ultrastructure, behavior, and algal flagellate affinities of the helioflagellate *Ciliophrys marina*, and the classification of the helioflagellates (Protista, Actinopoda, Heliozoea). *J. Protozool.* **29**: 19–29.
Deflandre, G. 1952. Classe des silicoflagellidés. In: Grassé, P.P. (ed.) *Traité de Zoologie*, vol. 1. Masson et Cie, Paris. pp. 425–38.
Fenchel, T. & Patterson, D.J. 1988. *Cafeteria roenbergensis* nov. gen., nov. sp., a heterotrophic microflagellate from marine plankton. *Marine Microbial Foodwebs* **3**: 9–19.
Fott, B. 1946. The planktonic species of the genus *Bicosoeca. Věst. Král. Spol. Nauk* **1944(19)**: 1–7.
Henriksen, P., Knipschildt, F., Moestrup, Ø. & Thomsen, H.A. 1993. Autecology, life history and toxicology of the silicoflagellate *Dictyocha speculum* (Silicoflagellata, Dictyochophyceae). *Phycologia* **32**: 29–39.
Hibberd, D.J. 1973. Observations on the ultrastructure of flagellar scales in the genus *Synura* (Chrysophyceae). *Arch. Mikrobiol.* **89**: 291–304.
Hibberd, D.J. 1976. The ultrastructure and taxonomy of the Chrysophyceae and Prymnesiophyceae (Haptophyceae): a survey with some new observations on the ultrastructure of the Chrysophyceae. *Bot. J. Linn. Soc.* **72**: 55–80.
Hibberd, D.J. 1978. The fine structure of *Synura sphagnicola* (Korsh.) Korsh. (Chrysophyceae). *Br. Phycol. J.* **13**: 403–12.
Hibberd, D.J. 1979. The structure and phylogenetic significance of the flagellar transition region in the chlorophyll *c*-containing algae. *BioSystems* **11**: 243–61.
Hibberd, D.J. 1986. Ultrastructure of the Chrysophyceae: phylogenetic implications and taxonomy. In: Kristiansen, J. & Andersen, R.A. (eds.) *Chrysophytes: Aspects and Problems.* Cambridge University Press, Cambridge, pp. 23–36.
Klug, G. 1936. Ein Beitrag zur Kenntnis von *Bicosoeca lacustris* J. Clark. *Arch. Protistenk.* **88**: 107–15.
Koutoulis, A., McFadden, G.I. & Wetherbee, R. 1988. Spine-scale reorientation in *Apedinella radians* (Pedinellales, Chrysophyceae): the microarchitecture and immunocytochemistry of the associated cytoskeleton. *Protoplasma* **147**: 25–41.
Kristiansen, J. 1986. The ultrastructural basis of chrysophyte systematics and phylogeny. *CRC Crit. Rev. Plant Sci.* **4**: 149–211.
Kristiansen, J. 1990. Phylum Chrysophyta. In: Margulis L., Corliss, J.,

90 *Chrysophyte algae*

Melkonian, M. & Chapman, D.J. (eds.) *Handbook of Protoctista*. Jones and Bartlett, Boston. pp. 438–53.

Larsen, J. 1985. Ultrastructure and taxonomy of *Actinomonas pusilla*, a heterotrophic member of the Pedinellales. *Br. Phycol. J.* **20**: 341–55.

Lee, J.J., Hutner, S.H. & Bovee, E.C. 1985 (eds.) *An Illustrated Guide to the Protozoa*. Society of Protozoologists, Lawrence, Kansas. 129pp.

Manton, I. 1955. Observations with the electron microscope on *Synura caroliniana* Whitford. *Proc. Leeds Philos. Lit. Soc., Sci. Sect.* **6**: 306–16.

Mignot, J.P. 1974 Etude ultrastructurale des *Bicosoeca*, protistes flagellés. *Protistologica* **10**: 543–65.

Mignot, J.-P. & Brugerolle, G. 1982. Scale formation in chrysomonad flagellates. *J. Ultrastruct. Res.* **81**: 13–26.

Moestrup, Ø. 1982. Flagellar structure in algae: a review, with new observations particularly on the Chrysophyceae, Phaeophyceae (Fucophyceae), Euglenophyceae, and *Reckertia*. *Phycologia* **21**: 427–528.

Moestrup, Ø. & Andersen, R.A. 1991. Organization of heterotrophic heterokonts. In: Patterson, D.J. & Larsen, J. (eds.) *The Biology of Free-Living Heterotrophic Flagellates*. Clarendon Press, Oxford. pp. 333–60.

Moestrup, Ø. & Thomsen, H.A. 1976. Fine structural studies on the flagellate genus *Bicosoeca*. I. *Bicosoeca maris* with particular emphasis on the flagellar apparatus. *Protistologica* **12**: 101–20.

Moestrup, Ø. & Thomsen, H.A. 1990. *Dictyocha speculum* (Silicoflagellata, Dictyochophyceae), studies on armoured and unarmoured stages. *K. vidensk. Selsk. Biol. Skr.* **37**: 1–57.

Parke, M. 1949. Studies on marine flagellates. *J. Mar. Biol. Assoc. U.K.* **28**: 255–86.

Pascher, A. 1910. Chrysomonaden aus dem Hirschberger Grossteiche. *Mon. Abh. Int. Rev. Ges. Hydrobiol. Hydrograph.* **1**: 1–66.

Pascher, A. 1914. Über Flagellaten und Algen. *Ber. Dtsch. Bot. Ges.* **32**: 136–60.

Patterson, D.J. & Fenchel, T. 1985. Insights into the evolution of heliozoa (Protozoa, Sarcodina) as provided by ultrastructural studies on a new species of flagellate from the genus *Pteridomonas*. *Biol. J. Linn. Soc.* **34**: 381–403.

Pedersen, S.M., Beech, P.L. & Thomsen, H.A. 1986. *Parapedinella reticulata* gen. et sp. nov. (Chrysophyceae) from Danish waters. *Nord. J. Bot.* **6**: 507–13.

Pienaar, R.N. 1980. Observations on the structure and composition of the cyst of *Prymnesium* (Prymnesiophyceae). *Proc. Electron Microsc. Soc. S. Africa* **10**: 73–4.

Pienaar, R.N. 1981. Ultrastructural studies on the cysts of *Prymnesium* (Prymnesiophyceae). *Phycologia* **20**: 112.

Pipes, L.D., Leedale, G.F. & Tyler, P.A. 1991. Ultrastructure of *Tessellaria volvocina* (Synurophyceae). *Br. Phycol. J.* **26**: 259–78.

Preisig, H., Vørs, N. & Hällfors, G. 1991. Diversity of heterotrophic heterokont flagellates. In: Patterson, D.J. & Larsen, J. (eds.) *The Biology of Free-Living Heterotrophic Flagellates*. Clarendon Press, Oxford. pp. 361–99.

Schnepf, E. & Deichgräber, G. 1969. Über die Feinstruktur von *Synura petersenii* unter besonderer Berücksichtigung der Morphogenese ihrer Kieselschuppen. *Protoplasma* **68**: 85–106.

Siver, P.A. 1991. *The Biology of Mallomonas: Morphology, Taxonomy and Ecology*. Kluwer, Dordrecht. 230pp.

Smith, R.McK. & Patterson, D.J. 1986. Analyses of heliozoan interrelationships: an example of the potentials and limitations of ultrastructural approaches to the study of protistan phylogeny. *Proc. R. Soc. Lond. B* **227**: 325–66.

Swale, E.M.F. 1969. A study of the nanoplankton flagellate *Pedinella hexacostata* Vȳsotskiĭ by light and electron microscopy. *Br. Phycol. J.* **4**: 65–86.

Tangen, K. 1974. Fytoplankton og planktoniske ciliater i en forurenset terskelfjord, Nordåsvatnet i Hordaland. PhD thesis, University of Oslo, 449pp.

Thomsen, H.A. 1988. Ultrastructural studies of the flagellate and cyst stages of *Pseudopedinella tricostata* (Pedinellales, Chrysophyceae). *Br. Phycol. J.* **23**: 1–16.

van Valkenburg, S.D. 1970. The ultrastructure of the silicoflagellate *Dictyocha fibula* Ehrenberg. PhD thesis, University of Washington, Seattle. 167pp.

van Valkenburg, S.D. & Norris, R.E. 1970. The growth and morphology of the silicoflagellate *Dictyocha fibula* Ehrenberg in culture. *J. Phycol.* **6**: 48–54.

Willén, T. 1963. Notes on Swedish plankton algae. *Nova Hedwigia* **5**: 39–56.

Zhang, X., Inouye, I. & Chihara, M. 1990. Taxonomy and ultrastructure of a freshwater scaly flagellate *Mallomonas tonsurata* var. *etortisetifera* var. nov. (Synurophyceae, Chromophyta). *Phycologia* **29**: 65–73.

Zimmermann, B., Moestrup, Ø. & Hällfors, G. 1984. Chrysophyte or heliozoon: ultrastructural studies on a cultured species of *Pseudopedinella* (Pedinellales ord. nov.), with comments on species taxonomy. *Protistologica* **20**: 591–612.

Part II
Development, physiology and nutrition

5

Comparative aspects of chrysophyte nutrition with emphasis on carbon, phosphorus and nitrogen

JOHN A. RAVEN

Introduction

Consideration of the nutrition of chrysophytes from a biochemical or biophysical viewpoint in comparison with that of other algal taxa has become relatively more difficult over the years. This is not necessarily a function of laxness in using presumptive chrysophytes for such work, but rather that many of the more widely used genera are now classified in other major taxa. The most obvious examples are the separation of the class Prymnesiophyceae, with such nutritionally well investigated genera as *Emiliania*, *Isochrysis* and *Pavlova*, and the recognition that *Olisthodiscus* (*Heterosigma*), a molecular-genetically and cell-biologically well understood organism, is probably a chloromonad (raphidophyte) (Heywood 1989; Patterson & van Valkenberg 1990). Even fewer data would be available if the chrysophytes *sensu stricto* were considered in the absence of the recently separated Synurophyceae. However, both of these classes (Chrysophyceae and Synurophyceae) are considered here, as are recently discovered marine picoplanktonic chrysophytes (Anderson 1987; Shapiro *et al.* 1989; Keller & Rice 1989), giving a range of organisms from marine picoplankton, through the more familiar freshwater unicells and colonies, to the large benthic freshwater *Hydrurus*.

To be more positive, there *is* a substantial body of data on various aspects of nutrition in the chrysophytes. We start by considering the evidence as to the ratio of major nutrients found in chrysophytes growing at their maximum specific growth rates, their qualitative allocation to different molecular species, and the mechanism of synthesis of these molecular species from such common intracellular substrates as hexose, ammonium and phosphate. The purpose of this analysis is to see whether resource requirements (energy, carbon, nitrogen, phosphorus, trace elements) for growth of chrysophytes are different from those of other algae.

We then turn to the forms in which the elements enter the cells and the entry mechanisms involved, and to the influence of the concentration of nutrients and the hydrodynamic regime in the medium on acquisition of elements. Finally, we consider how the mode of nutrition influences the maximum specific growth rate of chrysophytes and whether the occurrence of mixotrophy imposes a specific growth rate penalty relative to the rate attained by more nutritionally single-minded organisms.

Carbon nitrogen and phosphorus quota of chrysophytes in relation to growth rate

The comprehensive review by Sandgren (1988) suggests that the nitrogen and phosphorus requirements of photosynthetic chrysophytes are not significantly different from those of other (freshwater) planktophytes; this is shown for both Q_{max} and Q_0 for phosphorus in tables 2–3 of Sandgren (1988, as corrected by Sandgren, personal communication) with a molar C:P for Q_{max} in the range 43–144, and for Q_0 in the range 127–228 (see Turpin 1988). For the obligately chemoorganotrophic *Paraphysomonas imperforata* the optimal (for maximum specific growth rate) C:N:P molar ratio is 75:12:1 (Caron *et al.* 1990*a*). This can be compared with the Redfield ratio of 106:16:1 for photolithotrophic phytoplankton, and the C:N molar ratio of 7:1 for the photolithotrophic picoplankter *Aureococcus anophagefferens* (Dzurica *et al.* 1989).

While not considered in detail in this paper, data are also available on silicon (see Raven 1983) and on iron and copper (see Raven *et al.* 1989) requirements of chrysophytes. With respect to iron, chrysophytes, other Chromista, and red algae use only a soluble cytochrome c in electron transfer between cytochrome b_6-f and photosystem I in photosynthesis. This increases the iron requirement, but substantially decreases the copper requirement, relative to organisms (many Viridiplantae) which have only the cuproprotein plastocyanin as the redox agent mediating between the cytochrome b_6-f complex and photosystem I.

Metabolic pathways from hexose, ammonium and phosphate in chrysophytes

Taking triose/hexose, ammonium and phosphate as intracellular starting points for biosynthesis which are common to all trophic modes (photolithotrophy, mixotrophy, and the phagotrophic and saprotrophic versions of chemoorganotrophy exhibited by the chrysophytes), knowledge of the subsequent pathways is very patchy, even though the end-products are somewhat better characterized. Examples of end-products which have recently been reviewed or reported on include: lipids and sterols (Wood 1988; Bricelj *et al.* 1989; Patterson & van Valkenberg 1990), photosynthetic pigments (Jeffrey 1989; Bjornland & Liaaen-Jensen 1989; Bidigare 1989; Bidigare *et al.* 1990), the cellulosic component of the cell envelope (Haug 1974), the compatible solutes isofloridoside (Kauss 1967, 1974; Kremer 1980; Kremer & Kirst 1982) and β-dimethylsulphonium propionate (Keller *et al.* 1989), the vitamin ascorbic acid (which is synthesized via the pathway with inversion of the carbon chain of glucose such as occurs in animals, *Euglena* and the diatom *Cyclotella*: Helspar *et al.* 1982; Grün & Loewus 1984), and the photosynthetic enzyme ribulose bisphosphate carboxylase-oxygenase (Rubisco: Rothschild & Heywood 1987).

The extent to which there are distinctive chrysophyte lipids and photosynthetic pigments is not clear. For example, while *Ochromonas* was the first genus in which a chlorosulfolipid was found, similar compounds have subsequently been found elsewhere in algae (Wood 1988). In some cases, at least some enzymes of biosynthetic pathways have been characterized from chrysophytes (terpenoids: Coolbear & Threlfall 1989; isofloridoside: Kauss 1974); in another case (e.g. '$C_3 + C_1$' anaplerotic inorganic carbon fixation: Kauss & Kandler 1962), the nature of the enzymes involved was approached from the effects of vitamin deficiency on the auxotroph *Poterioochromas* (as *Ochromonas*) *malhamensis*.

Biochemistry of carbon incorporation

The chrysophytes have a range of processes by which exogenous carbon can be incorporated to produce the organic components of the organisms (see Holen & Boraas, this volume). It is possible that some chrysophytes are obligate photolithotrophs (i.e. cannot grow without light and carbon dioxide and have no stimulation of growth by organic compounds), although for the example quoted by Droop (1974: his table 19.5), i.e. *Mallomonas epithalatia*, the only two organic carbon sources tested were glucose and acetate: 500 mol m^{-3} glycerol was not tested (cf. Antia *et al.* 1969). Other photosynthetically competent chrysophytes, e.g. *Ochromonas danica* and *Poterioochromonas* (as *Ochromonas*) *malhamensis*, can grow in the dark on glucose (plus a few vitamins, etc.), i.e. as chemoorganotrophs. In addition to this saprotrophic growth, *Ochromonas* and *Poterioochromonas* spp. are also well-known phagotrophs (Myers & Graham 1956; Droop 1974; Andersson *et al.* 1989; Sanders 1991). Mixotrophy thus seems to be a major trophic mode in chrysophytes, although in the species which have been best examined the chemoorganotrophic mode dominates in terms of the achieved specific growth rate under conditions optimal for either chemoorganotrophy or photolithotrophy alone (see Myers & Graham 1956; Andersson *et al.* 1989; Caron *et al.* 1990*b*; Paran *et al.* 1991; Sanders 1991). Finally, there are non-photosynthetic chrysophytes which are obligate organotrophs, e.g. *Paraphysomonas* sp. (see Caron *et al.* 1990*a*).

   Photosynthetic incorporation of exogenous inorganic carbon has been investigated in *Poterioochromonas* (as *Ochromonas*) *malhamensis* (Kauss & Kandler 1963); short-term labelling with ^{14}C inorganic carbon shows that inorganic carbon is assimilated in the light via the photosynthetic carbon reduction cycle. Subsequent enzymatic evidence, again for *Ochromonas* spp., has demonstrated the presence of a Rubisco with a subunit composition (eight large, eight small) and subunit (53000 and 16500) and total (570000) molecular sizes (Rothschild & Heywood 1987) resembling those of other eukaryotes. It is not clear whether both subunits of chrysophyte Rubisco are plastid-encoded, as is the case in the (few) other chromophytes examined (Cattolico & Loiseaux-de Goër 1989; Kowallik 1989; also see Delancy *et al.*, this volume). Aside from a genetic approach to this question, it could also be addressed using the plastids competent in protein synthesis (but not in photosynthesis) recently isolated from *Poterioochromonas malhamensis* (Schurmann & Brandt 1991).

   As regards the thylakoid reactions and their energetic interactions with inorganic carbon fixation, 'state transitions' (the adjustment of excitation energy transfer to the two photoreaction centers as a function of the wavelength of photosynthetically active radiation: see Baker 1991) occur in *Ochromonas danica*. This is similar to findings in chlorophyll-*b*-containing organisms except certain micromonadophytes (probably including *Micromonas pusilla*, which was originally described as the chrysophyte *Chromulina pusilla*, with a chlorophyll-*c*-like pigment), in cyanobacteria and in red algae (Gibbs & Biggins 1989; Wilhelm 1990). However, other chromophyte algae tested (diatoms, prymnesiophytes, tribophytes and phaeophytes) either lack state transitions or, when they occur, they yield little in the way of changes in excitation energy delivery to the two photoreactions (see Gibbs & Biggins 1989; Raven *et al.* 1989; Wilhelm 1990; Fork *et al.* 1991). Lack of state transitions is presumably a function of the essentially identical transfers of excitation energy to the two photoreactions with static light-harvesting assemblies over the 400–700 nm range of excitation wavelengths (Gibbs & Biggins 1989; Raven *et al.* 1989). Why a chrysophyte should differ from the other, similarly pigmented chromophytes tested is not clear.

A final aspect of photosynthesis in chrysophytes that has been investigated involves the use of ^{18}O to separate oxygen uptake and oxygen evolution in illuminated photosynthetic cells. Weis & Brown (1959) used *Poterioochromonas* (as *Ochromonas*) *malhamensis* in establishing the now widely validated changes in oxygen uptake as a function of photon flux density that occurs in phototrophs with the C_3 pathway of inorganic carbon assimilation when Rubisco oxygenase is suppressed by high inorganic carbon concentrations. This involves a partial inhibition of the 'dark' respiration oxygen uptake as photon flux density increases from zero to $\sim 10\%$ of the photosynthesis-saturating value, and an increase in oxygen uptake to above the dark rate with further increases in photon flux density. This increase apparently involves the Mehler reaction in which oxygen uptake at the reducing end of photoreaction I is balanced by oxygen evolution at the oxidizing end of photoreaction II (see Badger 1985).

It is of interest to examine the stated motivation for using chrysophytes, and particularly *Poterioochromonas malhamensis*, in some of this work. The choice of *Poterioochromonas* (as *Ochromonas*) *malhamensis* by Weis & Brown (1959) was dictated by the intrinsically low ratio of light- and carbon-saturated photosynthesis to dark respiration of this organism, thus making effects on respiratory oxygen uptake more readily examined. Kauss & Kandler (1962, 1963) used *Poterioochromonas* (as *Ochromonas*) *malhamensis* as a convenient organism for studying the effects of thiamine and biotin deprivation on photosynthetic and non-photosynthetic inorganic carbon fixation; here it was the auxotrophic nature of *Poterioochromonas malhamensis* that attracted the investigators.

The biochemistry of photosynthesis in chrysophytes is thus not particularly remarkable, with the exception that excitation energy distribution between the photoreactions of the only chrysophyte tested (*Ochromonas danica*) differs from that of other chromophytes tested, and more closely resembles that of most chlorophyll-*b*-containing and phycobilin-containing phototrophs.

Turning to the alternative mode of carbon acquisition, i.e. chemoorganotrophy, the phagotrophic species can generally also have their organic carbon requirements met by soluble, low molecular weight organic carbon, i.e. saprotrophy (osmotrophy). The restrictions on the organic carbon compounds which can support saprotrophic growth (e.g. glucose but not acetate in *Poterioochromonas malhamensis* and *Ochromonas danica*) may reside in biochemistry (activation) constraints or (see below) limitations on entry across the plasmalemma. Similar possibilities, again untested, apply to the reason for obligate photolithotrophy of *Mallomonas epithalatia*.

Biochemistry of nitrogen incorporation

Allen (1969), Sandgren (1988) and Dzurica *et al.* (1989) have surveyed the nitrogen sources which can be used for growth of chrysophytes. The largely chemoorganotrophic *Poterioochromonas* (as *Ochromonas*) *malhamensis* uses amino acids and urea, while the more photolithotrophic chrysophytes can use nitrate as nitrogen source involving the use of nitrate and nitrite reductases, enzymes which have not yet been isolated from chrysophytes. The use of urea in *Poterioochromonas* (as *Ochromonas*) *malhamensis* involves secretion of urease (Sandgren 1988; see below).

Biochemistry of phosphate incorporation

While orthophosphate and various organic phosphates can serve as exogenous phosphorus sources for chrysophytes (Sandgren 1988; Dzurica *et al.* 1989), it is not clear whether phosphorus becomes available to intracellular enzymes uniquely as phosphate, or whether organic phosphates can enter the cells. We need to know more about the uptake of unhydrolyzed organic phosphates, and the role of the extracellular phosphatases of chrysophytes (Sandgren 1988).

Transport of carbon sources into chrysophytes

Little is known about the membrane transport processes involved in the uptake of either inorganic (Sandgren 1988; Veen 1991) or organic carbon sources by chrysophytes. For inorganic carbon, i.e. the $CO_2-HCO_3^--CO_3^{2-}$ in the medium supplying the CO_2 fixed in photosynthesis by Rubisco, the alternatives are diffusive entry of CO_2 through the lipid bilayer part of the membrane and (possibly) through protein channels (uniporters), and active entry via proteinaceous transporters which recognize and use an energy substrate (e.g. ATP or an ion gradient) to move CO_2 and/or HCO_3^- from the outside to the inside of the membrane against an electrochemical potential gradient for CO_2 and/or HCO_3^- (Raven 1991*a,b*). Few data are available for chrysophytes which permit a distinction between these various possibilities. Approaches which could be tried on chrysophytes include the following: gas-exchange correlates (inorganic carbon affinity, CO_2 compensation concentration, oxygen effects on photosynthesis) during steady-state photosynthesis; intracellular and extracellular carbonic anhydrase activity; comparison of rate of inorganic carbon assimilation from a medium with a high pH with a low equilibrium CO_2 concentration predicted from the known total inorganic carbon concentration via the Henderson–Hasselbalch equation, and not the higher dissolved CO_2 concentration predicted for air-equilibrium with the rate of uncatalyzed HCO_3^- to CO_2 conversion; isotope disequilibrium studies of CO_2 versus HCO_3^- as the species entering the cells in the natural or experimental absence of extracellular carbonic anhydrase (Raven 1991a,b).

In the absence of such data, we can follow Sandgren (1988) in suggesting that HCO_3^- use may not be present in freshwater chrysophytes which inhabit acidic water. Furthermore, we can appeal to comparative ecophysiology to suggest that the haptobenthic (equivalent to haptophytic *sensu* Luther 1949) *Hydrurus foetidus*, a large palmelloid chrysophyte living in fast-flowing cold waters enriched in CO_2 by chemoorganotrophic inputs to ground-water, has a purely diffusive CO_2 supply, as is the case for the freshwater red algae and bryophytes which live in similar habitats (Raven *et al.* 1986; MacFarlane & Raven 1985; 1989, 1990). However, measurements of the type outlined in the last paragraph are needed, not only because of the intrinsic interest of the mechanism of inorganic carbon entry, but also because the mechanism of CO_2 or HCO_3^- entry, and its implications for the functioning of Rubisco and the need for the glycolate-metabolizing pathways, can alter the photon, nitrogen, iron, manganese and molybdenum costs of growth (Raven 1991a,b).

Transport of nitrogen sources into chrysophytes

Little seems to be known of transport of nitrogen sources into chrysophytes (Raven 1980, 1986, 1990). However, data are available on the ammonium concentration needed to half-saturate ammonium net influx in the marine chrysophyte (class Dictyochophyceae, *sensu* Moestrup, this volume) unicell *Pseudopedinella*: 0.5 mmol m^{-3} (Harrison *et al.* 1989). Such low values are typical of small algal cells, and reflect both the occurrence of a high-affinity ammonium transport protein catalyzing an electrically driven ammonium uniport across the plasmalemma (perhaps an ammonium–sodium symporter in marine organisms), and small diffusion restrictions outside the cells (Raven 1980, 1984a, 1986).

Low molecular weight organic nitrogen compounds are being increasingly recognized as significant nitrogen sources for many microalgae in their natural habitats (Antia *et al.* 1991). In the case of urea, many microalgae have active transport mechanisms for urea in their plasmalemmas (Raven 1984a; Antia *et al.* 1991). While no data are available on such mechanisms in chrysophytes, it is of interest that extracellular urease is produced by *Poterioochromonas* (as *Ochromonas*) *malhamensis* (Sandgren 1988), so that the hydrolysis product ammonium could be the actual nitrogen species entering the cells. An analogous situation has been reported (alas, with no data on chrysophytes) for amino acids, where extracellular amino acid (and amine) oxidases yield ammonium, hydrogen peroxide and the deaminated oxo-analogue of the amino compound (Palenik & Morel 1990; Antia *et al.* 1991).

Clearly much work remains to be done on the role of extracellular hydrolytic and oxidative enzymes, and of specific transporters in the plasmalemma, in determining the kinds of nitrogen-containing molecules, and the range of their concentrations in the medium which supports chrysophyte growth. An interesting example of the latter is the coccoid marine picoplanktonic chrysophyte *Aureococcus anophagefferens* (Keller *et al.* 1989). Here the dominance of this 'brown tide' alga in mesocosms was relinquished in favor of algae with larger cells when the dissolved inorganic nitrogen concentration (ammonium plus nitrate) increased significantly above the < 1 mmol m^{-3} found in the brown tide in nature. As we shall see, small phytoplankton cells not only have a lower area-based requirement for influx of a given nutrient (e.g. NO_3^-) at the maximum specific growth rate, but also a smaller potential diffusive limitation of nutrient movement from the bulk phase to the cell surface (see Raven 1986).

It is also pertinent to note that the use of nitrates rather than ammonium as nitrogen sources increases the predicted iron, manganese and, especially, molybdenum cost of cell growth (see Raven 1991a,b).

Transport of phosphorus sources into chrysophytes

Inorganic phosphate is a good source of phosphorus for many chrysophytes, e.g. the freshwater species discussed by Sandgren (1988). Organic phosphates are also used by many chrysophytes (Sandgren 1988); in the marine picoplanktonic *Aureococcus anophagefferens* they can be used at higher external concentrations than can inorganic phosphate, and they offset the apparently toxic effects of high concentrations of inorganic phosphate (Dzurica *et al.* 1989). Some of the effects of organic phosphates on growth of *Aureococcus anophagefferens* could be a function of their action as chelating agents (Dzurica *et al.* 1989). Whether organic phosphates are necessarily hydrolyzed by the surface phosphatases commonly found in chrysophytes (Sandgren 1988) before phosphorus can be made available to the cytosol, i.e. with inorganic phosphate as the species entering the cell, is not clear. Certainly the half-saturation concentration of inorganic phosphate for growth (K_μ) of freshwater photolithotrophic planktonic chrysophytes (Sandgren 1988) in the range 0.005–0.025 μM for *Synura* and *Mallomonas*, and half-saturation concentrations for uptake (K_m) of 0.4–1.3 μM (Sandgren 1988), indicate a high-affinity uptake system for inorganic phosphate.

Transport of nutrient solutes from the bulk water phase to the surface of chrysophytes

During net uptake of nutrient solutes by an alga, the nutrient concentration at the surface of the organisms is less than that in even the most violently agitated and thus best-mixed bulk water phase because of the occurrence of a boundary or unstirred layer around the alga. This diffusion boundary layer is the distance through which diffusion alone (with no mass flow of nutrients and solvent together) is involved in solute transport, and, since net diffusive flux can only occur down a concentration gradient, the solute concentration at the cell surface must be less than that in the bulk medium.

To illustrate the factors determining the solute concentration gradient between the bulk medium and the cell surface, we use a version of Fick's law of diffusion in a planar system:

$$(C_b - C_s) = \frac{J \cdot L}{D}$$

where C_b = bulk phase concentration of solute (mol m^{-3})

C_s = cell surface concentration of solute (mol m^{-3})

J = net flux of solute from bulk phase to cell surface (mol m^{-2} s^{-1}) (=net flux of solute into the cell across the plasmalemma)

L = thickness of diffusion boundary layer (m)

D = diffusion coefficient of the solute in water (m^2 s^{-1})

With a constant value of D, $(C_b - C_s)$ increases if J is greater or if L is greater. We shall see that, since both J and L are lower in smaller phytoplankton cells, the $(C_b - C_s)$ value is also lower in these smaller cells, so that there is less likelihood of limitation of nutrient uptake by the diffusion boundary layer in the small cells.

For a phytoplankton cell or colony up to about 50 μm in diameter (i.e. the majority of planktonic chrysophytes), the thickness of the unstirred layer is numerically equal to the radius of the organism. This relationship holds for all of the naturally achievable velocities of movement of the organism relative to its immediate aqueous environment due to sinking of non-motile cells or swimming by flagellates, i.e. the unstirred layer thickness is not influenced by sinking or swimming. For nutrient-limited growth at a given fraction of their resource-saturated specific growth rate, smaller phytoplankton cells have smaller concentration differences between the bulk medium and the cell surface than do larger cells. This would occur even if the cell size dependence of specific growth rate in phytoplankton were such that the area-based influx of a given nutrient (mol m^{-2} plasmalemma per second) were independent of cell size, i.e. $\mu_m = ab^{-1/3}$ where μ_m is specific growth rate (m^3 m^{-3} s^{-1}), b is the cell volume (m^3) and a is a constant (m^{-2} s^{-1}). However, the best-investigated algae (diatoms and dinoflagellates) have exponent b values of between -0.12 and -0.16 rather than -0.33, indicating that the area-based nutrient influx is smaller at a given fraction of μ_{max} in smaller cells (Banse 1982; Raven 1986, 1987a). As we shall see below, there are insufficient data on chrysophytes to be sure that this larger (i.e. less negative) exponent of b is applicable to their phototrophic members. However, even with this uncertainty, the smaller unstirred layer thickness around smaller cells would still permit a given fraction of μ_m to be achieved by smaller cells with a smaller concentration difference between the bulk phase and the cell surface. This in turn means that, with identical transport characteristics of the plasmalemma (constant density of transporters, as moles transporter per square meter of membrane; constant specific reaction rate of transporters at substrate saturation, as moles solute transported per mole transporter per second; constant substrate affinity of the transporter for mediated entry; constant permeability coefficient and constant assimilatory enzyme properties (area-based content, specific reaction rate and affinity) for non-mediated entry), the smaller cell can grow at a given, nutrient-limited fraction of its specific growth rate with a lower steady-state bulk phase concentration of the limiting nutrient (cf. Keller *et al.* 1989).

Published values of bulk-phase solute concentration which yield half of the maximum specific growth rate (K_μ) for phosphorus (Sandgren 1988) and iron (Veen 1991), together with the μ_m value and cell size, suggest that external diffusion is not a major restraint on nutrient acquisition from low bulk-phase concentrations by planktonic chryso-phytes (see computations in Raven 1986, 1987*a*, 1988).

Turning to the haptobenthic (attached) phototrophic chrysophytes, the velocity of water movement relative to the organism can be much higher than the few hundred micrometers per second which characterize the environment of planktonic members. For attached algae, the velocity of water movement over the organism, together with the size, shape and flexibility of the organism and the extent to which it protrudes from the boundary layer around the substream to which it is attached, determine its effective diffusion boundary layer. An example of a haptobenthic chrysophyte which, by its life form and habitat, minimizes the diffusion boundary layer thickness over the thallus is *Hydrurus foetidus*. While no measurements of its specific growth rate, nutrient requirements and boundary layer thickness can be traced in the literature, the habitat and structure of *Hydrurus* (Parker *et al.* 1973; Vesk *et al.* 1984) suggest that the situation resembles that in red algae (e.g. *Lemanea*) and mosses (e.g. *Fontinalis*) growing in similar habitats (see MacFarlane & Raven 1985, 1989, 1990; Raven *et al.* 1986). The extent to which the benefits, in terms of improved carbon, nitrogen and phosphorus acquisition, of the com-bination of habitat (fast-flowing freshwater) and life form (branched elongate colony in a matrix) are offset by increased resource commitment to mechanical strengthening elements is unclear in *Hydrurus* or, indeed, in the red algae and mosses (see Raven 1989; Kraemer & Chapman 1991). An intriguing point about *Hydrurus* is that, despite the obvious structural capacity of the matrix material in resisting tension imposed by the drag of water moving at $> 1\,\mathrm{m\,s^{-1}}$ over the thallus, this tension-resistance is not employed in regulating the volume of the cells with their contents hyperosmolar to the freshwater medium in the manner of the turgor-resisting walls of plants (see Raven 1982, 1984*a*). Instead, the cells all have contractile vacuoles, the volume-regulating mechanism employed by freshwater organisms without a complete, turgor-containing wall (e.g. flagellate or amoeboid cells: Raven 1982, 1984*a*). The use of contractile vacuoles is energetically more economical than cell wall production during rapid growth, but walls are energetically better during slow, or no, growth (Raven 1982). How the energetic balance is struck in *Hydrurus*, which in any case has a tension-resistant matrix (see above), requires further investigation.

Interception of food particles by phagotrophic chrysophytes

Our consideration of the uptake of soluble nutrients showed that the motility so common in planktonic chrysophytes has little or no influence, beneficial or detrimental, on this process in a homogeneous environment, i.e. one with no gradients of nutrient solute concentration in the bulk phase. We shall see later that flagellar motility could help to optimize acquisition of light and of nutrients when the gradients of energy and chemical resources are, as is often the case, inverse (see Raven & Richardson 1984, for discussion in the context of dinoflagellates).

By contrast, motility is definitely an advantage for acquisition of particles by phagotrophic phytoplankton such as many chrysophytes (Fenchel 1982a,b; Veen 1991), because they employ the 'direct interception' (raptorial) capture method as defined by Fenchel (1986). We note that the 'filtering' protozoans also require relative water motion (swimming for planktonic organisms), while 'diffusion' feeding does not. Fenchel (1986) notes that raptorial feeding is favored if the food particle to predator size ratio exceeds 1:10, so that only flagellates less than 10 μm in diameter are effective bacterivores (food particles < 1 μm). Filter feeders are favoured by lower ratios of food to predator size (Fenchel 1986). Fenchel (1987) notes that filter feeding is enhanced by attachment of the predator to an object over which the water flow rate is greater than could be achieved in swimming by a free-living flagellate. This recalls arguments used earlier about the boundary layer thickness over phototrophs when free-floating (or swimming) or attached. We note that swimming, whether in phagotrophic or phototrophic flagellates, consumes only approximately 1% of the total energy transformed by rapidly growing cells (Fenchel & Finlay 1983; Raven 1982, 1984a, Raven & Richardson 1984).

Phagotrophic nutrition in obligately chemoorganotrophic chrysophytes in nature is normally the major source of all their nutrients; saprotrophy is much less important (Fenchel 1987). For mixotrophic chrysophytes, the extent of carbon acquisition by phagotrophy ranges from less than 10% at μ_m for *Dinobryon divergens* (Veen 1991) to a dominant role in many *Ochromonas* and *Poterioochromonas* species (e.g. Myers & Graham 1956; Andersson *et al.* 1989; Caron *et al.* 1990b; Sanders *et al.* 1990; also see Holen & Boraas, this volume). Phagotrophy apparently has a uniformly more significant role in nitrogen, phosphorus and iron supply to mixotrophic chrysophytes (Caron *et al.* 1990b; Veen 1991). However, comparison of the ratio of carbon to nitrogen, or phosphorus, or iron, in prey and predator (see above) shows that acquisition of significant quantities of elements other than carbon by phagotrophy also necessitates ingestion of significant, perhaps excessive, quantities of carbon (Raven, unpublished data).

The roles of flagella in chrysophyte nutrition

We have already seen that flagellar activity is not significant in the acquisition of soluble nutrients from a homogeneous bulk phase, but is important for phagotrophy. Raven & Richardson (1984) showed that phototrophic flagellates exhibiting diel migrations in a stratified water column between an upper, well-illuminated but nutrient-poor zone in the daytime and a lower, nutrient-rich zone at night can benefit in both energetic and nutritional terms over a stationary cell. Such vertical diel migrations are known for dinophyte, cryptomonad, chloromonad and chlorophyte flagellates (Watanabe *et al.* 1983; Raven & Richardson 1984; Salonen *et al.* 1984; Sommer & Gliwicz 1986) and have been reported for the chrysophytes *Mallomonas* and *Synura* (Jones 1988).

Another possible use of flagella in phototrophic phytoplankton organisms, including chrysophytes, relates to light absorption. Although backscatter in water bodies leads to a significant contribution of upwelling radiation, the major component is downwelling. This means that the part of a phytoplankton organism which faces the downwelling radiation has a greater incident photon flux density than does the part which faces the upwelling radiation. Depending on the size and the concentration of chromophores in the organism (as moles chromophore per cubic meter organism volume), the absorptance (fractional absorption) of vector (entirely downwelling) radiation is in the range 0.1–0.99 in the 400–700 nm wavelength range (Raven 1984*a*,*b*). This means that there is a substantial gradient in the rate of excitation of photosynthetic pigments for the larger chrysophytes between the illuminated side and the opposite side for stationary cells in a vector radiation field. Rotation about the appropriate axis (i.e. one at right angles to the incident radiation) would lead to temporal averaging of excitation of chromophores. This could, with an appropriate rate of rotation, increase photosynthesis under light-limited or light-saturated (but not photoinhibitory) conditions, and perhaps reduce photoinhibition (Raven 1984*a*,*b*; Laws *et al.* 1986; Osborne & Raven 1986; Greene & Gerard 1990; see also Pearcy 1990). The data on which these assertions are based involve alterations in pigment excitation frequency induced other than by rotation, and can involve alterations in the ratios of photosynthetic catalysts (Greene & Gerard 1990).

The possible relevance of these phenomena to flagellar motility is that many flagellates, including chrysophytes, rotate as they swim. Pienaar (1980) quotes rotation times of 0.25–1.25 s for pigmented chrysophytes; such times are within the range in which constructive effects of alterations of incident photon flux densities can be observed (Greene & Gerard 1990). Whether such gyrations (Pienaar 1980) have the roles suggested in maximizing photosynthetic gain, and the extent to which rotation by small-scale turbulence interferes with these roles, need further investigation. A start has been made by investigating the effect of externally imposed (i.e. additional to effects imposed by cell movements) fluctuating irradiance regimes in cultures of the non-motile *Aureococcus* (Milligan & Cosper 1991). Studies on the effects of small-scale turbulence on growth and photosynthesis of dinoflagellates (Thomas & Gibson 1990; Thomas *et al.* 1991) are also of relevance.

Aside from the role of gyrations in the swimming of flagellates (which, since it occurs in non-photosynthetic chrysophytes, means that the putative roles suggested above are 'emergent properties'), the occurrence of flagella and/or of phagotrophy in freshwater flagellates requires volume regulation by contractile vacuoles. While walled cells of some higher plants can regularly obtain free-living microorganisms to become endosymbionts (e.g. legumes internalizing rhizobia), a direct nutritional role of phagocytosis, as in chrysophytes, is inconsistent with the presence of walls. Raven (1982) has considered the energetics of growth at various specific growth rates for freshwater microalgae relying on contractile vacuoles or on cell walls for their volume regulation (see also the discussion of *Hydrurus* above, and compare motile vegetative cells of chrysophytes with stomatocysts).

Kinetic and energetic aspects of photolithotrophy and phagotrophy: are there disadvantages of mixotrophy?

Raven (1984*a,b*, 1986, 1987*a*) considered the restrictions on maximum specific growth rates which are imposed by the occurrence of a photolitho-trophic nutrition by organisms. The conclusion from these analyses is that additional catalytic and structural machinery is involved in generating reduced organic carbon from inorganic carbon using light energy relative to that which is needed for the uptake of soluble low molecular weight organic sources of carbon and energy for saprotrophs; this additional material which must be synthesized in each cell doubling, imposes a constraint on specific growth rate. The quantitative predictions are in general accord with the observed maximum specific growth rates of photolithotrophs and saprotrophs (Raven 1984*a,b*, 1986, 1987*a*), i.e. higher maximum specific growth rates for saprotrophic microorganisms than for similar-sized photolithotrophs when both are growing under optimal conditions (cf. Bell 1985, who presents a less specific analysis). Within each main trophic mode, reductions in maximum specific growth rate have been noted if resources which require more manipulation (e.g. nitrate rather than ammonium as nitrogen source) are supplied (Stacey *et al.* 1977; Turpin *et al.* 1985; Raven *et al.* 1992). These effects are most obvious for organisms with high maximum specific growth rates, i.e. those with least 'room' to house extra catalysts in cells already packed with essential catalysts.

It is likely that phagotrophy involves more additional catalysts than the simplest form of saprotrophy considered by Raven (1987*a*), since the use of exogenous low molecular weight chemoorganotrophic substrate requires only transport systems at the plasmalemma and a minimal set of assimilatory enzymes. The phagotroph must also have a range of transport and assimilatory catalysts which deal with the products of digestion within intracellular food vacuoles. Digestion presupposes digestive enzymes; such enzymes are, of course, produced by many saprotrophs (e.g. those using macromolecular, including solid, substrates), but in saprobes they are secreted into the growth medium rather than into the vacuoles (spaces which are, topologically, identical with respect to relation to the cytosol). At all events, the *minimum* nitrogen cost of digestive enzymes in a phagotroph growing at a specific growth rate of $10^{-4}\,\mathrm{s}^{-1}$ has been computed to be less than 10^{-5} of the total cell nitrogen (Raven, unpublished data). The other components of the phagotrophic apparatus are the flagella and the ingestion (phagocytotic) apparatus. The cost of flagella is small in terms of a reduction in the maximum specific growth rate (Raven & Richardson 1984), i.e. only a few percent at most. Furthermore, as we saw earlier, the flagella which are commonly found in obligately photolithotrophic chrysophytes as well as in those which are mixotrophic can have important functions in all three nutritional types, so that it is not strictly possible to say that the costs of flagella are only chargeable to phagotrophic and mixotrophic chrysophytes. The final cost is that of the ingestion apparatus. Clearly, if we are to believe the endosymbiotic hypothesis of the origin of plastids and mitochondria, the ancestors of even obligately photolithotrophic chrysophytes must have had a phagocytotic apparatus. Whether this apparatus, or at least information for it, is found in organisms which have no overt phagocytotic apparatus is not clear. At all events, the fraction of cellular resources which is devoted to the phagocytotic apparatus in obligate chemoorgano-trophs (phagotrophs) in excess of that in obligate photolithotrophs is only of the order of 5% as compared with the 50% or so devoted to the photosynthetic apparatus in obligate photolithotrophs (Raven 1984*a,b*, 1987*b*, and unpublished data).

Overall, the increment of resource inputs needed for phagotrophy rather than saprotrophy is much less than that for photolithotrophy compared with saprotrophy. This is in agreement with the observed differences in observed maximum specific growth between obligately chemoorgano-trophic (normally, mainly phagotrophic) and obligately photolithotrophic chrysophytes of similar size (Fig. 5.1). The fastest-growing *Paraphysomonas*

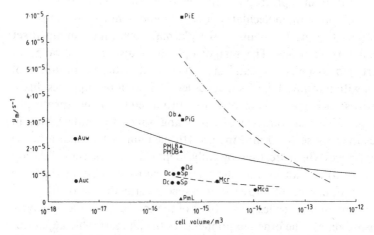

Figure 5.1. Specific growth rates of chrysophytes, normalized where necessary to 20 °C values assuming a Q_{10} of 2.0, as a function of cell size and nutritional mode. For comparison, mean values for photolithotrophic growth of diatoms (———) and dinoflagellates (- - - - -), and of chemoorganotrophic growth of unicells (mainly phagotrophically nourished cilates) (—·—·—) are also shown; all three regressions are from Banse (1982).

●, Predominantly or obligately photolithotrophic organisms growing photo-lithotrophically. Dd, *Dinobryon divergens* (Veen, 1991); Dc, *Dinobryon cylindricum* (two strains) (Sandgren 1988); Sp, *Synura petersonii* (two strains) (Sandgren 1988); Mcr, *Mallomonas cratis* (Sandgren 1988); Mca, *Mallomonas caudata* (Sandgren 1988); Auc, *Aureococcus anophagefferens* (cultured) (Dzurica *et al.* 1989); Auw, *Aureococcus anophagefferens* (wild population)* (Dzurica *et al.* 1989).

■, Obligate chemoorganotrophs growing phagotrophically. PiG, *Paraphyso-monas imperforata* (Goldman *et al.* 1987); PiE, *Paraphysomonas imperforata* (Edwards *et al.* 1989).

▲, Mixotrophic organisms. PMLB, *Poterioochromonas malhamensis* grown in light with bacteria (Caron *et al.* 1990*b*); PMDB, *Poterioochromonas malhamensis* compares well with fast-growing amoebae and ciliates (Banse 1982; see Fig. 5.1). However, none of the photolithotrophically growing chryso-phytes grow as fast as diatoms of a similar size. At least one of the mixotrophic organisms grows faster than any of the photolithotrophically growing cells, although the reported growth rate is achieved in the presence of bacteria as a phagotrophic food source with or without light, and growth in the light (or dark) in inorganic media lacking bacteria is negligible (Caron *et al.* 1990*b*).

The occurrence of the genetic information for the plastid, and its expression (in part, at least, even during dark chemoorganotrophic growth), may account for the lower μ_m values for mixotrophic chrysophytes than for obligately chemoorganotrophic ones (Fig. 5.1). The available data do not permit an unequivocal answer to the question of whether apparently obligate photolithotrophic chrysophytes such as *Mallomonas* have higher specific growth rates than mixotrophs such as *Dinobryon*, which (unlike the *Poterioochromonas* and *Ochromonas* spp. in Fig. 5.1) can readily grow photolithotrophically with a phagotrophic component of carbon supply of less than 10% at μ_m under mixotrophic growth conditions (Veen 1991). It is important to note that it is the photolitho-trophic growth rates of mixotrophs which are cited in Fig. 5.1. It is of interest that the fastest-growing photolithotrophic chrysophytes grow less rapidly than the fastest-growing microalgae of all/any taxa, i.e. diatoms and green algae (Fig. 5.1: compare with Raven 1986, 1987*a*). The photolithotrophic growth data for chrysophytes are quite close to the regression of Banse (1982) for dinoflagellates (Fig. 5.1).

Summary

This chapter surveys knowledge of chrysophyte carbon, nitrogen and phosphorus nutrition, and leads to the following conclusions and questions. With regard to carbon, photosynthetic carbon assimilation in chrysophytes uses the C_3 pathway. It is not clear whether chrysophytes: (1) can use exogenous HCO_3^-; (2) have CO_2 concentrating mechanisms such as

---

grown in dark with bacteria (Caron *et al.* 1990*b*); PML, *Poterioochromonas malhamensis* 'grown' in light without bacteria (Caron *et al.* 1990*b*); Ob, *Ochromonas* sp. from Baltic grown in light or dark with bacteria (Andersson *et al.* 1989).

* The photolithotrophic growth rate of the wild population of *Aureococcus* could be an overestimate as a result of hydrodynamic concentration of cells and/or chemoorganotrophy.

commonly occur in algae; or (3) have many obligately photolithotrophic species. Concerning nitrogen, NH_4^+ and NO_3^- can be used by phototrophically growing chrysophytes, and at least those capable of phagotrophy can use amino acids. What is the metabolic basis for, and ecological significance of, preferences for particular nitrogen sources? Regarding phosphorus, some chrysophytes can use inorganic *and* organic phosphorus. How does this capacity relate to extracellular phosphatases?

The greater maximum specific growth rate of obligate chemoorganotrophic (phagotrophic) than of photolithotrophic chrysophytes can theoretically be accounted for by the costs of producing the two sorts of metabolic machinery. Does mixotrophy reduce maximum specific growth rate relative to either nutritional mode alone? What happens to the organic carbon acquired phagotrophically with nitrogen, phosphorus and iron in mixotrophs which depend mainly on photosynthesis for their carbon input? Increased understanding of the ecology and paleoecology of chrysophytes requires further investigations of the physiology and biochemistry of their nutrition. In addition to such studies, which would elevate knowledge of chrysophyte nutrition to the level achieved for the best-known algae, the nutritional versatility found among the chrysophytes makes these algae particularly useful in dissecting the implications for growth rates of protists of the nutritional mode(s) which they employ.

Acknowledgments

Work on algal nutrition in the author's laboratory is supported by the Natural Environment Research Council. Stimulatory and moderating influences of past and present colleagues in Dundee, and of Dr Craig Sandgren, are gratefully acknowledged. Ideas on the role of turbulence and of rotation during flagellar motility on photon supply to plankton cells were initiated in discussions at the 1982 Woods Hole course on Energetics and Transport in Aquatic Plants.

References

Allen, M.B. 1969. Structure, physiology and biochemistry of the Chrysophyceae. *Annu. Rev. Microbiol.* **23**: 29–46.
Andersen, R.A. 1987. Synurophyceae *classis nov.*, a new class of algae. *Am. J. Bot.* **74**: 337–53.
Andersson, A., Falk, S., Samuelsson, G. & Hägstrom, A. 1989. Nutritional characteristics of a mixotrophic nanoflagellate, *Ochromonas* sp. *Microb. Ecol.* **17**: 251–62.

Antia, N.J., Cheng, J.Y. & Taylor, F.J.R. 1969. The heterotrophic growth of a marine photosynthetic cryptomonad (*Chroomonas salina*). In: Margalef, R. (Ed.) *Proceedings of the Sixth International Seaweed Symposium.* Subsecretaria De La Marina Mercante, Madrid. pp. 17–29.

Antia, N.J., Harrison, P.J. & Oliveira, L. 1991. The role of dissolved organic nitrogen in phytoplankton nutrition, cell biology and ecology. *Phycologia* **30**: 1–89.

Badger, M.R. 1985, Photosynthetic oxygen exchange. *Annu. Rev. Plant. Physiol.* **36**: 27–53.

Baker, N.R. 1991. A possible role for photosystem II in environmental perturbations of photosynthesis. *Physiol. Plant.* **81**: 563–70.

Banse, K. 1982. Cell volumes, maximal growth rates of unicellular algae and ciliates, and the role of ciliates in the marine pelagial. *Limnol. Oceanogr.* **27**: 1059–71.

Bell, G. 1985. The origin and early evolution of germ cells as illustrated by the Volvocales. In: Halvorson, H. & Monroy, A. (eds.) *The Origin and Evolution of Sex.* A.R. Liss, New York, pp. 221–56.

Bidigare, R.R. 1989. Photosynthetic pigment composition of the brown tide algae: unique chlorophyll and carotenoid derivatives. In: Cosper, E.M., Bricelj, V.M. & Carpenter, E.M. (eds) *Novel Phytoplankton Blooms: Causes and Impact of Recurrent Brown Tides and Other Unusual Blooms.* Coastal and Estuarine Studies 35. Springer, Berlin. pp. 57–75.

Bidigare, R.R., Kennicutt, M.C. II, Ordrusek, M.E., Keller, M.D. & Guillard, R.R.L. 1990. Novel chlorophyll-related compounds in marine phytoplankton: distributions and geochemical implications. *Energy Fuels* **4**: 653–7.

Bjørnland, T. & Liaaen-Jensen, S. 1989. Distribution patterns of carotenoids in relation to chromophyte phylogeny and systematics. in: Green, J.C., Diver, W.L. & Leadbeater, B.S.C. (eds) *The Chromophyte Algae: Problems and Perspectives.* Oxford University Press, Oxford, pp. 37–60.

Bricelj, V.M. Fischer, N.S., Guckert, J.B. & Chu, F.-L. E. 1989. Lipid composition and nutritional value of the brown tide alga *Aureococcus anophagefferens.* In: Cosper, E.M., Bricelj, V.M. & Carpenter, E.M. (eds) *Novel Phytoplankton Blooms: Causes and Impact of Recurrent Brown Tides and Other Unusual Blooms,* Coastal and Estuarine Studies 35. Springer, Berlin. pp. 85–99.

Caron, D.A., Goldman, J.C. & Dennett, M.R. 1990*a*. Carbon utilization by the omnivorous flagellate *Paraphysomonas imperforata. Limnol. Oceanogr.* **35**: 192–201.

Caron, D.A., Porter, K.G. & Sanders, R.W. 1990*b*. Carbon, nitrogen and phosphorus budgets for the mixotrophic phytoflagellate *Poteriochromonas malhamensis* (Chrysophyceae) during bacterial growth. *Limnol. Oceanogr.* **35**: 433–43.

Cattolico, R.A. & Loiseaux-de Goër, S. 1989. Analysis of chloroplast evolution and phylogeny: a molecular approach. In: Green, J.C., Diver, W.L. & Leadbeater, B.S.C. (eds.) *The Chromophyte Algae: Problems and Perspectives.* Oxford University Press, Oxford, pp. 85–100.

Coolbear, T. & Threlfall, D.R. 1989. Biosynthesis of terpenoid lipids. In: Ratledge, C. & Wilkinson, S.G. (eds.) *Microbial Lipids,* vol. 2. Academic Press, London, pp. 114–254.

Droop, M.R. 1974. Heterotrophy of carbon. In: Stewart, W.D.P. (ed.) *Algal Physiology and Biochemistry.* Blackwell Scientific, Oxford. pp. 530–59.

Chrysophyte algae

Dzurica, S., Lee, C., Cosper, E.M. & Carpenter, E.J. 1989. Roles of environmental variables, specifically organic compounds and micronutrients, in the growth of chrysophyte *Aureococcus anaphagefferens*. In: Cosper, E.M., Bricelj, V.M. and Carpenter, E.M. (eds) *Novel Phytoplankton Blooms, Causes and Impact of Recurrent Brown Tides and Other Unusual Blooms.* Coastal and Estuarine Studies 35. Springer, Berlin. pp. 229–52.

Edwards, A.M., Leadbeater, B.S.C. & Greene, J.C. 1989. Predatory chrysophytes: their role in carbon and energy turnover in the sea. *Br. Phycol. J.* **24**: 303.

Fenchel, T. 1982*a*. Ecology of heterotrophic microflagellates. I. Some important forms and their functional morphology. *Mar. Ecol. Prog. Ser.* **8**: 211–223.

Fenchel, T. 1982*b*. Ecology of heterotrophic microflagellates. II. Bioenergetics and growth. *Mar. Ecol. Prog. Ser.* **8**: 225–231.

Fenchel, T. 1986. Protozoan filter feeding. *Prog. Protistol.* **1**: 65–114.

Fenchel, T. 1987. Ecology of Protozoa. Springer, Berlin.

Fenchel, T. & Finlay, B.J. 1983. Respiration rates in heterotrophic, free-living protozoa. *Microb. Ecol.* **9**: 99–122.

Fork, D.C., Herbert, S.K. & Malkin, S. 1991. Light energy distribution in the brown algae *Macrocystis pyrifera* (giant kelp). *Plant physiol.* **95**: 731–9.

Gibbs, P.B. & Biggins, J. 1989. Regulation of distribution of excitation energy in *Ochromonas danica*, an organism containing a chlorophyll a/c/carotenoid light harvesting antenna. *Photosynthesis Res.* **21**: 81–91.

Goldman, J.C., Caron, D.A. & Dennett, M.R. 1987. Nutrient cycling in a microflagellate food chain. IV. Phytoplankton–microflagellate interactions. *Limnol. Oceanogr.* **32**: 1239–52.

Greene, R.M. & Gerard, V.A. 1990. Effects of high-frequency light fluctuations on growth and photoacclimation of the red alga *Chondrus crispus. Mar. Biol.* **105**: 337–44.

Grün, M. & Loewus, F.A. 1984. L-Ascorbic acid biosynthesis in the euryhaline diatom *Cyclotella cryptica. Planta* **160**: 6–11.

Harrison, P.J., Parslow, J.S. & Conway, H.L. 1989. Determination of nutrient uptake kinetic parameters: a comparison of methods. *Mar. Ecol. Prog. Ser.* **52**: 301–12.

Haug, A. 1974. Chemistry and biochemistry of algal cell-wall polysaccharides. In: Northcote, D.H. (ed.) *MTP International Reviews of Science, Biochemistry*, series 1, vol. II, *Plant Biochemistry*, Butterworth, London. pp. 51–88.

Helspar, J.P., Kagan, L., Hilby, C.L., Maynard, T.M. & Loewus, F.A. 1982. L-Ascorbic acid biosynthesis in *Ochromonas danica. Plant Physiol.* **69**: 465–8.

Heywood, P. 1989. Some affinities of the Rhaphidophyceae with other chromophyte algae. In: Green, J.C., Diver, W.L. & Leadbeater, B.S.C. (eds.) *The Chromophyte Algae: Problems and Perspectives.* Oxford University Press, Oxford, pp. 279–93.

Jeffrey, S.W. 1989. Chlorophyll c pigments and their distribution in the chromophyte algae. In: Green, J.C., Diver, W.L. & Leadbeater, B.S.C. (eds.) *The Chromophyte Algae: Problems and Perspectives.* Oxford Press, Oxford. pp. 13–36.

Jones, R.J. 1988. Vertical distribution and diel migration of flagellated phytoplankton in a small humic lake. *Hydrobiologia* **16**: 75–87.

Kauss, H. 1967. Isofloridosid und osmoregulation bei *Ochromonas malhamensis*. *Z. Pflanzen Physiol.* **56**: 453–65.

Kauss, H. 1974 Osmoregulation in *Ochromonas*. In: Zimmerman, U. & Dainty, J. (eds.) *Membrane Transport in Plants*. Springer, Berlin. pp. 90–4.

Kauss, H. & Kandler, O. 1962. Die Kohlensaureassimilation von *Ochromonas malhamensis* bei Thiamin und Biotinmangel. I. Die heterotrophe Kohlensaureassimilation. *Arch. Mikrobiol.* **42**: 204–18.

Kauss, H. & Kandler, O. 1963. Die Kohlensaureassimilation von *Ochromonas malhamensis* bei Thiamin- und Biotinmangel. II. Photosynthese. *Arch. Mikrobiol.* **44**: 406–20.

Keller, A.A. & Rice, R.L. 1989. Effect of nutrient enrichment on natural populations of the brown tide phytoplankton *Aureococcus anophagefferens* (Chrysophyceae). *J. Phycol.* **25**: 636–46.

Keller, M.D., Bellows, W.K. & Guillard, R.R.C. 1989. Dimethylsulfide production and marine phytoplankton: an additional impact of unusual blooms. In: Cosper, E.M., Bricelj, V.M. & Carpenter, E.J. (eds.) *Novel Phytoplankton Blooms. Causes and Impact of Recurrent Brown Tides and other Unusual Blooms*. Coastal and Estuarine Studies 35. Springer, Berlin. pp. 101–15.

Kowallik, K.V. 1989. Molecular aspects and phylogenetic implications of plastid genomes of certain chromophytes. In: Green, J.C., Diver, W.L. & Leadbeater, B.S.C. (eds.) *The Chromophyte Algae: Problems and Perspectives*. Oxford University Press, Oxford. pp. 101–24.

Kraemer, G.P. & Chapman, D.J. 1991. Effects of tensile force and nutrient availability on carbon uptake and cell wall synthesis in bladders of juvenile *Egregia menziesii* (Turn.) Aresch. (*Phaeophyta*). *J. Exp. Mar. Biol. Ecol.* **149**: 267–77.

Kremer, B.P. 1980. Taxonomic implications of algal photoassimilate patterns. *Br. Phycol. J.* **15**: 399–409.

Kremer, B.P. & Kirst, G.O. 1982. Biosynthesis of photosynthates and taxonomy of algae. *Z. Naturforsch.* **73C**: 761–71.

Laws, E.A., Taguchi, S., Hirata, J. & Pang, L. 1986. High algal production rates achieved in a shallow outdoor flume. *Biotechnol. Bioeng.* **28**: 191–7.

Luther, H. 1949. Vorschlag zu einer ökologischen Grundeinleitung der Hydrophyten. *Act. Bot. Fenn.* **49**: 1–15.

MacFarlane, J.J. & Raven, J.A. 1985. External and internal CO_2 transport in *Lemanea*: interactions with the kinetics of ribulose bisphosphate carboxylase. *J. Exp. Bot.* **36**: 610–22.

MacFarlane, J.J. & Raven, J.A. 1989. Quantitative determinations of unstirred layer permeability and kinetic parameters of RUBISCO in *Lemanea mamillosa*. *J. Exp. Bot.* **40**: 321–7.

MacFarlane, J.J. & Raven, J.A. 1990. C, N and P nutrition of *Lemanea mamillosa* Kutz, (Batrachospermales, Rhodophyta) in the Dighty Burn, Angus, Scotland. *Plant Cell Environm.* **13**: 1–13.

Milligan, A.J. & Cosper, E.M. 1991. Photosynthetic and growth efficiency of the brown tide alga *Aureococcus anophagefferens* in fluctuating light. *J. Phycol.* **27**: 50s.

Myers, J. & Graham, J.-R. 1956. The role of photosynthesis in the physiology of *Ochromonas*. *J. Cell Comp. Physiol.* **47**: 397–414.

Osborne, B.A. & Raven, J.A. 1986. Light absorption by plants and its implications for photosynthesis. *Biol. Rev.* **51**: 1–61.

Palenik, B. & Morel, F.M.M. 1990. Amino acid utilization by marine phytoplankton: a novel mechanism. *Limnol. Oceanogr.* **35**: 260–9.

Paran, N., Budinsky, Z. & Berman, T. 1991. Interactions between mixotrophic flagellates and bacteria in aquatic ecosystems. *Symbiosis* **10**: 219–31.

Parker, B.C., Samsel, G.L. & Prescott, G.W. 1973. Comparison of microhabitats of microscopic subalpine stream algae. *Am. Midland Nat.* **90**: 143–53.

Patterson, G.W & Van Valkenberg, S.D. 1990. Sterols of *Dictyocha fibula* (Chrysophyceae) and *Olisthodiscus luteus* (Raphidophyceae). *J. Phycol.* **26**: 484–89.

Pearcy, R.W. 1990. Sunflecks and photosynthesis in plant canopies. *Annu. Rev. Plant Physiol. Plant Mol. Biol.* **41**: 421–53.

Peinaar, R.N. 1980. Chrysophytes. In: Cox, E.R. (ed.) *Phytoflagellates.* Elsevier/North-Holland, New York. pp. 213–42.

Raven, J.A. 1980. Nutrient transport in microalgae. *Adv. Microb. Physiol.* **21**: 47–226.

Raven, J.A. 1982. The energetics of freshwater algae: energy requirements for biosynthesis and volume regulation. *New Phytol.* **92**: 1–20.

Raven, J.A. 1983. The transport and function of silicon in plants. *Biol. Rev.* **58**: 179–207.

Raven, J.A. 1984a. *Energetics and Transport in Aquatic Plants.* A.R. Liss, New York.

Raven, J.A. 1984b. A cost–benefit analysis of photon absorption by photosynthetic unicells. *New Phytol.* **98**: 593–625.

Raven, J.A. 1986. Physiological consequences of extremely small size for autotrophic organisms in the sea. In: Platt, T. & Li, W.K.W. (eds.) *Photosynthetic Picoplankton. Can. Bull. Fish Aq. Sci.* **214**: 1–70.

Raven, J.A. 1987a. Limits to growth. In: Borowitzka, M.A. & Borowitzki, L.J. (eds). *Microalgal Biotechnology.* Cambridge University Press, Cambridge. pp. 331–56.

Raven, J.A. 1987b. The role of vacuoles. *New Phytol.* **106**: 357–422.

Raven, J.A. 1988. Algae. In: Baker, D.A. & Hall, J.L. (eds.) *Solute Transport in Plant Cells and Tissues.* Longman, Harlow. pp. 166–219.

Raven, J.A. 1989. Algae on the move. *Trans. Bot. Soc. Edinb.* **45**: 167–86.

Raven, J.A. 1990. Transport systems in algae and bryophytes: an overview. *Enzymol.* **174**: 366–89.

Raven, J.A. 1991a. Implications of inorganic C utilization: ecology, evolution and geochemistry. *Can. J. Bot.* **69**: 908–24.

Raven, J.A. 1991b. Physiology of inorganic C acquisition and implications for resource use efficiency by marine phytoplankton: relation to increased CO_2 and temperature. *Plant Cell Environm.* **14**: 779–94.

Raven, J.A. & Richardson, K. 1984. Dinophyte flagella: a cost–benefit analysis. *New Phytol.* **98**: 259–76.

Raven, J.A., MacFarlane, J.J. & Griffiths, H. 1986. The application of carbon isotope techniques. In: Crawford, R.M.M. (ed.) *Plant Life in Aquatic and Amphibious Habitats. British Ecological Society Special Symposium.* Blackwell Scientific, Oxford, pp. 129–49.

Raven, J.A., Johnston, A.M. & Surif, M.B. 1989. The photosynthetic apparatus as a phyletic character. In: Green, J.C., Diver, W.L. & Leadbeater, B.S.C. (eds.) *The Chromophyte Algae: Problems and Perspectives.* Oxford University Press, Oxford. pp. 41–60.

Raven, J.A., Wollenweber, B. & Handley, L.L. 1992. A comparison of ammonium and nitrate as nitrogen sources for photolithotrophs. *New Phytol.* **13**: 11–32.

Rothschild, L.J. & Heywood, P. 1987. Protistan phylogeny and chloroplast evolution: conflicts and congruence. In: Corliss, J.O. & Patterson, D.J. (eds.) *Progress in Protistology*, vol. 2. Biopress, Bristol, pp. 1–68.

Salonen, K., Jones, R.I. & Arrola, L. 1984. Hypolimnetic phosphorus retrieval by diel vertical migrations of lake phytoplankton. *Freshwater Biol.* **14**: 431–8.

Sanders, R.W. 1991. Mixotrophic protists in marine and freshwater ecosystems. *J. Protozool.* **38**: 76–81.

Sanders, R.W., Porter, K.G. & Caron, D.A. 1990. Relationship between phototrophy and phagotrophy in the mixotrophic chrysophyte *Poteriochromonas malhamensis*. *Microb. Ecol.* **19**: 97–109.

Sandgren, C.D. 1988. The ecology of chrysophyte flagellates: their growth and perennation strategies as freshwater phytoplankton. In: Sandgren, C.D. (ed.) *Growth and Reproductive Strategies of Freshwater Phytoplankton.* Cambridge University Press, Cambridge. pp. 9–104.

Schürmann, A. & Brant, P. 1991. Isolation of chloroplasts from *Poteriochromonas malhamensis* with the capacity for translation. *Plant Cell Physiol.* **32**: 533–40.

Shapiro. L.P., Haugen, E.M., Keller, M.D., Bidigare, R.R., Campbell, L. & Guillard, R.R.L. 1989. Taxonomic affinities of marine coccoid ultraphytoplankton: a comparison of immunochemical surface antigen cross-reactions and HPLC chloroplast pigment signatures. *J. Phycol.* **25**: 794–7.

Sommer, U. & Gliwicz, Z.M. 1986. Long range vertical migration of *Volvox* in tropical Lake Cahora Basa (Mozambique). *Limnol. Oceanogr.* **31**: 650–3.

Stacey, G., van Baaten, C. & Tabita, F.R. 1977. Isolation and characterization of a marine *Anabaena* sp. capable of rapid growth on molecular nitrogen. *Arch. Microbiol.* **114**: 197–201.

Thomas, W.H. & Gibson, C.H. 1990. Effects of small-scale turbulence in microalgae. *J. Appl. Phycol.* **2**: 71–8.

Thomas, W.H., Gibson, C.H. & Vernet, M. 1991. Inhibition of *Gonyaulax polyedra* Stein by quantified small-scale turbulence. *J. Phycol.* **27**: 72s.

Turpin, D.H. 1988. Physiological mechanisms in phytoplankton resource competition. In: Sandgren, C.D. (ed.) *Growth and Reproductive Strategies of Freshwater Phytoplakton.* Cambridge University Press, Cambridge, pp. 316–68.

Turpin. D.H., Layzell, D.B. & Elrifi, I.R. 1985. Modelling the C economy of *Anabaena flos-aquae*: estimates of establishment, maintenance and active costs associated with growth on NH_4^+, NO_3^- and N_2. *Plant Physiol.* **78**: 746–52.

Veen, A. 1991. Ecophysiological studies on the phagotrophic phytoflagellate *Dinobryon divergens* Imhof. PhD thesis, University of Amsterdam. 125pp.

Vesk, M., Hoffman, L.R. & Pickett-Heaps, J.D. 1984. Mitotis and cell division in *Hydrurus foetidus* (Chrysophyceae). *J. Phycol.* **20**: 461–470.

Watanabe, M.M. Nakamura, Y. & Kohata, K. 1983. Diurnal vertical migration and dark uptake of nitrate and phosphate of the red tide flagellates, *Heterosigma akashiwa* Hada and *Chattonella antiqua* (Hada) Ono (Raphidophyceae). *Jpn. J. Phycol.* **31**: 161–6.

Weis, D. & Brown, A.M. 1959. Kinetic relationship between photosynthesis and

118   *Chrysophyte algae*

respiration in the algal flagellate, *Ochromonas malhamensis. Plant Physiol.* **34**: 235–9.

Wilhem, C. 1990. The biochemistry and physiology of light-harvesting processes in chlorophyll *b*- and chlorophyll *c*-containing algae. *Plant Physiol. Biochem.* **28**: 293–306.

Wood, B.J.B. 1988. Lipids of algae and protozoa. In: Ratledge, C. & Wilkinson, S.G. (eds.) *Microbial Lipids*, vol. 1. Academic Press, London, pp. 807–67.

6

Mixotrophy in chrysophytes

DALE A. HOLEN & MARTIN E. BORAAS

Microbial food web

Within the past decade there have been significant changes in the perception of the interactions among trophic levels in aquatic environments. The once-traditional phytoplankton–macrozooplankton–fish food web concepts have been revised to more fully accommodate the effects of both prokaryotes and nutritionally diverse protists. For instance, about 40–60% of the carbon fixed by all phytoplankton is thought to pass through the heterotrophic bacterial community (Cole *et al.* 1988) rather than into zooplankton directly. Furthermore, oligotrophic marine and freshwaters are now characterized by the preponderance of an assemblage of small phytoplankton less than 2 μm in diameter called picoplankton. This assemblage is composed primarily of chroococcoid cyanobacteria such as *Synechococcus* and *Chlorella*-like eukaryotes, both of which contribute significantly to overall primary production (Platt *et al.* 1983; Stockner & Antia 1986; Stockner 1988). Because these tiny cells are poorly grazed by macrozooplankton, they contribute little to the traditional food web.

Much of this bacterial and picoplanktonic carbon is transferred to higher trophic levels via diverse protozoa. In a series of seminal papers (Fenchel 1982*a, b, c, d*), small flagellates < 20 μm in diameter were reported to be efficient bacterivores with relatively high particle ingestion rates. Given the observed densities of these small protozoa in natural systems together with the calculated grazing rates, Fenchel concluded that such protozoa could have a significant impact on natural bacterial abundances. Subsequent research regarding the grazing habits and feeding efficiencies of pelagic protozoans has implicated small flagellates and ciliates as the primary grazers of picoplankton, i.e. bacteria and algae, in

most freshwater and marine pelagic environments. Cladoceran macrozooplankton, however, have also been shown to have an impact on bacterial
mortality in some freshwater systems (Güde 1988; Vaqué & Pace 1992).
Bacterivorous protozoa are now thought to control pelagic bacterial
population densities, and have an impact on both bacterial species
composition and size distribution through the preferential ingestion of
large bacterial cells (Güde 1979; Gonzalez *et al.* 1990; Holen & Boraas
1991). As protozoa, in turn, can be a food source for macrozooplankton,
bacterial carbon may be shunted into metazoans via picoplankton–
protozoan–macrozooplankton coupling (Turner *et al.* 1988; Sanders &
Porter 1990; Carrick & Fahnenstiel 1991). Given the high metabolic rates
(Fenchel & Finlay 1983) and rapid nutrient turnover rates (Johannes
1964) of very small heterotrophs (see Raven, this volume), these protists
can also have a major influence on energy fluxes and nutrient cycling in
these systems (Caron *et al.* 1988).

The rather circuitous route of primary production from phototrophic
picoplankton to bacterioplankton to protozoa, facilitated via remineralization and predation, has been called the 'microbial loop' (Azam *et al.*
1983) (Fig. 6.1). In oligotrophic pelagic environments, where small size is
advantageous due to narrower boundary layers around small cells (Raven,
this volume) ecosystem metabolism is often dominated by autotrophic
and heterotrophic picoplankton. The microbial loop, with its relatively
inefficient transfer of carbon to higher trophic levels, strongly influences
ecosystem productivity in these pelagic environments. In contrast, more
eutrophic environments are characterized by larger phytoplankton and
denser metazoan populations, facilitating the more direct transfer of
carbon via the classical food web. Bacterial populations in these more
productive environments appear to be regulated by other factors in
addition to grazing loss, i.e., nutrient concentrations (Berninger *et al.* 1991;
Gasol & Vaqué 1993). Furthermore, inputs of allochthonous materials
in these environments often reduces the impact of nutrient remineralization
by protozoan predators.

Mixotrophy

Algal protists with chloroplasts are typically viewed as primary producers.
Many photosynthetic algae are also known to supplement growth
osmotrophically with dissolved organic carbon. Digestive vacuoles in
some algae, however, suggest that they may also use phagotrophy to
supplement phototrophic growth. It is now apparent that phagotrophic

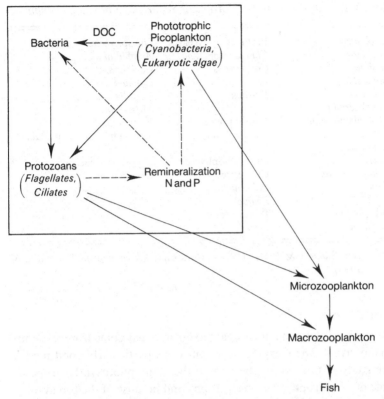

Figure 6.1. Simplified diagram of the major pathways for carbon and nutrient transfer in the microbial food web. Continuous arrows represent carbon transfer via phagotrophy and dashed arrows represent remineralization processes. DOC, dissolved organic carbon.

algae are common in both freshwater and oceanic habitats, can attain relatively high abundances in these systems, and can have a major impact on total bacterivory (Sanders *et al.* 1989; Sanders 1991; Berninger *et al.* 1992). At least seven algal classes – Chrysophyceae, Cryptophyceae, Dictyochophyceae, Dinophyceae, Raphidophyceae, Xanthophyceae and Prymnesiophyceae – have been reported to contain mixotrophic protists (Table 6.1). In addition, the discovery of ciliates that harbor functional chloroplasts sequestered from algal prey further exemplifies the abundance of organisms that utilize mixotrophy in both marine and freshwater ecosystems (Stoecker *et al.* 1987, 1988; Rogerson *et al.* 1989). Furthermore, many animal–algal symbioses with similar characteristics have been described, particularly in marine environments (Smith & Douglas 1987).

Table 6.1. *Flagellates with chloroplasts reported to be phagotrophic*

Chrysophyceae	Prymnesiophyceae	Dinophyceae
Catenochrysis	*Chrysochromulina*	*Amphidinium*
Chromulina	*Coccolithus*	*Ceratium*
Chrysamoeba	*Prymnesium*	*Gymnodinium*
Chrysococcus	*Pavlova*	*Gyrodinium*
Chrysosphaerella		*Massartia*
Chrysostephanosphaera	Xanthophyceae	*Peridinium*
Dinobryon	*Chlorochromonas*	*Alexandrium* (= *Gonyaulax*)
Ochromonas		
Phaeaster	Dictyochophyceae	Cryptophyceae
Poterioochromonas	*Cyrtophora*	*Cryptomonas*
Uroglena	*Palatinella*	
	Pedinella	
Raphidophyceae	*Pseudopedinella*	
Heterosigma		

*Modified from Sanders & Porter (1988), Moestrup (this volume) and Nygaard &
Tobiesen (1993).*

Chrysophyte mixotrophy

Much of what is known about algal mixotrophy has come from the study
of chrysophytes. The Chrysophyceae provides particularly good models
for the study of mixotrophy because of the many photosynthetic species
capable of osmotrophy and phagotrophy and because of the broad range
of photosynthetic capacity that they exhibit (Table 6.2; Raven, this
volume). Mixotrophy in chrysophytes was observed as early as the 1890s
for a species of *Ochromonas* (see Pringsheim 1952) and Pascher (1943) first
described the uptake of small particles by *Dinobryon*. However, the
ecological significance of this mode of nutrition was not appreciated until
Bird & Kalff (1986) reported that populations of *Dinobryon* could be
quantitatively important in lakes as bacterial consumers during times of
peak algal abundance. More recently, Berninger *et al.* (1992) found that
48% of the phototrophic nanoplankton (2–20 µm in diameter) in Lake
Lacawac (Pennsylvania, USA) were mixotrophs which accounted for at
least 90% of the bacterivory. The majority of mixotrophs in this lake were
small unicellular algae, possibly chrysophytes and *Dinobryon* spp. A
significant proportion of mixotrophic flagellates in other systems are
chrysophytes as well (Bird & Kalff 1987; Bennett *et al.* 1990). Phototrophic
chrysophytes can have ingestion rates comparable to their strictly heterotrophic counterparts (Nygaard & Hessen 1990; Porter, 1988), underscoring
their potential importance as bacterial and picoplankton consumers.

Table 6.2. *Spectrum of nutritional capabilities for representative chrysophytes*

Strict phototrophy:	*Mallamonas, Synura*
Mixotrophy:	
Obligate phototrophy and facultative phagotrophy	*Dinobryon divergens* *Dinobryon cylindricum*
Obligate phototrophy and obligate phagotrophy:	*Uroglena americana* *Ochromonas* spp. *Chrysamoeba*
Facultative phototrophy and facultative phagotropy + osmotrophy	*Ochromonas danica* *Poterioochromonas malhamensis*
Strict osmotrophy:	*Paraphysomonas*
Strict phagotrophy:	*Spumella* spp. *Ochromonas vallesciaca*

Types of mixotrophy

The dependency on particulate food by chrysophytes varies from an obligate to facultative requirement among genera, and seems to depict nutritional stages in an adaptive gradient from strict phototrophy to strict phagotrophy (Table 6.2). The nutritional types pertaining to mixotrophy are discussed below together with related experimental evidence.

Obligate mixotrophy

There are few documented instances of obligate mixotrophy in chrysophytes but the colonial chrysophyte *Uroglena americana* appears to be an example. Ishida & Kimura (1986) reported a requirement for both photosynthesis and a bacterial food source for growth of this flagellate. They determined that whole bacteria or bacterial lysates provided essential phospholipids that the algae were unable to synthesize. In addition, undescribed species of *Ochromonas* and *Chrysamoeba*, originally isolated from the western North Atlantic, appeared to be obligate mixotrophs by virtue of their inability to be cultured axenically and because growth on bacteria would occur only in illuminated cultures (Estep *et al.* 1984).

Predominant phagotrophic mixotrophy

Mixotrophs, such as photosynthetic species of *Ochromonas* and *Poterioochromonas malhamensis*, are efficient phagotrophs which can achieve their maximum specific growth rate, μ_{max}, only when they are provided with particulate prey. *P. malhamensis* growth rates in batch culture ranged from 1.5 to 2.2 d^{-1} with measured ingestion rates of 60 to 90 bacteria per flagellate per hour when bacteria were not limiting (Sanders *et al.* 1990; Caron *et al.* 1990). Bacterial abundance, rather than light intensity, appeared to limit the ingestion and growth rates of this flagellate. Moreover, the addition of NH_4^+ and PO_4^{3-}, either singly or combined, to flagellate–bacterial cultures did not affect bactivory or flagellate generation time, suggesting that bactivory was the dominant means of nutrition. Using a continuous culture with an initial bacterial concentration of 4×10^7 cells ml^{-1}, we observed a growth rate for this flagellate of 2.1 d^{-1} with a mean ingestion rate of 108 bacteria per flagellate per hour in our laboratory (Fig. 6.2). *P. malhamensis* appears equally adept at utilizing dissolved organic sources of carbon for growth, as we observed a growth rate of 1.3 d^{-1} when cultured on an enriched organic medium (*Ochromonas* medium; Starr 1978). Also, Lewitus & Caron (1991*a*) reported growth rates for this flagellate of 0.49 d^{-1} on 0.083 M ethanol and 0.59 d^{-1} and 0.72 d^{-1} on 0.055 and 0.029 M glycerol, respectively.

Phototrophic growth in *P. malhamensis*, in contrast, is significantly slower. Sanders *et al.* (1990) observed growth of *P. malhamensis* at <0.1 d^{-1} when cultured on a basal salts medium at 200 µEin m^{-2} s^{-1}. Moreover, photosynthesis contributed only approximately 7% of the total carbon budget of this flagellate when cultured on bacteria at a concentration of 5.4×10^7 bacteria ml^{-1}, suggesting the relatively minor contribution of photosynthesis to growth when abundant bacteria are available. In our laboratory, we cultured *P. malhamensis* on a defined inorganic medium (DYIV: modified by C. Sandgren (unpublished) from DYIII: Lehman 1976) at a light intensity of 200 µEin m^{-2} s^{-1} and achieved a growth rate of 0.068 d^{-1}. Myers & Graham (1956) reported a phototrophic growth rate for this flagellate of about 0.2 d^{-1}. Optimal phototrophic growth, however, may be pH dependent, as Lewitus & Caron (1991*a*) cultured the flagellate at a growth rate of about 0.3 d^{-1} at a pH of 4.8.

Many species of *Ochromonas* appear to be similar to *P. malhamensis* with regard to phagotrophic capabilities. Andersson *et al.* (1989) reported a μ_{max} for an *Ochromonas* sp. isolated from the Baltic Sea of 3.2 d^{-1} when cultured on *Escherichia coli* minicells in the light and the dark. As with *P. malhamensis*, growth rate was strongly correlated with bacterial concentration. As growth rates were independent of light conditions, photosynthesis appears to be negligible for this flagellate when it is actively growing on bacteria. Furthermore, axenic cultures of this isolate

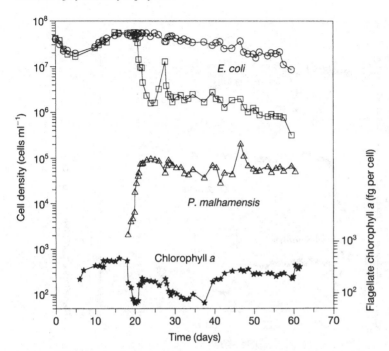

Figure 6.2. Growth of *Poterioochromonas malhamensis* in a two-stage chemostat culture on *E. coli*. Densities of *E. coli* in the first and second stages of chemostat are represented by circles and squares, respectively. The chemostat, set at a dilution rate of $1.0\,\mathrm{d}^{-1}$, was illuminated continuously with cool-white fluorescent bulbs ($200\,\mu\mathrm{Ein\,m^{-2}\,s^{-1}}$) at 25 °C. The flagellate was grown on mineral medium until day 18, when it was inoculated into the second stage. In batch culture the chlorophyll *a* content increased up to *c*. 600 fg per cell. After exposure to bacteria, the cell-specific chlorophyll *a* declined to a basal value of 60 fg per cell and then increased to *c*. 300 fg per cell. The flagellate entered and maintained a steady state thereafter. Cellular chlorophyll *a* appeared to have little effect on standing crop.

cultured on different inorganic media at $100\,\mu\mathrm{Ein\,m^{-2}\,s^{-1}}$ showed no significant increase in density even after 5 weeks of incubation.

In continuous culture, we observed a $\mu_{\max}$ for *O. danica* of $2.4\,\mathrm{d}^{-1}$ with a mean ingestion rate of 182 bacteria per flagellate per hour when cultured on *E. coli* at an initial concentration of 3×10^7 cells ml^{-1} (Fig. 6.3). Fenchel (1982*b*) reported a maximum uptake rate for an *Ochromonas* sp. of 190 bacteria per flagellate per hour. Although growth rates were not reported, *O. minima* ingested isotope-labelled and DTAF-stained bacteria at a rate of 36 and 29 bacteria per flagellate per hour, respectively (Nygaard & Hessen 1990). The observed ingestion and growth rates of

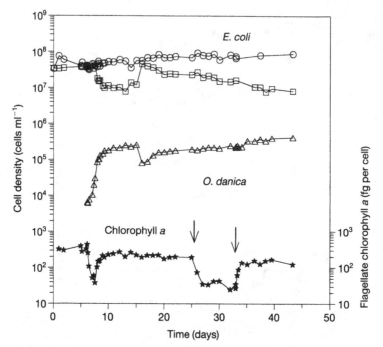

Figure 6.3. Growth of *Ochromonas danica* in a two-stage chemostat on *E. coli*. Circles and squares represent bacterial densities in the first and second stages of the chemostat, respectively. The flagellate was grown phototrophically for 6 days prior to inoculation into the second stage. Flagellate cell-specific chlorophyll *a* concentration decreased from 400 fg per cell to 40 fg per cell during growth on bacteria. The dilution rate was increased from $0.5 \, d^{-1}$ to $1.7 \, d^{-1}$ at day 14. After 25 days the lights were turned off (first arrow) and the cell-specific chlorophyll *a* decreased from 300 fg per cell to 30 fg per cell. After illumination was restored at day 33 (second arrow), the chlorophyll *a* returned to 150 fg per cell.

these mixotrophs, which are not dissimilar to colorless flagellates, reflect their preference for phagotrophy. Because phototrophic growth is slow, these chrysophytes may utilize photosynthesis as a survival mechanism during periods of low abundance of particulate food (Sanders *et al.* 1990).

An example of obligate phagotrophy in a mixotroph may be the flagellate *O. vallesciaca*, which appeared as a contaminant in a two-stage algal–rotifer chemostat (Boraas *et al.* 1988). This chrysophyte appears to have an active chloroplast in the light, as evidenced by red fluorescence under blue-light excitation, but does not demonstrate ^{14}C uptake (Lu 1988). Maintained in a chemostat on a diet of the coccoid green alga *Chlorella pyrenoidosa*, it appears to be selective for food particles $>3 \, \mu m$. Smaller prey, e.g. the bacterium *Xenorhabdus luminescens*, which can sustain growth of other flagellate species, cannot sustain growth of *O. vallesciaca* (Seale *et al.* 1990). Generally, however, *Ochromonas* spp. are versatile in their preference for food as they have been shown to ingest bacteria (Fenchel 1982*b*), yeast (Aaronson 1974; Lu 1988), cyanobacteria (Cole & Wynne 1974; Daley *et al.* 1973), green algae (Boraas *et al.* 1988) and other flagellates (Boraas *et al.* 1992).

Whereas prey availability has been shown to have a direct influence on the specific growth rate of predominant phagotrophic mixotrophs, the factors responsible for the relatively slow μ_{max} during phototrophic growth remain unclear. Myers & Graham (1956) suggested that the low concentration of cellular chlorophyll in *P. malhamensis*, in comparison with similar-sized algae, may restrict its photosynthetic potential. Andersson *et al.* (1989) calculated a cell carbon-to-chlorophyll-*a* ratio of 77 to 250 for *Ochromonas*. They speculate that the high carbon-to-chlorophyll ratio, relative to other marine phytoplankton (typically 25 to 100), would result in a lower photosynthetic capacity in comparison with other eukaryotic algae. Raven (this volume) compares growth rates of chrysophytes with different modes of nutrition. The higher specific growth rate of colorless versus mixotrophic species, when cultured heterotrophically, may be a result of the higher 'cost' of maintaining a photosynthetic apparatus. This would suggest that mixotrophy, although seemingly advantageous for the organism, probably involves an evolutionary trade-off or metabolic cost (see below).

Predominantly phototrophic mixotrophy

Although species of *Dinobryon* at times of peak abundance have been shown to contribute significantly to bacterial grazing in some lakes (Bird & Kalff 1986), phototrophy appears to be the dominant mode of nutrition in this flagellate. Under certain conditions, however, phagotrophy may contribute significantly to the total carbon uptake. Bird & Kalff (1989) reported that a *Dinobryon* and *Ochromonas* assemblage, localized in a metalimnetic stratum at 7 m depth. (≈ 5 μEin m^{-2} s^{-1} light level) obtained 79% of its total carbon from phagotrophy. Caron *et al.* (1993) also concluded that *Dinobryon cylindricum*, growing in bacterized cultures, may have obtained 25% of its acquired carbon from bacterial ingestion, the remaining 75% being attributed to photosynthesis.

An obligate requirement for light was demonstrated for both *D. cylindricum* (Caron *et al.* 1993) and *D. divergens* (Veen 1991). *D. divergens* was able to achieve a μ_{max} of about $1.0\,d^{-1}$ when cultured photosynthetically and showed no enhancement of growth when bacteria were added at saturating food densities. The addition of bacteria to light-limited cultures, however, stimulated the growth rate from 0.56 to $0.88\,d^{-1}$. Under these conditions the maximum ingestion rates measured, using 0.57 µm fluorescently tagged microspheres (FM), was about 210 FM $cell^{-1}\,h^{-1}$ at a concentration exceeding 10^7 FM ml^{-1}. Even under light-limited conditions, *D. divergens* obtained $< 10\%$ of its carbon requirement from bacteria.

Lehman (1976) was able to culture species of *Dinobryon* axenically, achieving growth rates of $0.69\,d^{-1}$ for *D. cylindricum* and *D. sociale* and $0.35\,d^{-1}$ for *D. sertularia*. This concurs with Sandgren (1986) who reported phototrophic growth rates for non-axenic *D. cylindricum* ranging from 0.30 to $0.50\,d^{-1}$ over 10–20 °C. Additionally, maximum specific growth rate estimates for four clones of *D. cylindricum*, cultured on mineral media in non-axenic semi-continuous culture, were 0.9, 0.5, 0.6 and $0.75\,d^{-1}$ (Sandgren 1988). Caron *et al.* (1993), however, observed poor growth of *D. cylindricum* in axenic cultures but achieved a μ_{max} of $0.7\,d^{-1}$ with culture on bacteria at a concentration of 4×10^7 bacteria ml^{-1}. Light intensity had a marked effect on ingestion rate and survival of this flagellate. Ingestion rates ranged from 5–10 bacteria per flagellate per hour at ≥ 150 µEin $m^{-2}\,s^{-1}$ to 1.8 bacteria per flagellate per hour at 30 µEin $m^{-2}\,s^{-1}$. Ingestion rate and growth ceased for cultures incubated in continuous darkness, indicating an obligate requirement for both light and particulate food.

Obligate phototrophy with facultative phagotrophy has also been observed in the cryptomonads *Cryptomonas ovata* and *C. erosa*. Observed ingestion rates ranged from 0.7 to 1.7 bacteria per flagellate per hour and contributed about 2% of the organism's daily carbon budget (Tranvik *et al.* 1989). Using fluorescently labeled beads, Porter (1988) reported similar ingestion rates for *C. ovata* and *C. erosa*, ranging from 0.3 to 2.6 beads per flagellate per hour. As photosynthesis appears to be the dominant mode of carbon acquisition in these algae, the ingestion of bacteria may be a means of obtaining limiting nutrients or other factors for augmenting phototrophic growth.

Nutrient acquisition via phagotrophy

It is clear that in laboratory environments, mixotrophs such as *P. malhamensis* and *Ochromonas* can obtain all of the carbon and nutrients needed for maintenance and growth from an enriched organic medium or the ingestion of bacteria. Likewise, species which are predominantly phototrophic but facultatively phagotrophic may utilize ingested prey to supply nutrients which are limiting to photosynthesis. In natural environments, however, where bacterial densities are typically limiting, the contribution of phagotrophy and osmotrophy to the nutrition of mixotrophic algae is not clear. Although a supplementary dual osmotrophic–phagotrophic nutritional mode seems likely, the uptake of dissolved sources of nutrients may be of minor importance, at least in oligotrophic systems. Capriulo (1990) suggests that because of the low concentrations of dissolved organics typical of most marine ecosystems, uptake by flagellates should be diffusion limited and proportional to the surface area of the cell. The relatively large boundary layer surrounding these cells, in comparison with bacteria, would suggest that mixotrophic algae are poor competitors for dissolved nutrients.

The effect of nutrient limitation, particularly phosphorus, on algal growth has been well documented. Phagotrophy in algae, as an alternative source of nutrients, has the potential for alleviating this constraint on growth. Such supplementary sources could be particularly important in oligotrophic environments or under conditions of light limitation where autotrophic contributions are reduced. Because most chrysophytes have been shown to be photoauxotrophs that require several B vitamins, this obligate requirement could also be satisfied through prey ingestion. Aaronson & Baker (1959) demonstrated that both *O. danica* and *P. malhamensis* were able to obtain their requirement for biotin by ingesting bacteria in a biotin-free medium. As previously mentioned, Ishida & Kimura (1986) demonstrated that phototrophic growth of the mixotroph *Uroglena* required phospholipids which could be supplied by feeding them bacteria. Given the observed ingestion rates of some mixotrophic species, which are not dissimilar to the grazing rates of obligate heterotrophs (Fenchel 1982b; Porter 1988; Nygaard & Hessen 1990), and the nutritional composition of bacteria (Herbert 1976), phagotrophy can potentially meet the many nutrient requirements of mixotrophs.

Factors affecting ingestion

Laboratory feeding studies with *P. malhamensis* have indicated that prey abundance is important in regulating flagellate ingestion rate. The ingestion rate of this flagellate increases with increasing prey abundance and probably exhibits the characteristic Holling type II functional response that has been reported for other bactivorous protozoa (Holen & Boraas 1991). In laboratory cultures, *P. malhamensis* has been observed to graze bacteria down to a 'threshold' concentration of about 1×10^6 cells ml^{-1}. As particulate food becomes limiting, flagellate densities in illuminated cultures tend to stabilize as photosynthesis becomes the dominant nutritional mode; in dark cultures, on the other hand, the flagellate population decreases. Presumably, bacteria present at densities below this minimum concentration are not encountered with sufficient frequency by this raptorial feeder for positive growth of the flagellate to occur by phagotrophic means. Factors other than prey abundance, however, may influence ingestion rate as well. While Sanders *et al.* (1990) and Caron *et al.* (1990) observed no effects of light and nutrient concentration on flagellate grazing rates, Porter (1988) observed an inverse relationship between both light intensity and ingestion rate and light intensity and glucose assimilation for *P. malhamensis*. Ingestion rate decreased from 1.3 to 0.5 beads per flagellate per hour with increasing light intensity and decreased from 1.4 to *c*. 0 beads per flagellate per hour with increasing glucose concentration. However, it should be noted that Porter's data are questionable because they represent ingestion rates that are almost two orders of magnitude lower than other reported ingestion rates for this species. In addition, Veen (1991) suggests that the treatment performed by Porter, i.e. centrifugation of cells, may have negatively influenced the uptake rate of *P. malhamensis*.

The concentration of dissolved organic carbon was shown to affect the ingestion rate of *Ochromonas* sp. (Sibbald & Albright 1991). In treatments where the dissolved organic carbon (DOC) concentration was limiting, the flagellate had an ingestion rate that was 2.5 times higher than equivalent treatments where the DOC concentration was in excess. Limiting nitrogen, phosphorus and vitamins in dark cultures also stimulated phagotrophy. The addition of 3-(3,4)-dichlorophenyl-*1,1*-dimethylurea (DCMU), an inhibitor of photosystem II, however, resulted in no significant difference in ingestion rates for the flagellate in the light and the dark. Thus, light was not shown to affect ingestion.

Loss of cellular chlorophyll

One of the interesting phenomena associated with the change in nutritional mode in some mixotrophic chrysophytes is the decrease in cellular chlorophyll *a* that occurs when the phototrophically-reared flagellates are added to organically enriched or bacterized cultures. This reduction in chlorophyll *a* concentration was observed in *P. malhamensis* when the flagellates were cultured on an abundance of organic food, both in the light and in the dark (Myers & Graham 1956; Handa *et al.* 1981; Sanders *et al.* 1990; Lewitus & Caron 1991*a*).

Using the design of Boraas (1983), we cultured both *P. malhamensis* and *O. danica* in two-stage chemostats on the bacterium *E. coli* to test the effect of phagotrophic nutrition on chlorophyll *a* content. We initially cultured the flagellates in illuminated batch cultures on an inorganic medium to maximize cellular chlorophyll *a* concentrations, and then inoculated the flagellates into the second stage of separate two-stage chemostats containing *E. coli* at a concentration of about 4×10^7 bacteria ml^{-1} (Figs. 6.2, 6.3). There was a dramatic decrease in flagellate chlorophyll *a* per cell concomitant with an exponential increase in flagellate densities. *P. malhamensis* chlorophyll *a* decreased from 600 fg chl *a* $cell^{-1}$ to 60 fg chl *a* $cell^{-1}$. The chlorophyll *a* in *O. danica* decreased from 400 fg chl *a* $cell^{-1}$ to 40 fg chl *a* $cell^{-1}$. The decrease in chlorophyll *a* was correlated with a reduction in chlorophyll autofluorescence as detected by epifluorescence microscopy. As bacterial abundance became limiting to flagellate growth, the cell-specific chlorophyll *a* content in both *P. malhamensis* and *O. danica* increased, but to levels less than those observed before phagotrophy was stimulated, and then entered into a steady state at a final concentration of about 200 fg per flagellate (Figs. 6.2 and 6.3).

The reduction in cellular chlorophyll in *P. malhamensis* appears to be the result of simple dilution by growth as opposed to active degradation (data not shown). The inverse relationship between cell-specific chlorophyll concentration and organic food availability would suggest that a transition in nutritional strategies occurs when these mixotrophs switch from phototrophic to phagotrophic growth. As organic food, i.e. bacteria, becomes available, the cell-specific chlorophyll *a* concentration of the flagellate declines and photosynthesis is reduced (Holen, unpublished data). Thus, there appears to be a transition in dependency from photosynthesis to phagotrophy during growth. Although this transitional period has not been fully investigated, in *P. malhamensis* phagotrophy and phototrophy were observed to occur together (Sanders *et al.* 1990).

The chlorophyll *a* concentration in the cell appears to be directly related to particulate food availability and its subsequent effect on flagellate growth rate. But light is also required for chlorophyll concentration above a basal cell-specific concentration. When growing in a chemostat with a continuous supply of *E. coli*, the chlorophyll *a* concentration in *O. danica* decreased from 200 to 50 fg per cell when the lights were turned off.

Restoring illumination resulted in an increase in cellular chlorophyll *a* back to the original concentration (Fig. 6.3). During the 190 h time interval there was no change in flagellate density, supporting the suggestion that photosynthesis plays a minor role in their metabolism when abundant particulate food is available.

Changes in cell-specific chlorophyll concentration measured via epifluorescence microscopy or direct fluorometric determination are often used as an index of changing nutritional mode in these mixotrophic flagellates. However, an increase in chlorophyll concentration is not necessarily correlated with an increase in photosynthetic rate. The reduction in chlorophyll *a* concentration to 50 fg per cell in *O. danica*, when cultured on *E. coli* in the dark (Fig. 6.3), resulted in the cessation of photosynthesis as evidenced by the lack of ^{14}C uptake when incubated in an artificial light incubator for 1 h at 200 μEin m^{-2} s^{-1}. The ^{14}C uptake measurements during the increase in *O. danica* chlorophyll after illumination was resumed, indicated that photosynthesis did not become active until the flagellate had achieved a chlorophyll *a* concentration of 150 fg per cell (Fig. 6.4). In addition, *O. vallesciaca*, as previously mentioned, contains chlorophyll, as evidenced by chlorophyll autofluorescence, but does not appear to have a photosynthetically competent chloroplast, as evidenced by the lack of autotrophic ^{14}C incorporation in the light. This suggests that the use of chlorophyll autofluorescence via epifluorescence microscopy as an index of phototrophic activity in mixotrophs, particularly for whole water samples, should be used with caution.

A loss of activity in the photosynthetic apparatus as a result of heterotrophic growth is not limited to chrysophytes. A similar response was seen in the marine cryptophyte *Pyrenomonas salina* (Lewitus & Caron 1991*b*). The addition of glycerol to light-limited cultures resulted in an increase in growth rate, a reduced phycoerythrin content and a decrease in thylakoid number. The latter two indices imply a decrease in photosynthetic capacity. The change in the photosynthetic apparatus that was observed in these cryptophytes suggests that heterotrophic nutrition ensues at the expense of photosynthetic capability. The relationship between cell-specific chlorophyll concentration and food supply in chrysophytes may not be as direct as it appears. The increase in cell-specific chlorophyll *a* during heterotrophic growth is presumably a response to food limitation. Lewitus & Caron (1991*a*), however, observed an increase in cellular chlorophyll *a* in *P. malhamensis*, cultured on glucose, when the substrate concentration was relatively high and flagellate growth was still in mid-exponential phase. The authors suggest that cyclic AMP, which is

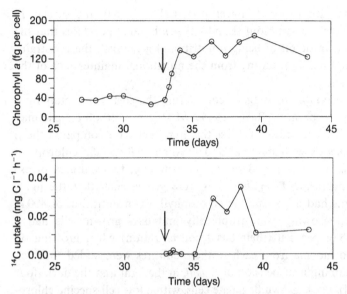

Figure 6.4. Effect of exposure to light on *O. danica* chlorophyll *a* concentration and production in a two-stage chemostat (from Fig. 6.3). The population was maintained at a dilution rate of $1.7 \, d^{-1}$ on *E. coli* in the dark. At the arrow, the culture was illuminated. Cell-specific chlorophyll *a* increased immediately after the lights were turned on. However, photosynthesis (measured by P–I curves) was delayed for almost 3 days in continuous light.

released by the flagellate extracellularly during growth on glucose, conditioned the medium and facilitated the initiation of chlorophyll synthesis. This is supported by the similar response observed by Berchtold & Bachofen (1977) for *Chlorella fusca*.

Trade-offs between phagotrophy and phototrophy

The ecological role of mixotrophy is very probably underappreciated at the moment. With the multiple nutritional pathways of these organisms and the ubiquity of bacteria as potential prey, why are mixotrophs not the usual dominant organisms in lakes? Is there an inherent trade-off between phagotrophy and phototrophy that compensates for the apparent advantage? Alternatively, is mixotrophy a specialization that either renders the nutritional generalist competitively inferior to phagotrophic and phototrophic specialists or renders the generalist more susceptible to mortality losses, e.g. by predation? It has been demonstrated that the ciliate *Paramecium bursaria*, a mixotroph by virtue of its *Chlorella* symbionts, has about a 17% lower potential μ_{max} than the similar *P. aurelia* which lacks a symbiont (Landis 1986). Landis speculates that the lower μ_{max} is a trade-off of being a nutritional generalist, the constraint to growth presumably resulting from the nutritional maintenance of the algal symbionts.

It seems reasonable to suggest, considering the cost of maintaining a photosynthetic apparatus (Raven 1984a, b), that mixotrophic algae may exhibit a similar constraint. To test this hypothesis we compared the μ for *P. malhamensis* with minimal and maximal cell-specific chlorophyll concentrations when cultured on *E. coli*. Initially, two cultures of exponentially increasing *P. malhamensis* were grown such that the light-grown cultures had a cell-specific chlorophyll *a* concentration of 800 fg per cell when growing phototrophically and dark-grown cells had a content of 48 fg per cell (their basal concentration) when growing on bacteria. Two inocula from each of these cultures were added to batch cultures containing a suspension of *E. coli*, in the light and the dark (four cultures total). Dark-grown flagellate cells, with a low cell-specific chlorophyll *a* concentration, had a statistically significantly higher μ than light-grown cells, with a high cell-specific chlorophyll *a* concentration, when cultured on *E. coli*, irrespective of light conditions (Fig. 6.5). These data suggest that a trade-off to mixotrophy for *P. malhamensis* when maintaining maximum chlorophyll may be a constraint on μ when an abundance of organic food is available. Therefore the ability of mixotrophs to reduce their chlorophyll 'load' when growing heterotrophically would have a selective advantage by virtue of an increase in the cell-specific growth rate. In addition, we have found in a comparison between light and dark bacterized cultures containing *P. malhamensis* that the flagellate yields in the dark cultures are consistently about twice those in the illuminated cultures (Holen, unpublished data). This would be consistent with a hypothesis suggesting that chlorophyll synthesis and maintenance represents a metabolic cost to the organism. In dark cultures which preclude chlorophyll synthesis above a basal concentration, flagellates continue through additional divisions after bacteria have been grazed to threshold densities (Fenchel 1982c). In illuminated cultures, however, as bacteria become limiting, chlorophyll synthesis proceeds, utilizing resources that could be used for maintenance and growth, and subsequently results in a lower flagellate yield.

The serial-endosymbiotic theory for the origin of eukaryotic cells suggests that phagotrophy in protists was the ancestral mode of nutrition and that photosynthetic capability arose as a result of the integration of an

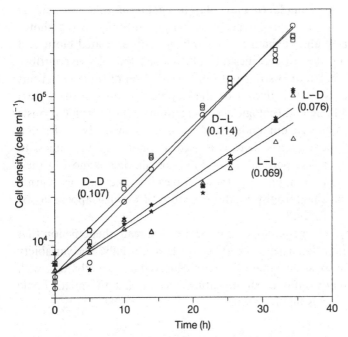

Figure 6.5. The effect of chlorophyll concentration on growth of *P. malhamensis* on a bacterial suspension of *E. coli* in batch cultures. See text for experimental design. The four treatments include flagellate cells with maximal chlorophyll incubated in the light (L–L) and the dark (L–D), and flagellate cells with minimal chlorophyll incubated in the light (D–L) and the dark (D–D). Flagellate cells with minimal chlorophyll *a* per cell had significantly higher growth rates than flagellates with maximal chlorophyll *a* per cell, irrespective of light condition. The numbers in parentheses indicate the slope of the line (i.e., the growth rate in hours).

algal symbiont with heterotrophic host (Margulis 1981). The variability in the contribution of phototrophic versus phagotrophic nutrition that has been observed in chrysophytes suggests that some species, e.g. *Ochromonas* sp. and *P. malhamensis*, may have a high probability of the secondary loss of chloroplast function. This is further evidenced by the finding that *Spumella*, a small colorless chrysophyte, has a rudimentary chloroplast (Mignot 1977; Moestrup, this volume). Whether other colorless chrysophytes represent organisms which have completely lost their plastids, or in contrast never acquired a plastid, and thus diverged from plastidic chrysophytes at an early age in evolution (Round 1980), is unclear.

It has become increasingly apparent that mixotrophic chrysophytes are a common component of many freshwater and marine communities, and in times of peak abundance may play a significant role in bacterial mortality. Given the varying dependence on phagotrophy versus photo-trophy that these algae exhibit and the influence of associated biotic and environmental factors, it becomes difficult to assess the relative contribu-tions of either nutritional strategy to the metabolism of these organisms *in situ*, as well as to their impact on the trophic dynamics of the system as a whole. Although autecological investigations of mixotrophic protists have greatly enhanced our understanding of the growth dynamics and physiology of these species there is a need for more *in situ* studies. In aquatic systems where particulate food may be limiting, is most of their nutrition derived from photosynthesis? Likewise, what is the diurnal contribution of phagotrophy to those mixotrophs which appear to be primarily phototrophic?

Mixotrophy, in general, appears to be a widespread phenomenon that enables these nutritional generalists to compete successfully. The variability in mixotrophic constructs that have been observed and the ubiquity with which these innovations occur in nature suggest that the mixotrophic mode of nutrition is highly favored.

Summary

Recent investigations into the dynamics of the microbial food web have brought about fundamental changes in our understanding of its structure and its impact on higher trophic levels. Particularly in oligotrophic systems, small organisms less than 20 µm, composed primarily of eukaryotic yalgae, cyanobacteria, planktonic bacteria and protozoa, are responsible for most of the energy and nutrient flux that occurs. The transfer of these nutrients and carbon from the microbial food web to higher trophic levels is facilitated by the grazing on protozoa by larger predators. An integral part of this protozoan assemblage is the mixotrophic flagellates which utilize both photosynthesis and phagotrophy for growth. Chrysophytes exhibit a range in mixotrophy from obligate phagotrophy to obligate phototrophy. The chrysophytes *Poterioochromonas malhamensis* and species of *Ochromonas* are predominantly phagotrophic but can use photosyn-thesis as a means of survival during periods of low food abundance. Species of *Dinobryon*, in contrast, are primarily phototrophic but can augment growth with particulate food. Although mixotrophy would appear to be nutritionally advantageous to these algae, constraints on growth due to a dual nutritional capability may exist. The ubiquity with which these and other mixotrophs are found in nature, together with the high abundances that they can achieve, suggests that they should be considered an important component of the microbial food web.

Acknowledgments

The authors gratefully acknowledge Craig Sandgren and Robert Sanders for their critical review of an earlier draft of this chapter. Their input substantially improved its content.

References

Aaronson, S. 1974. The biology and ultrastructure of phagotrophy in *Ochromonas danica* (Chrysophyceae: Chrysomonadida). *J. Gen. Microbiol.* **83**: 21–9.

Aaronson, S. & Baker, H. 1959. A comparative biochemical study of two species of *Ochromonas, J. Protozool.* **6**: 282–4.

Andersson, S., Larsson, U. & Hagström, Å. 1989. Nutritional characteristics of a mixotrophic nanoflagellate *Ochromonas* sp. *Microb. Ecol.* **17**: 252–62.

Azam, F., Fenchel, T., Field, J.G., Gray, J.S., Meyer-Reil, L.A. & Thingstad, F. 1983. The ecological role of water-column microbes in the sea. *Mar. Ecol. Prog. Ser.* **1**: 257-63.

Bennett, S., Sanders, R.W. & Porter, K.G. 1990. Heterotrophic, autotrophic and mixotrophic nanoflagellates: seasonal abundances and bacterivory in a eutrophic lake. *Limnol. Oceanogr.* **35**: 1821–32.

Berchtold, M. & Bachofen, R. 1977. Possible role of cyclic AMP in the synthesis of chlorophyll in *Chlorella fusca. Arch. Microbiol.* **112**: 173–7.

Berninger, U.-G., Finlay, B.J. & Kuuppo-Leinikki, P. 1991. Protozoan control of bacterial abundances in freshwater. *Limnol. Oceanogr.* **36**: 139–46.

Berninger, U.-G., Caron, D.A. & Sanders, R.W. 1992. Mixotrophic algae in three ice-covered lakes of the Pocono Mountains, USA. *Freshwater Biol.* **28**: 263–72.

Bird, D.F. & Kalff, J. 1986. Bacterial grazing by planktonic lake algae. *Science* **231**: 493–5.

Bird, D.F. & Kalff, J. 1987. Algal phagotrophy: regulating factors and importance relative to photosynthesis in *Dinobryon* (Chrysophyceae). *Limnol. Oceanogr.* **32**: 277–84.

Bird, D.F. & Kalff, J. 1989. Phagotrophic sustenance of a metalimnetic phytoplankton peak. *Limnol. Oceanogr.* **34**: 155–62.

Boraas, M.E. 1983. Population dynamics of food-limited rotifers in two-stage chemostat culture. *Limnol. Oceanogr.* **28**: 546–63.

Boraas, M.E., Estep, K.W., Johnson, P.W. & McN. Sieburth, J. 1988. Phagotrophic phototrophs: the ecological significance of mixotrophy. *J. Protozool.* **35**: 249–52.

Boraas, M.E., Seale, D.B. & Holen, D.A. 1992. Predatory behavior of *Ochromonas* analyzed with video microscopy. *Arch. Hydrobiol.* **123**: 459–68.

Capriulo, G.M. 1990. Feeding-related ecology of marine protozoa. In: Capriulo, G.M. (ed.) *Ecology of Marine Protozoa.* Oxford University Press, New York. pp. 186–259.

Caron, D.A., Goldman, J.C. & Dennett, M.R. 1988. Experimental demonstration of the roles of bacteria and bacterivorous protozoa in plankton nutrient cycles. *Hydrobiologia* **15**: 27–40.

Caron, D.A., Porter, K.G. & Sanders, R.W. 1990. Carbon, nitrogen and phosphorus budgets for the mixotrophic phytoflagellate *Poterioochromonas malhamensis* (Chrysophyceae) during bacterial ingestion. *Limnol. Oceanogr.* **35**: 433–43.

Caron, D.A., Sanders, R.W., Lim, E.L., Marrasé, C., Amaral, L.A., Whitney, S., Aoki, R.B. & Porter, K.G. 1993. Light-dependent phagotrophy in the freshwater mixotrophic chrysophyte *Dinobryon cylindricum. Microb. Ecol.* **25**: 93–111.

Carrick, H.J. & Fahnenstiel, G.L. 1991. The importance of zooplankton–protozoan couplings in Lake Michigan. *Limnol. Oceanogr.* **36**: 1335–45.

Cole, G.T. & Wynne, M.J. 1974. Endocytosis of *Microcystis aeruginosa* by *Ochromonas danica. J. Phycol.* **10**: 397–410.

Cole, J.J., Finlay, S. & Pace, M.L. 1988. Bacterial production in fresh and saltwater ecosystems: a cross-system overview. *Mar. Ecol. Prog. Ser.* **43**: 1–10.

Daley, R.J., Morris, G.P. & Brown, S.R. 1973. Phagotrophic ingestion of a blue-green alga by *Ochromonas. J. Protozool.* **20**: 58–61.

Estep, K.W., Davis, P.G., Hargraves, P.E. & McN. Sieburth, J. 1984. Chloroplast containing microflagellates in natural populations of North Atlantic nanoplankton: their identification and distribution; including a description of five new species of *Chrysochromulina* (Prymnesiophyceae). *Prostistologica* **20**: 613–34.

Fenchel, T. 1982*a*. Ecology of heterotrophic microflagellates. I. Some important forms and their functional morphology. *Mar. Ecol. Prog. Ser.* **8**: 211–23.

Fenchel, T. 1982*b*. Ecology of heterotrophic microflagellates. II. Bioenergetics and growth. *Mar. Ecol. Prog. Ser.* **8**: 225–31.

Fenchel, T. 1982*c*. Ecology of heterotrophic microflagellates. III. Adaptations to heterogeneous environments. *Mar. Ecol. Prog. Ser.* **9**: 25–33.

Fenchel, T. 1982*d*. Ecology of heterotrophic microflagellates. IV. Quantitative occurrence and importance as bacterial consumers. *Mar. Ecol. Prog. Ser.* **9**: 35–42.

Fenchel, T. & Finlay, B.J. 1983. Respiration rates in heterotrophic, free-living protozoa. *Microb. Ecol.* **9**: 99–122.

Gasol, J.M. & Vaqué, D. 1993. Lack of coupling between heterotrophic nanoflagellates and bacteria: a general phenomenon across aquatic systems. *Limnol. Oceanogr.* **38**: 657–65.

Gonzales, J.M., Sherr, E.B. & Sherr, B.F. 1990. Size-selective grazing on bacteria by natural assemblages of estuarine flagellates and ciliates. *Appl. Envir. Microbiol.* **56**: 583–9.

Güde, H. 1979. Grazing by protozoa as selection factor for activated sludge bacteria. *Microbiol. Ecol.* **5**: 225–37.

Güde, H. 1988. Direct and indirect influences of crustacean zooplankton on bacterioplankton of Lake Constance. *Hydrobiologia* **159**: 63–73.

Handa, A.K., Bressan, R.A., Quader, H. & Filner, P. 1981. Association of formation and release of cyclic AMP with glucose depletion and onset of chlorophyll synthesis in *Poterioochromonas malhamensis, Plant Physiol.* **68**: 460–3.

Herbert, D. 1976. Stoichiometric aspects of microbial growth. In: Dean, A.C.R., Ellwood, D.C., Evans, C.G.T. & Melling, J. (eds.) *Continuous Culture*: 6. *Applications and New Fields*. Ellis Horwood, Chichester, England. pp. 1–30.

Holen, D.A. & Boraas, M.E. 1991. The feeding behavior of *Spumella* sp. as a function of particle size: implications for bacterial size in pelagic systems. *Hydrobiologia* **220**: 73–88.

Ishida, Y. & Kimura, B. 1986. Photosynthetic phagotrophy of Chrysophyceae: evolutionary aspects. *Microbiol. Sci.* **3**: 132–5.

Johannes, R.E. 1964. Phosphorus excretion and body size in marine animals: microzooplankton and nutrient regeneration. *Science* **146**: 923–4.

Landis, W.G. 1986. The interplay among ecology, breeding systems and genetics in the *Paramecium aurelia* and *Paramecium bursaria* complexes. *Prog. Protistol.* **1**: 287–307.

Lehman, J.T. 1976. Ecological and nutritional studies on *Dinobryon* Ehrenb.: seasonal periodicity and the phosphate toxicity problem. *Limnol. Oceanogr.* **21**: 646–58.

Lewitus, A.J. & Caron, D.A. 1991a. Physiological responses of photoflagellates to dissolved organic substrate additions. I. Dominant role of heterotrophic nutrition in *Poterioochromonas malhamensis* (Chrysophyceae). *Plant. Cell. Physiol.* **32**: 671–80.

Lewitus, A.J. & Caron, D.A. 1991b. Physiological responses of phytoflagellates to dissolved organic substrate additions. II. Dominant role of autotrophic nutrition in *Pyrenomonas salina* (Cryptophyceae). *Plant Cell Physiol.* **32**: 791–801.

Lu, T. 1988. A study on the growth and nutrition of a fresh-water species of *Ochromonas*. Master's thesis, University of Wisconsin-Milwaukee.

Margulis, L. 1981. *Symbiosis in Cell Evolution.* Freeman, San Franciso.

Mignot, J.P. 1977. Etude ultrastructurale d'un flagellé du genre *Spumella* Cienk (= *Heterochromonas* Pascher = *Monas* O.F. Muller), Chrysomonadine Leucoplastidiée. *Protistologica* **13**: 219–31.

Myers J. & Graham, J. 1956. The role of photosynthesis in the physiology of *Ochromonas*. *J. Cell Comp. Physiol.* **47**: 397–414.

Nygaard, K. & Hessen, D.O. 1990. Use of ^{14}C-protein-labeled bacteria for estimating clearance rates by heterotrophic and mixotrophic flagellates. *Mar. Ecol. Prog. Ser.* **68**: 7–114.

Nygaard, K. & Tobiesen, A. 1993. Bacterivory in algae: a survival strategy during nutrient limitation. *Limnol. Oceanogr.* **38**: 273–9.

Pascher, A. 1943. Zur Kenntnis verschiedener Ausbilkungen der planktonischer *Dinobryon*. *Int. Rev. Ges. Hydrobiol.* **43**: 110–23.

Platt, T., Subba Rao, D.V. & Irwin, B. 1983. Photosynthesis of picoplankton in the oligotrophic ocean. *Nature* **301**: 702–4.

Porter, K.G. 1988. Phagotrophic phytoflagellates in microbial food webs. *Hydrobiologia* **159**: 80–97.

Pringsheim, E.G. 1952. On the nutrition of *Ochromonas*. *Q. J. Microsc. Sci.* **93**: 71–96.

Raven, J.A. 1984a. *Energetics and Transport in Aquatic Plants.* A.R. Liss, New York.

Raven, J.A. 1984b. A cost-benefit analysis of photon absorption by photosynthetic unicells. *New Phytol.* **98**: 593–625.

Rogerson, A., Finlay, B.J. & Berninger, U.G. 1989. Sequestered chloroplasts in the freshwater ciliate *Strombidium viridae* (Ciliphora: Oligotrichida). *Trans. Am. Microsc. Soc.* **108**: 117–26.

Round, F.E. 1980. The evolution of pigmented and unpigmented unicells: a reconsideration of the protista. *Biosystems* **12**: 61–9.

Sanders, R.W. 1991. Mixotrophic protists in marine and freshwater ecosystems. *J. Protozool.* **38**: 76–81.

Sanders, R.W. & Porter, K.G. 1988. Phagotrophic phytoflagellates. In: Marshall, K.C. (ed.) *Advances in Microbial Ecology*. Plenum Publishing, New York. pp. 167–92.

Sanders, R.W. & Porter, K.G. 1990. Bacterivorous flagellates as food resources for the freshwater crustacean zooplankter *Daphnia ambigua*. *Limnol. Oceanogr.* **35**: 188–91.

Sanders, R.W., Porter, K.G., Bennett, S.J. & DeBiase, A.E. 1989. Seasonal patterns of bacterivory by flagellates, ciliates, rotifers and cladocerans. *Limnol. Oceanogr.* **34**: 673–87.

Sanders, R.W., Porter, K.G. & Caron, D.A. 1990. Relationship between photo-trophy and phagotrophy in the mixotrophic chrysophyte *Poterioochromonas malhamensis*. *Microb. Ecol.* **19**: 97–109.

Sandgren, C.D. 1986. Effects of environmental temperature on the vegetative growth and sexual life history of *Dinobryon cylindricum* Imhof. In: Kristiansen, J. & Anderson, R.A. (eds.) *Chrysophytes: Aspects and Problems*. Cambridge University Press, Cambridge. pp. 207–25.

Sandgren, C.D. 1988. The ecology of chrysophyte flagellates: their growth and perennation strategies as freshwater phytoplankton. In: Sandgren, C.D. (ed.) *Growth and Reproductive Strategies of Freshwater Phytoplankton*. Cambridge University Press, Cambridge. pp. 9–104.

Seale, D.B., Boraas, M.E., Holen, D.A. & Nealson, K.H. 1990. Use of bioluminescent bacteria, *Xenorhabdus luminescens*, to measure predation on bacteria by freshwater microflagellates. *FEMS Microbiol. Ecol.* **73**: 31–40.

Sibbald, M.J. & Albright, L.J. 1991. The influence of light and nutrients on phagotrophy by the mixotrophic nanoflagellate *Ochromonas* sp. *Mar. Microbial Food Webs* **5**: 39–47.

Smith, D.C. & Douglas, A.E. 1987. *The Biology of Symbiosis*. Edward Arnold, London. 302pp.

Starr, R.C. 1978. The culture collection of algae at the University of Texas at Austin. *J. Phycol.* **14** (Suppl.): 47–100.

Stockner, J.G. 1988. Phototrophic picoplankton: an overview from marine and freshwater ecosystems. *Limnol. Oceanogr.* **33**: 765–75.

Stockner, J.C. & Antia, N.J. 1986. Algal picoplankton from marine and freshwater ecosystems: a multidisciplinary perspective. *Can. J. Fish. Aquat. Sci.* **43**: 2472–503.

Stoecker, D.K., Michaels, A.E. & Davis, L.H. 1987. Larger proportion of marine planktonic ciliates found to contain functional chloroplasts. *Nature* **326**: 790–2.

Stoecker, D.K., Silver, M.W., Michaels, A.E. & Davis, L.H. 1988. Obligate mixotrophy in *Laboea strobila*, a ciliate which retains chloroplasts. *Mar. Biol.* **99**: 415–23.

Tranvik, L.J., Porter, K.G. & McN. Sieburth, J. 1989. Occurrence of bacterivory in *Cryptomonas*, a common freshwater phytoplankter. *Oecologia* **78**: 473–6.

Turner, J.T., Tester, P.A. & Ferguson, R.L. 1988. The marine cladoceran *Penilla avisostris* and the 'microbial loop' of pelagic food webs. *Limnol. Oceanogr.* **33**: 245–55.

Vaqué, D. & Pace, M.L. 1992. Grazing on bacteria by flagellates and cladocer-ans in lakes of contrasting food-web structure. *J. Plankton Res.* **14**: 307–21.

Veen, A. 1991. Ecophysiological studies on the phagotrophic phytoflagellate *Dinobryon divergens* Imhof. PhD thesis, University of Amsterdam, The Netherlands.

7

Biomineralization and scale production in the Chrysophyta

BARRY S.C. LEADBEATER &
DAVID A.N. BARKER

Introduction

Biomineralization is the process by which living organisms assemble structures from naturally occurring inorganic compounds. Most groups of living organisms have members that deposit minerals and in many instances the mineralized structures provide skeletal support and protection for softer organic parts. Within the Chrysophyta a diverse range of biomineralized structures are produced (Preisig 1986). The biogenic material most extensively deposited by chrysophytes is silica, which occurs almost universally in stomatocysts and is commonly present in mineralized scales. Other minerals are deposited by members of the Chrysophyceae as well. For instance, iron and manganese mineralized material occurs in the brown-colored stalks of *Anthophysa vegetans*, the brown-colored loricae of *Pseudokephyrion pseudospirale*, and the gelatinous holdfasts of *Phaeothamnion articulata* (Preisig 1986). Calcareous deposits are rare in the Chrysophyta, being limited to the pseudocysts of a few species of the Sarcinochrysidales and the mucilage of some species, such as *Celloniella*, which form gelatinous colonies (Preisig 1986; see Moestrup, this volume).

This review is concerned with the utilization of silica by chrysophytes. The information will be presented in the overall context of the biogeochemical cycling of silica.

Biogeochemical cycling of silica

Silicon is the second most abundant element in the Earth's crust, accounting for some 28% of its mass. Invariably it occurs in chemical composition with oxygen, as in the crystalline and cryptocrystalline polymers of silica (SiO_2). Common naturally-occurring substances in-

141

volving additional elements include: kaolinite, feldspars, micas, and the so-called clay minerals (Stumm & Morgan 1970). Available measurements confirm that all of these minerals are only sparingly soluble in water. Below pH 9.0 silicon is generally released as weak monosilicic acid and it is in this form that silicon is generally utilized by protistan organisms. The silicon content of marine and freshwater habitats varies considerably, but, in general, concentrations of reactive silicate are higher in freshwaters than in the sea.

Planktonic cells take up monosilicic acid by an active transport mechanism which remains effective at low external concentrations of silicic acid. Such organisms are capable of reducing the external concentration of soluble reactive silicate to barely detectable levels. Intracellular deposition of polymerized silica characteristically occurs within membrane-bounded vesicles known as silica deposition vesicles (SDVs). When the silica structure is mature it is then extruded to the outer surface of the protoplast. This is achieved either by exocytosis or by withdrawal of the surrounding protoplast from the outer surface of the silicified structure.

Once in contact with the aquatic medium silica structures dissolve. In marine habitats, where the water column may be up to 6 km in depth, re-cycling of silicon is very nearly complete. However, most freshwater lakes lack sufficient depth of water in which sedimenting silica particles can dissolve. Once at the sediment–water interface or within the surface sediment layers, amorphous silica structures continue to dissolve with the result that 95% or more of silicon may be returned to the water column. Silica particles that do not dissolve may eventually become incorporated into sediments and ultimately they may become lithified to form rock. Of the siliceous remains of protista that reach the sediment surface in marine environments only about 2% or less are preserved in fossil deposits (Hurd 1973; Heath 1974). In freshwater lakes, the percentage of biogenic silica that is incorporated into the permanent sediments may be higher (Reynolds 1986).

Background to studies on the utilization of silica by chrysophytes

Recognition that chrysophytes produce silicified structures came relatively early in the history of chrysophyte studies. One of the earliest records of silicified stomatocysts was that of Cienkowski (1870) for *Chromulina nebulosa*, and for silicified scales there is Iwanoff's (1899) record for *Mallomonas acaroides*. Until 1983, all chrysophyte taxa bearing silica scales were classified exclusively within the single family Mallomonadaceae,

Table 7.1. *Genera within Mallomonadaceae at time of Bourrelly's (1981) review and subsequent changes to classification in accordance with Preisig & Hibberd (1982a, 1983), Andersen (1987) and Preisig (this volume)*

After Bourrelly (1981)	Classification 1993
Chrysophyceae	Synurophyceae
Mallomonadaceae	Synuraceae
Synura Ehr.	*Synura* Ehr.
Mallomonas Perty	*Tessellaria*
Mallomonopsis Matvienko	Mallomonadaceae
Paraphysomonas (Stokes) de Saedeleer	*Mallomonas* Perty
Spiniferomonas Takahashi	*Mallomonopsis* Matvienko
Chrysosphaerella Lauterborn	Chrysophyceae
Conradiella Pascher	Paraphysomonadaceae
Microglena Ehrenberg	*Paraphysomonas* (Stokes) de Saedeleer
Catenochrysis Perman	*Spiniferomonas* (Takahashi) Nicholls
	Chrysosphaerella Lauterborn
	Polylepidomonas Preisig & Hibberd
	Conradiella Pascher
	Microglena Ehrenberg
	Catenochrysis Perman

which at the time of Bourrelly's (1981) survey contained nine genera (Table 7.1). Since 1981, a number of important changes to the classification of these genera have taken place (Table 7.1), with the result that silica-bearing chrysophytes are no longer considered to form a homogeneous grouping. As a result of an extensive study of *Paraphysomonas* and related genera, Preisig & Hibberd (1982a, b, 1983) came to the conclusion that, in terms of cell structure, *Paraphysomonas*, *Spiniferomonas*, *Chrysosphaerella* and *Polylepidomonas* were more closely related to members of the Ochromonadaceae than to members of the Mallomonadaceae. They resolved this matter taxonomically by creating a new family, Paraphysomonadaceae, to include Chrysophyceae with an *Ochromonas*-type of cell structure but which produces scales (Preisig & Hibberd 1983).

Within the Paraphysomonadaceae, a number of taxonomic readjustments became necessary as a result of Preisig & Hibberd's (1982a, b, 1983) work (see Preisig, this volume) The type species of *Spiniferomonas* and the only species of *Lepidochromas* were found to be colorless and to have scales similar to *Paraphysomonas*. The subsequent transfer of these species to *Paraphysomonas* invalidated both generic names. To accommodate the six remaining pigmented species previously ascribed to *Spiniferomonas* a new genus, *Chromophysomonas*, was therefore created (Preisig & Hibberd 1982b). In addition, another new genus, *Polylepidomonas*, was created for *Paraphysomonas vacuolata*, which was shown to be pigmented (Preisig & Hibberd 1983). More recently Nicholls (1985) and Skogstad & Reymond (1989) have shown that cells identified as *Spiniferomonas bourrellyi*, the type species of *Spiniferomonas*, have chloroplasts. Nicholls' (1985) observation therefore revalidates the genus *Spiniferomonas*. The systematic position of the three other genera attributed to the Mallomonadaceae, namely *Conradiella*, *Microglena* and *Catenochrysis*, remains uncertain until they can be cultured and studied with electron microscopy.

Based on an accumulation of evidence, Andersen (1987) came to the conclusion that *Synura*, *Mallomonas* and *Mallomonopsis* were sufficiently dissimilar from the remainder of the Chrysophyceae to warrant their inclusion in a new class, Synurophyceae Andersen. This move follows up Preisig & Hibberd's (1983) conclusion that members of the Mallomonadaceae are not closely related to the Paraphysomonadaceae. Recently, on the basis of an ultrastructural study, *Tessellaria*, a colonial chrysophyte from Tasmania with silica scales, has been ascribed to the Synurophyceae (Tyler *et al.* 1989; Pipes *et al.* 1991; Pipes & Leedale 1992).

Microanatomy of siliceous structures

Scales

Scale morphology has been and remains an important character in chrysophyte taxonomy (Kristiansen 1986a, b). However, in contrast to the Prymnesiophyceae in which there is an underlying consistency to the microfibrillar composition of scales, chrysophyte scales do not have an obvious universal substructure. The existence of an organic component within scales has only been demonstrated for *Synura petersenii* (McGrory & Leadbeater 1981).

Whilst silica structures will inevitably increase the density of cells, the overall form and mechanical strength of scales is generally achieved with minimal use of silica. Thus, baseplates may be perforated as in *Paraphysomonas foraminifera* (Lucas 1967), *Tessellaria volvocina* (Tyler *et al.* 1989) and *Synura petersenii* (Fig. 7.5), or of a meshwork type such as those of *Paraphysomonas homolepis* (Preisig & Hibberd 1982b). Long spines may be constructed of an open meshwork as in *Paraphysomonas eiffelii*, or may be very thin in which case they may be supported by engaged buttresses as in *P. caelifrica* and *Chromophysomonas trioralis*, or may be hollow as in *Paraphysomonas vestita* (Preisig & Hibberd 1982a,

1983). In *P. vestita* the hollow shaft bears annular or helical markings (Fig. 7.6), a characteristic way of providing extra strength to the wall of a cylinder. In *Synura*, as exemplified by *S. petersenii*, transverse supporting struts on the distal surface of the baseplate extend from the cylinder to the scale periphery, thereby supporting the central cylinder and ensuring that the relatively thin baseplate remains flat (Fig. 7.6). The rim, although appearing substantial, is not solid when observed in vertical section but has a central layer of small spherical chambers which lack silica (Lead-beater 1986). Similarly, each of the massive silica scales of *Mallomonas splendens* consists of an outer shell of silica supported internally by a framework of silica partitions (Beech *et al.* 1990).

In addition to a layer of scales, many species of *Mallomonas* bear a partial or complete covering of bristles (Kristiansen 1986a). As in *Mallomonas splendens*, the base of the bristle articulates with the proximal surface of the dome of the accompanying baseplate-scale. Bristles may be sculpted in various ways, including longitudinal grooves, lateral serrations or distal hooks. In chrysophytes, spines and bristles may be arranged in a close-packed array over the entire surface of the cell as in *Paraphysomonas vestita* and *P. imperforata*; they may be generally distributed over the surface of the cell but interspersed with plate-scales as in *Spiniferomonas bourrellyi* (Preisig & Hibberd 1982a); or they may be located at the two poles as in *Mallomonas splendens* (Beech *et al.* 1990).

Scale case structure

On the protoplast surface, scales are usually combined to form a case (Figs. 7.1 and 7.3). Scales may be arranged in one distinctive layer, as in *Synura* (Fig. 7.15) and *Mallomonas*, but often, as in species of *Paraphysomonas*, overlap between scales may be such as to give the impression of a multilayered case (Fig. 7.2). However, with few exceptions there is no clear distinction between underlying and overlying layers of scales as there is within the Prymnesiophyceae. Instead, overlapping plate-scales in a layer are mixed with spine- or crown-scales which are themselves in direct contact with the protoplast surface (Preisig & Hibberd 1982a). *Tessellaria volvocina*, a colonial synurophyte from Tasmania, produces plate- and spine-scales which are not associated with individual cells but are located in a multilayered investment on the surface of the colony (Tyler *et al.* 1989; Pipes & Leedale 1992).

A character which distinctively separates *Synura* and *Mallomonas* from other silica-scale-bearing chrysophytes is the construction of the scale

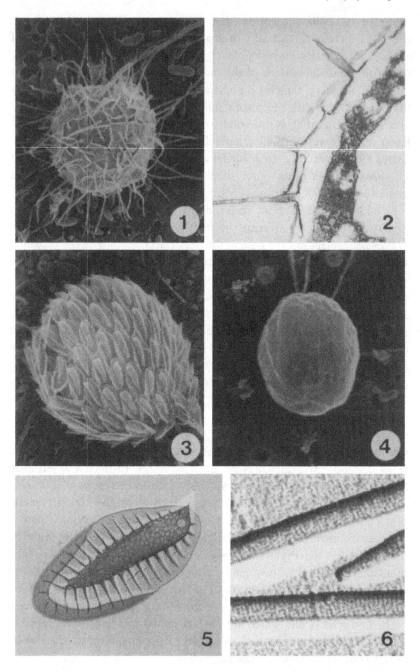

case. The arrangement has been most thoroughly worked out for *Synura petersenii*, which has a case made up of precisely arranged, overlapping scales (Leadbeater, 1984, 1986, 1990). The distal surface of the rim of each scale is covered by a layer of diffuse material which serves as an adhesive for holding scales together (see Wetherbee *et al.*, this volume) In an undisturbed scale case, the imbrication of the scales is so precise that the rims are exactly covered and cannot be seen when viewed from the outside (Fig. 7.3). The number of scales per case varies according to the size of the cell (see Smol, this volume). For the strain used extensively for culture work in Birmingham, large cells contain approximately 90 scales (Leadbeater 1990). When a *Synura* or *Mallomonas* cell divides, approximately half the parent scale case is bequeathed to each of the two daughter cells (Beech *et al.* 1990), so that during the subsequent interphase each of the daughter cells adds scales to the existing case until a larger case with 90 scales is produced. Thus, during interphase enough silica is taken up from the medium and deposited to produce this complement of scales. *Mallomonas splendens* has a basically similar arrangement of body scales around the protoplast although the morphology and patterning of the distal scale surface differs from that of *S. petersenii* (Beech *et al.* 1990).

Cysts

Whilst there is an extensive literature (see Sandgren 1983; Cronberg, 1986; Smol, this volume) relating to ecological and paleolimnological aspects of chrysophycean cysts, knowledge regarding the factors responsible for triggering encystment and excystment and the role of cysts in the life history of chrysophytes is far from complete (see Sandgren 1991). There is general agreement that cysts are perennating structures and in this context their thick siliceous walls are well designed to withstand mechanical damage and dissolution, thereby protecting the enclosed protoplast. There are a limited number of studies relating to the development of cysts,

---

Figure 7.1. *Paraphysomonas vestita.* Scanning electron micrograph (SEM) of whole cell with covering of scales. × 5000.
Figure 7.2. *P. vestita.* Vertical section through scale case showing imbrication of scales. × 25 000.
Figure 7.3. *Synura petersenii.* SEM of silica-replete cell covered with scales. × 2500.
Figure 7.4. *S. petersenii.* SEM of naked, silica-impoverished cell. × 2000.
Figure 7.5. Scale case of *S. petersenii* showing perforated baseplate and central cylinder, transverse supporting struts and rim. × 11 000.
Figure 7.6. *P. vestita.* Shadowcast whole mount of part of a spine shaft showing transverse markings. × 120 000.

particularly the ultrastructure of silica deposition (Hibberd 1977; Sandgren 1980*a*, *b*, 1983, 1989; Skogstad & Reymond 1989), but no information is currently available on the silicon content or solubility of cysts.

The cellular silicon content of chrysophytes

The only chrysophytes in which the cellular silicon content has been studied in detail are *Synura petersenii* (Sandgren & Barlow 1989; Saxby 1990; P.C. Towlson, unpublished observations), *Paraphysomonas vestita* (Leadbeater, unpublished observations) and *P. imperforata* (Barker, unpublished observations). For all taxa, the determination of cellular silicon content was based on batch culture growth experiments in which growth and silica uptake were measured (Table 7.2).

When cellular silicon content is expressed in relation to cell surface area, the values obtained for *Synura petersenii* are of the same order of magnitude as those obtained for several diatoms (Table 7.2) (Klaveness & Guillard 1975; Saxby 1990). The values obtained for *Paraphysomonas vestita* and *P. imperforata* are lower.

Requirement of chrysophytes for silicon

Klaveness & Guillard (1975) investigated growth and silica uptake kinetics of *Synura petersenii* at a range of initial silicate concentrations. Using inocula containing low concentrations of *Synura* colonies they measured growth in batch culture. They found that below a concentration of 1.0 µM reactive silicate the rate of growth of *Synura* was considerably decreased and that half of the maximum growth rate was achieved at a concentration of only 0.23 µM silicon. The growth rate, once set to a constant rate by initial conditions, remained unchanged even beyond the point where the silicate concentration in the medium started to drop measurably. Thus, it appears that there is not an immediate relationship between the growth rate of *S. petersenii* and the silicon concentration of the medium. These results do not confirm an absolute requirement for silicon but they do show that *S. petersenii* requires small amounts of silicate to grow well and that this species is capable of depleting the medium of silicate. Experiments carried out by C. Sandgren & S. Barlow (personal communication) using a continuous culture system also demonstrated that below a threshold concentration of dissolved silicate the rate of growth was proportionately related to the silicon content of the medium.

Table 7.2. *Silicon content of* Synura petersenii, Paraphysomonas vestita *and* P. imperforata *relative to cell surface area. Comparable data are given for a selection of diatoms and the collared flagellate* Stephanoeca diplocostata

Species	Silicon per cell (pg Si per cell)	Cell surface area (μm^2)	Silicon unit surface (pg Si μm^{-2})	Reference
Synura petersenii	22–24	500	0.046	Klaveness & Guillard (1975)
	26–29	308	0.09	Saxby (1990)
Paraphysomonas vestita	0.84	85	0.01	Leadbeater (unpublished)
Paraphysomonas imperforata	0.42	61	0.007	Barker (unpublished)
Asterionella formosa	60.6	860	0.07	Reynolds (1984)
Fragillaria crotonensis	117	1058	0.11	Reynolds (1984)
Stephanodiscus hantzschii	16	404	0.04	Reynolds (1984)
Thalassiosira decipiens	250	1800	0.13	Paasche (1973)
T. pseudonana			0.019	Klaveness & Guillard (1975)
Stephanoeca diplocostata	2.1	80	0.026	Leadbeater & Davies (1984)

However, in both Klaveness & Guillard's (1975) and Sandgren & Barlow's (1989) experiments, silica impoverishment affected scale production well before growth was severely affected. In silica-impoverished media (<0.33 µM Si), *Synura* initially produces less well silicified scales and ultimately no scales at all, with the result that eventually colonies of 'naked' cells develop (Fig. 7.4) (Sandgren & Barlow 1989; Saxby 1990; see also Sandgren & Walton, this volume). Cultures of scaleless cells will grow indefinitely in medium with undetectable amounts of silica provided they are illuminated and supplied with other nutrients. Similar results were obtained for *Paraphysomonas* except, of course, that this taxon is phagotrophic and not photoautotrophic. Batch cultures of *P. imperforata* and *P. vestita* grown at unmeasurable levels of silicon eventually produce completely scaleless cells.

The response to silica impoverishment described for *Synura* and *Paraphysomonas* is in contrast to that demonstrated by diatoms, where depletion of silica ultimately leads to a cessation of growth. Diatoms have an absolute requirement for silicon at two levels. They require silicon in minute amounts for two enzymes critical to DNA synthesis, namely DNA polymerase and thymidilate kinase, and for synthesis of a number of DNA-binding proteins (Ludwig & Volcani 1986). In addition they have a major silicon requirement for valve production during cell division. In the absence of silicon new valves cannot be deposited, with the result that, following nuclear division, diatoms are unable to undergo cytokinesis (Busby & Lewin 1967).

Germanium dioxide is considered to be a specific inhibitor of silicate metabolism and, for those organisms considered to have an absolute silicon requirement, treatment with germanium dioxide would be expected to inhibit growth (Lewin 1966). The degree of inhibition is dependent on the atomic ratio between germanium and silicon present in the medium. For *Synura petersenii* and *Paraphysomonas* addition of germanium dioxide at low concentrations inhibited growth (Klaveness & Guillard 1975; Lee 1978). However, the results are not sufficiently comprehensive to conclude that these organisms have an absolute requirement for silicon.

Silica replenishment of silica-impoverished cells

When silicon is resupplied to actively growing silicon-impoverished *Synura petersenii* cells without scales the overall result is that within about 24 h most cells have achieved a complete covering of new scales (Sandgren & Barlow 1989; Saxby 1990; P.C. Towlson, unpublished observations).

There are conflicting accounts relating to the effects of silicon replenishment on growth. Saxby (1990) found that, in all experiments and at two light intensities used (16 and 243 μmol m^{-2} s^{-1}), reactive silicate addition to silicon-deprived batch cultures yielded an immediate increase in growth rate comparable to that observed for silicon-replete cells in silicon-rich medium. On the other hand, Sandgren & Barlow (1989) found that replenishment of silicon to chemostat cells grown at a light intensity of 250 μmol m^{-2} s^{-1} initially inhibited growth only to be followed by an enhanced rate of growth for the period 24–48 h after silicon addition. The results of P.C. Towlson (unpublished observations) are more or less in agreement with those of Sandgren & Barlow (1989).

For silicon-impoverished naked cells to make up a deficit of silica scales when supplied with silicon, the normal closely coupled relationship between the rates of silicon uptake and growth must in some way be modified. Instead of producing half a mature case of scales during interphase as would normally be required for silicon-replete cells, silicon-replenished naked cells must produce a whole case of scales. This catching up process could be achieved in a number of ways:

(1) The duration of interphase may remain the same but the rate of silicon uptake is increased.
(2) The duration of interphase may remain the same but the period of silicon uptake within interphase is extended if possible.
(3) The duration of interphase is lengthened but the rate of silicon uptake remains as normal.
(4) Various combinations of the above three conditions may occur.

The practical effects of (1) and (2) would result in an increased quantity of silicon being taken up during a period of normal growth. To distinguish between options (1) and (2) would require a method of determining the period during interphase in which silicon is taken up and in which scales are produced. The latter could be determined by careful study of sectioned material prepared for electron microscopy. In this way it might be possible to determine whether or not all cells in interphase contain developing scales. The practical effect of (3) would be a reduction in the rate of growth immediately after supply of silicon to silicon-impoverished cells. The results obtained by Sandgren & Barlow (1989) and Towlson (unpublished observations) for *S. petersenii* could be explained by option (3). The results obtained by Saxby (1990) could be explained by either options (1) or (2). However, the differences in these results are not irreconcilable since the different options need not be mutually exclusive, and the manner in which

Table 7.3. *Paraphysomonas imperforata growth and silicon uptake rates for cells grown at 20 °C in silica-replete medium, silica-impoverished medium and for silicon-impoverished cells after addition of silicon*

	Doubling time (h)	Rate of silicon uptake (10^9 µg Si h^{-1} per cell)
Silica-replete culture	4.1	35
Silica-impoverished cells in silica-depleted medium	4.1	–
Silica-depleted cells replenished with silica	4.1	(i)[a] 224 (ii)[b] 82

[a] For 1.5 h immediately after silicon addition.
[b] For the following 4 h.

cells respond to silicon replenishment probably depends on the physiological condition of the cells at the time of silicon resupply.

In *Paraphysomonas*, replenishment of silicon to silicon-starved cells also results in each cell producing a full complement of scales. In experiments carried out on *P. imperforata*, growth was not affected but the rate of silicon uptake increased for 1.5 h immediately after spiking with silicon and continued for the next 4 h at an enhanced rate (Table 7.3; Barker unpublished observations).

Intracellular deposition of silica

A universal feature of the intracellular deposition of silica by protista is that the polymerized silica is deposited within membrane-bounded silica deposition vesicles (SDVs). The origin of SDVs in chrysophytes has been the source of much speculation, with the Golgi apparatus and endoplasmic reticulum (ER) being most frequently cited (Mignot & Brugerolle 1982; Lee 1978). Whereas non-mineralized scales, such as the flagellar scales of *Synura petersenii* (Fig. 7.7) and the flagellar and body scales of *Sphaleromantis tetragona*, develop within the dictyosome (Fig. 7.8; Manton & Harris 1966), the source of SDVs in chrysophytes remains unclear. Occasionally in *Synura petersenii*, large cisternae similar to SDVs are observed near to the dictyosome (Fig. 7.9, arrows), but such appearances are relatively uncommon. In both the Paraphysomonadaceae and the Synurophyceae, silica scales are deposited during interphase, and exocystosis of scales probably occurs throughout this period of the cell cycle.

In *Paraphysomonas vestita*, which has a covering of siliceous spines, SDVs are first recognized at right angles to the dictyosome (Fig. 7.11). They can be distinguished by their distinctive flattened form and the associated layer of darker staining, diffuse material. A flattened cisterna of ER, also at right angles to the dictyosome and closely appressed to the SDV, always intervenes between the dictyosome and the SDV (Fig. 7.11). Direct contact between the dictyosome and a SDV has never been observed. SDVs containing developing scales migrate from a position near to the dictyosome into the peripheral cytoplasm; the ER remains associated with the proximal surface of the SDV (Fig. 7.12). As silica is deposited, the area of ER appressed to the surface of a SDV reduces, until only the central region remains in contact. Scale formation in other species of *Paraphysomonas*, *Chromophysomonas*, *Polylepidomonas* and *Chrysosphaerella* is more or less identical to that in *Paraphysomonas vestita*. In all taxa there is close association between a SDV and the subtending ER. The latter may serve as a template for shaping a scale (Lee 1978; Preisig & Hibberd 1983), and this is particularly apparent in *P. quadrispina* which has tower-like spines on the distal surface of some scales. So far, neither microtubules nor microfilaments have been observed in association with developing scales.

In *Synura*, SDVs are positioned on the distal surface of the periplastidial ER surrounding one of the two chloroplasts (Fig. 7.10; Mignot & Brugerolle 1982; Leadbeater 1984, 1986, 1990). On the basis of the similarity in cytochemical staining of dictyosome vesicles and SDVs, Mignot & Brugerolle (1982) suggested that SDVs originated from dictyosome cisternae or dictyosome-derived vesicles. Once associated with the distal surface of the periplastidial ER, each SDV becomes associated with an extensive array of actin microfilaments and overlying microtubules (Brugerolle & Bricheux 1984; Leadbeater 1986). The microtubules are probably responsible for positioning SDVs and the microfilaments for shaping the rims and baseplates. A diverticulum of the periplastidial ER projects outwards and forms a mould around which the central hollow cylinder of the scale forms. The scale matures by deposition of silica, and as the scale becomes more rigid the microtubules and microfilaments disappear (Leadbeater 1986). SDV's containing mature scales subsequently migrate towards the plasmalemma. In *Mallomonas* spp. bristles are deposited in separate SDVs which are independent of the baseplate scales (Wujek & Kristiansen 1978; Mignot & Brugerolle 1982; Beech *et al.* 1990). As with the body scales, SDVs containing bristles are associated with chloroplast ER. In *Tessellaria volvocina*, ascribed by Pipes *et al.* (1991) to

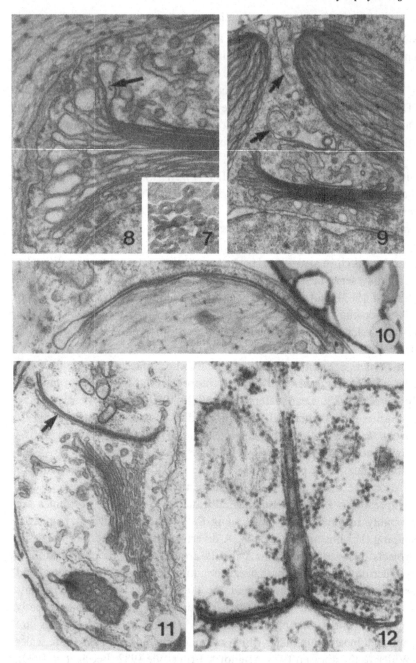

the Synurophyceae, SDVs originate at the anterior end of the cell and apparently grow by addition of coated vesicles from the dictyosome. Microfilaments and microtubules are associated with SDVs during silica deposition (Pipes & Leedale 1992).

Two features associated with scale production, namely close association between ER and SDVs and the shaping role of the ER, appear to be universal to members of the Chrysophyceae and Synurophyceae. It is likely that microfilaments will also be shown to be universal to these two groups. However, so far microtubules have only been observed in association with SDVs in members of the Synurophyceae.

Experiments involving microtubule inhibitors, such as colchicine, podo-phyllotoxin and nocodazole, applied at concentrations that inhibit cell division, also affect the thickness and shape of scales (Figs. 7.13, 7.14; Leadbeater 1984; Sandgren & Barlow 1989). In colchicine-treated cells, mature SDVs that have moved away from the chloroplast contain thin scales and in some sections microtubules close to a SDV appear to remain intact. Whether or not scale deformities resulting from treatment can be directly related to microtubule malfunction is debatable. Colchicine treatment of *Paraphysomonas* cells, which apparently do not have microtubules associated with their SDVs, also results in the production of thinner scales.

When mature, silicified scales are extruded on to the surface of the plasmalemma. This involves confluence of the plasmalemma with the membrane of the SDV (Figs. 7.17, 7.18). Once extruded, scales become incorporated into the overlying scale case. In *Synura* the specific manner in which SDVs develop on one chloroplast only and the helicoidal disposition of developing SDVs has been interpreted by Leadbeater (1990) as indicating that scales develop within the protoplast in a position close to where they will be incorporated in the overlying scale case. This interpretation would appear to explain how scales can be inserted in a very precise manner into the scale case without disturbing the overall

---

Figure 7.7–7.10. *Synura petersenii*. Figure 7.7. Shadowcast whole mount of organic flagellar scales. × 25 000. Figure 7.8. Dictyosome with flagellar scale in cisterna (arrow). × 30 000. Figure 7.9. Dictyosome with nearby large flattened cisternae (arrows) located between chloroplasts. × 15 000. Figure 7.10. Immature silica deposition vesicle (SDV) on surface of chloroplast. × 25 000.
Figures 7.11, 7.12. *Paraphysomonas vestita*. Figure 7.11. SDV and underlying endoplasmic reticulum (arrow) perpendicular to the long axis of the adjacent dictyosome. × 27 000. Figure 7.12. Late stage in spine development showing fully formed scale in SDV. × 50 000.

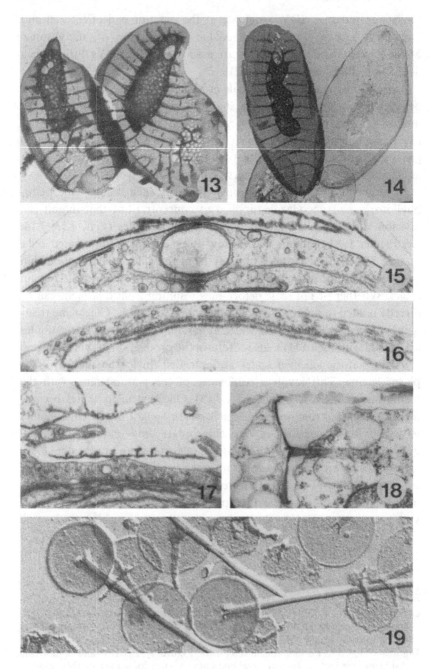

arrangement. However, when naked *Synura* cells produce scales as a result of silica replenishment, scales do not emerge in rows and there is some evidence that they can be moved around on the surface of the plasmalemma (Sandgren & Barlow 1989; P.C. Towlson, unpublished observations). Siver & Glew (1990) have suggested that naked cells form a scale case by producing horizontal ranks of scales from the rear end of the cell. As additional ranks of scales are produced, previously extruded scales are moved anteriorly. Whilst this description may be suitable for explaining how a naked cell assembles a scale case it does not explain how cells with an existing scale case insert additional scales. Beech *et al.* (1990) have observed the deployment of bristles at the posterior end of cells of *Mallomonas splendens*. Deployment involves a complex series of movements with the bristles being inverted by a protoplasmic protuberance outside the protoplast. When in position at the posterior end of the cell the subtending baseplates are extruded. In *Paraphysomonas* species with spines, it is not clear how scales with long projecting spine shafts are extruded and added to the scale case. The few sections that are available showing emergence of a scale suggest that the spine emerges sideways.

The ultrastructural events involved in the deposition of silica in cysts have been reviewed elsewhere (Sandgren 1983). Two matters are of significance here. Firstly, in cyst wall development dictyosome vesicles are closely associated with the SDV although no direct connection between the membrane systems has been observed. Secondly, silica is deposited as the SDV expands, whereas in scale formation the SDV achieves its shape prior to silica deposition.

Dissolution of silica

Batch culture experiments, involving chrysophytes with silica scales in which growth and silica content of the medium have been monitored,

---

Figure 7.13–7.17. *Synura petersenii.* Figures 7.13, 7.14. Deformed and weakly silicified scales produced by colchicine-treated cells. × 10 000. Figures 7.15. Longitudinal section of colchicine-treated cell showing a SDV containing a weakly silicified scale. × 20 000. Figure 7.16. Longitudinal section of a colchicine-treated cell showing a row of microtubules close to the distal surface of a SDV. × 55 000. Figure 7.17. Stage in exocytosis of a plate scale. × 15 000.
Figure 7.18. *Paraphysomonas vestita.* Stage in exocytosis of spine scale. × 27 000.
Figure 7.19. *P. vestita.* Shadowcast whole mount of scales, some partially dissolved. × 18 000.

Table 7.4. *Silicon content per cell, approximate surface area of scales, silicon per unit scale surface area and calculated time for complete dissolution for fully silicified scales of three chrysophytes*

	Silicon content per cell (pg Si per cell)	Approximate surface area of silica scales (μm^2)	Silicon content per unit scale surface area (pg Si μm^{-2})	Time for complete dissolution (days)
Paraphysomonas imperforata	0.42	141	0.0029	11
P. vestita	0.84	171	0.0049	19
Synura petersenii	26	500	0.052	200

See text for further details.

reveal that at some time during the stationary phase of growth, reactive silicate in the medium starts to increase as a result of net silica loss from the scales. Subsequent values for silica dissolution, although net, must approximate to gross since there will be minimal uptake of silicon from the medium.

The rate of dissolution of silica particles in water is dependent on: (1) the absolute dissolution rate of the silica species, (2) the surface area in contact with water, (3) the concentration of soluble silicate present in the water, (4) the presence of organic material on or within the siliceous structure, (5) environmental factors including temperature, salinity, pH of medium, water currents and turbulence. Most protistan siliceous structures are heterogeneous in composition and do not dissolve uniformly. By reference to data obtained for diatoms, the rate of leaching of silicon from biogenic silica is unlikely to exceed 0.1×10^{-9} mol m^{-2} s^{-1}, that is, $< 3 \times 10^{-9}$ pg Si μm^{-2} s^{-1} of silicified surface (Reynolds 1986). Given the initial quantity of silicon per unit area of scale, the minimum time required for its complete solution may be derived (Table 7.4). However, on actively growing cells silica structures generally take longer to dissolve. This is borne out by laboratory experiments at 20 °C in which scales of *Paraphysomonas vestita* took 35 days to dissolve completely. It is also unusual to observe living cells with partially dissolved scales or spines, which suggests that dissolution is slow in these conditions.

Conditions in surface waters are often favorable for silica dissolution since they are usually, warm, turbulent and undersaturated with respect to silicon. Individual scales in surface waters, especially in seawater, are likely to dissolve completely before they settle out of the water column. However, once out of surface water and into colder strata, rates of silica dissolution are slower. The fate of silica particles will depend on the time they remain surrounded by water undersaturated in silicon. In marine environments whole cells, scale cases and single scales may become incorporated into aggregates of debris or marine snow and will fall rapidly to the benthos. Silica particles grazed by zooplankton and subsequently incorporated into fecal pellets will also have an accelerated descent to the benthos.

When silica particles reach the sediment surface dissolution continues, but once immobilized and buried, siliceous material dissolving into near-static interstitial water rapidly saturates the concentration gradient and approaches equilibrium with reprecipitation. Solution times may then be increased by several orders of magnitude to tens of thousands of years. It is this inhibited dissolution which enables siliceous oozes to form at the bottoms of lakes and in parts of the ocean (Berner 1980). Silica turnover in lakes and oceans varies, but for a freshwater lake such as Lake Michigan rates may be greater than 95% (Parker *et al.* 1977). Nevertheless, in smaller, shallower lakes there are extensive records of chrysophyte cysts and scales preserved in sediments (Cronberg 1986; Smol, this volume).

Concluding comments

In this discussion, two silica-producing groups of pigmented algae, namely the silicoflagellates (known variously as the Dictyochales Haeckel, Silicoflagellata Borgert or Dictyochophyceae Silva) and the Parmales Booth and Marchant, at various times included within the Chrysophyceae, have been omitted. The silicoflagellates (see Moestrup, this volume) make a significant contribution to silicon fluxes in the oceans (Leadbeater 1991). Members of the poorly understood order Parmales are abundant in polar and subpolar seas (Mann & Marchant 1989; also see Preisig, this volume) but little is known about their silicon requirements. Life histories of members of both groups remain enigmatic but their contribution to the cycling of silicon within the oceans may be of significance.

In comparison with diatoms, chrysophytes probably play only a minor part in the global cycling of silicon in aquatic ecosystems. In marine environments, where species of *Paraphysomonas* are the only important chrysophyte members of the flora, the contribution made by chrysophytes to silica cycling must be minimal. In freshwater systems the impact of

Table 7.5. *Differences in silicon uptake and deposition by diatoms and chrysophytes*

Diatoms	Chrysophytes
Cells cannot divide without formation of new silica valves	Cells can divide without covering of silica scales
Absolute requirement for silicon demonstrated	Absolute requirement for silicon not demonstrated so far
Silicon taken up during about one tenth of cell cycle immediately after nuclear division	Silicon taken up throughout interphase
Silica valves produced as one event prior to cytokinesis	Silica scales deposited throughout interphase and added to existing scale case
Silicon replenishment to silicon-depleted cells results in return to normal growth and silicon uptake rates (growth and silicon uptake rates usually closely coupled)	Silicon replenishment to silicon-depleted cells results in alteration of growth rate and an enhancement of silicon uptake rates (although normally coupled rates can alter with respect to one another)

chrysophytes on silica cycling may be considerable especially in water bodies that support large populations of *Synura*, *Mallomonas* or *Chrysosphaerella*.

Comparison of the silicon requirements of scale-covered chrysophytes with that of diatoms is inevitable because both utilize silicon for cell coverings and both are found in similar habitats. Some of the more important points of contrast between these two groups are noted in Table 7.5. Results obtained so far indicate that chrysophytes utilize silicon in a manner similar to that of loricate choanoflagellates and probably silica-depositing heliozoa. However, many questions remain unanswered, including such disparate matters as: whether silicon has an obligate role in chrysophyte cell metabolism; what triggers silica cyst formation; what are the functions of scales and scale coverings, and how scales are inserted into an existing scale case.

Summary

Members of the Paraphysomonadaceae and Synurophyceae bear one or more extracellular layers of silicified scales. In the Paraphysomonadaceae, scale form varies from the long-shafted spines in *Paraphysomonas vestita* to meshwork scales in *P. homeolepis* and *P. eiffelii*. In the Synurophyceae,

Tessellaria volvocina has flat perforated plate-scales whereas the scales of *Synura* and *Mallomonas* spp. are highly modified with a posterior rim that allows for their characteristic imbrication within a scale case. In planktonic species, where the additional density of mineralized scales will increase the chances of a cell sinking out of the water column, the form and mechanical strength of scales is achieved with the minimal use of silica. The occurrence of silica within cyst walls is an important diagnostic character of the Chrysophyta. Thick siliceous-walled cysts protect the enclosed protoplast during periods of adversity and their resistance to decay and dissolution makes them ideal structures for preservation within the bottom sediments of water bodies, thereby ultimately contributing to the fossil record.

Scale-depositing chrysophytes typically do not have an obligate require-ment for silica in large quantities, since in the absence of soluble reactive silicate in the medium 'naked' cells without scale cases are produced. Resupply of silica to silica-starved cells results in the production of a complete scale case within 24 h in laboratory conditions. Silica deposition in chrysophytes occurs intracellularly within membrane-bounded silica deposition vesicles (SDVs). Mature scales are extruded on to the surface of cells where they become incorporated into the overlying scale case. Once in contact with the medium scales slowly dissolve, although in cold acid freshwaters dissolution is so slow that some scales ultimately become preserved in benthic sediments thereby providing another valuable ecological marker in the fossil record.

References

Andersen, R.A. 1987. Synurophyceae classis nov.: a new class of algae. *Am. J. Bot.* **74**: 337–53.

Beech, P.L. Wetherbee, R. & Pickett-Heaps, J.D. 1990. Secretion and deployment of bristles in *Mallomonas splendens* (Synurophyceae), *J. Phycol.* **26**: 112–22.

Berner, R.A. 1980. *Early Diagenesis: A Theoretical Approach*. Princeton University Press, Princeton, N.J.

Bourrelly. P. 1981. *Les Algues d'Eau Douce*, vol. 2, *Les Algues Jaunes et Brunes* (revised and enlarged edition). Société Nouvelle des Éditions Boubée, Paris.

Brugerolle, G. & Bricheux, G. 1984. Actin microfilaments are involved in scale formation of the chrysomonad cell *Synura. Protoplasma* **123**: 202–12.

Busby, W.F. & Lewin, J. 1967. Silicate uptake and silica shell formation by synchronously dividing cells of the diatom *Navicula pelliculosa* (Breb) Hilse, *J. Phycol.* **3**: 127–31.

162   *Chrysophyte algae*

Cienkowski, L. 1870. Über Palmellaceen und einige Flagellaten. *Arch. Mikrosk. Anat.* **6**: 421–38.
Cronberg, G. 1986. Chrysophycean cysts and scales in lake sediments. In: Kristiansen, J. & Andersen, R.A. (eds.) *Chrysophytes: Aspects and Problems.* Cambridge University Press, Cambridge. pp. 281–315.
Heath, G.R. 1974. Dissolved silica and deep sea sediments. *Soc. Econ. Palaeo. Min. Special Publ.* **20**: 77–93.
Hibberd, 1977. Ultrastructure of cyst formation in *Ochromonas tuberculata* (Chrysophyceae). *J. Phycol.* **13**: 309–20.
Hurd, D.C. 1973. Interactions of biogenic opal sediment and seawater in the Central Equatorial Pacific. *Geochim. Cosmochim. Acta* **37**: 2257–82.
Iwanoff, L. 1899. Beitrag zur Kenntnis der Morphologie und Systematik der Chrysomonaden. *Bull. Ac. Imp. Sc. St. Petersburg* **11**: 247–62.
Klaveness, D. & Guillard, R.R.L. 1975. The requirement for silicon in *Synura petersenii. J. Phycol.* **11**: 349–55.
Kristiansen, J. 1986*a*. The ultrastructural bases of chrysophyte systematics and phylogeny. *CRC Crit. Rev. Plant Sci.* **4**: 149–211.
Kristiansen, J. 1986*b*. Identification, ecology, and distribution of silica scale-bearing Chrysophyceae, a critical approach. In: Kristiansen, J. & Andersen, R.A. (eds.) *Chrysophytes: aspects and problems.* Cambridge University Press, Cambridge. pp. 229–39.
Leadbeater, B.S.C. 1984. Silicification of 'cell walls' of certain protistan flagellates. *Proc. Trans. R. Soc. Lond. B* **304**: 529–36.
Leadbeater, B.S.C. 1986. Scale case construction in *Synura petersenii* Korsch (Chrysophyceae). In: Kristiansen, J. & Andersen, R.A. (eds.) *Chrysophytes: Aspects and Problems.* Cambridge University Press, Cambridge. pp. 121–31.
Leadbeater, B.S.C. 1990. Ultrastructure and assembly of the scale case in *Synura* (Synurophyceae Andersen). *Br. Phycol. J.* **25**: 117–32.
Leadbeater, B.S.C. 1991. Protista and mineral cycling in the sea. In: Reid, P.C., Turley, C.M. & Burkill, P.H. (eds.) *Protozoa and their Role in Marine Processes.* Springer, Berlin. pp. 361–85.
Leadbeater, B.S.C. & Davies, M.E. 1984. Developmental studies on the loricate choanoflagellate *Stephanoeca diplocostata* Ellis. III. Growth and turnover of silica: preliminary observations. *J. Exp. Mar. Biol. Ecol.* **81**: 251–68.
Lee, R.E. 1978. Formation of scales in *Paraphysomonas vestita* and the inhibition of growth by germanium dioxide. *J. Protozool.* **25**: 163–6.
Lewin, J.C. 1966. Silicon metabolism in diatoms. V. Germanium dioxide, a specific inhibitor of diatom growth. *Phycologia* **2**: 1–12.
Lucas, I.A.N. 1967. Two new marine species of *Paraphysomonas. J. Mar. Biol. Assoc. U.K.* **47**: 329–34.
Ludwig, J.R. & Volcani, B.E. 1986. A molecular biology approach to understanding silicon metabolism in diatoms. In: Leadbeater, B.S.C. & Riding, R. (eds.) *Biomineralization in Lower Plants and Animals.* Systematics Association Special Volume 30. Clarendon Press, Oxford. pp. 315–25.
McGrory, C.B. & Leadbeater, B.S.C. 1981. Ultrastructure and deposition of silica in the Chrysophyceae. In: Simpson, T.L. & Volcani, B.E. (eds.) *Silicon and Siliceous Structures in Biological Systems.* Springer, New York. pp. 201–30.
Mann, D.G. & Marchant, H.J. 1989. The origins of the diatom and its life cycle. In: Green, J.C., Leadbeater, B.S.C. & Diver, W.L. (eds.) *The*

Chromophyte Algae: Problems and Perspectives. Systematics Association Special Volume 38. Clarendon Press, Oxford. pp. 307–23.

Manton, I. & Harris, K. 1966. Observations on the microanatomy of the brown flagellate *Sphaleromantis tetragona* Skuja with special reference to the flagellar apparatus and scales. *J. Linn. Soc (Bot.)* **59**: 397–403.

Mignot, J.P. & Brugerolle, G. 1982. Scale formation in chrysomonad flagellates. *J. Ultrastruct. Res.* **81**: 13–26.

Nicholls, K. 1985. The validity of the genus *Spiniferomonas* (Chrysophyceae). *Nord. J. Bot.* **5**: 403–6.

Paasche, E. 1973. The influence of cell size on growth rate, silica content, and some other properties of four marine diatom species. *Norw. J. Bot.* **20**: 197–204.

Parker, J.I., Conway, H.L. & Yaguchi, E.M. 1977. Dissolution of diatom frustules and dissolution of amorphous silicon in Lake Michigan. *J. Fish. Res. Bd. Can.* **34**: 545–51.

Pipes, L.D. & Leedale, G.F. 1992. Scale formation in *Tessellaria volvocina*. (Synurophyceae). *Br. phycol. J.* **27**: 1–19.

Pipes, L.D., Leedale, G.F. & Tyler, P.A. 1991. The ultrastructure of *Tessellaria volvocina* (Synurophyceae). *Br. Phycol. J.* **26**: 259–78.

Preisig, H.R. 1986. Biomineralization in the Chrysophyceae. In: Kristiansen, J. & Andersen, R.A. (eds.) *Chrysophytes: Aspects and Problems*. Cambridge University Press, Cambridge, pp. 71–4.

Preisig, H.R. & Hibberd, D.J. 1982*a*. Ultrastructure and taxonomy of *Paraphysomonas* (Chrysophyceae) and related genera: I. *Nord. J. Bot.* **2**: 397–420.

Preisig, H.R. & Hibberd, D.J. 1982*b*. Ultrastructure and taxonomy of *Paraphysomonas* (Chrysophyceae) and related genera: II. *Nord. J. Bot.* **2**: 601–38.

Preisig, H.R. & Hibberd, D.J. 1983. Ultrastructure and taxonomy of *Paraphysomonas* (Chrysophyceae) and related genera: III. *Nord. J. Bot.* **4**: 279–85.

Reynolds, C.S. 1984. *The Ecology of Freshwater Phytoplankton*. Cambridge University Press, Cambridge. 384pp.

Reynolds, C.S. 1986. Diatoms and the geochemical cycling of silicon. In: Leadbeater, B.S.C. & Riding, R. (eds) *Biomineralization in Lower plants and Animals*. Systematics Association Special Volume 30. Clarendon Press, Oxford. pp. 269–89.

Sandgren, C.D. 1980*a*. An ultrastructural investigation of resting cyst formation in *Dinobryon cylindricum* Imhof (Chrysophyceae, Chrysophycota). *Protistologica* **16**: 259–75.

Sandgren, C.D. 1980*b*. Resting cyst formation in selected chrysophyte flagellates: an ultrastructural survey including a proposal for the phylogenetic significance of interspecific variations in the encystment process. *Protistologica* **16**: 289–303.

Sandgren, C.D. 1983. Survival strategies of chrysophycean flagellates: reproduction and the formation of resistant resting cysts. In: Fryxell, G.A. (ed.) *Survival Strategies of the Algae*. Cambridge University Press, Cambridge. pp. 23–48.

Sandgren, C.D. 1989. SEM investigations of statospore (stomatocysts) development in diverse members of the Chrysophyceae and Synurophyceae. *Nova Hedwigia* **95**: 45–69.

Sandgren, C.D. 1991. Chrysophyte reproduction and resting cysts: a paleolimnologist's primer. *J. Paleolimnol.* **5**: 1–9.

Sandgren, C.D. & Barlow, S.B. 1989. Siliceous scale production in chrysophyte algae. II. SEM observations regarding the effects of metabolic inhibitors on scale regeneration in a laboratory population of scale-free *Synura petersenii* cells. *Nova Hedwigia* **95**: 27–44.

Saxby, K.J. 1990. The physiological ecology of freshwater chrysophytes with special reference to *Synura petersenii*. PhD thesis, Birmingham University, UK.

Siver, P.A. & Glew, J.R. 1990. The arrangement of scales and bristles on *Mallomonas* (Chrysophyceae): a proposed mechanism for the formation of the cell covering. *Can. J. Bot.* **68**: 374–80.

Skogstad, A. & Reymond, O.L. 1989. An ultrastructural study of vegetative cells, encystment, and mature statospores in *Spiniferomonas bourrellyi* (Chrysophyceae). *Nova Hedwigia* **95**: 71–9.

Stumm, W. & Morgan, J.J. 1970. *Aquatic Chemistry*. Wiley, New York.

Tyler, P.A., Pipes, L.D., Croome, R.L. & Leedale, G.F. 1989. *Tessellaria volvocina* rediscovered. *Br. Phycol. J.* **24**: 329–37.

Wujeck, D.E. & Kristiansen, J. 1978. Observations on bristle- and scale-production in *Mallomonas caudata* (Chrysophyceae) *Arch. Protistenk.* **120**: 213–21.

8

Immunological and ultrastructural studies of scale development and deployment in *Mallomonas* and *Apedinella*

RICHARD WETHERBEE, MARTHA LUDWIG & ANTHONY KOUTOULIS

Introduction

The diversity and morphology of scales and scale-like structures (e.g., spines, spine-scales, bristles) are remarkable among the different protistan groups, and the mechanism of their assembly and deployment can vary considerably (for review, see Romanovicz 1981). Arguably the most spectacular scale-bearing algae are found in the division Chrysophyta, which includes the organisms under investigation here: *Mallomonas splendens* (G.S. West) Playfair em. Croome, Dürrschmidt & Tyler (Synurophyceae) and *Apedinella radians* (Lohmann) Campbell (Pedinellophyceae). These two species collectively exhibit a wide range of surface features, some of them unique, and are excellent experimental systems for studying the development of scales and scale cases.

A number of cytological techniques have been used to investigate scale formation and development, most notably scanning and transmission electron microscopy. The formation of synurophycean scales (including the scale-like 'bristles' of *Mallomonas*) has been followed in several species at the ultrastructural level (e.g., Mignot & Brugerolle 1982; Brugerolle & Bricheux 1984). The exact manner of scale deployment onto the surface, however, is unknown, although two possible mechanisms have been put forward (Leadbeater 1990; Siver & Glew 1990). Recently, *in vivo* observations of bristle secretion and deployment in *M. splendens* have been made using image-enhanced video microscopy, and corroborated ultrastructurally with thin-sectioned material (Beech *et al.* 1990).

The development of immunocytochemical techniques has extended our knowledge of the cytoskeletal components active in these processes as well as the nature and role of surface molecules associated with the scale layer. Microtubules and actin filaments have been identified ultrastructurally and/or with commercially available antibodies and appear to be involved in aspects of scale morphogenesis and deployment (Mignot & Brugerolle 1982; Brugerolle & Bricheux 1984) and in spine-scale reorientation (Koutoulis *et al.* 1988). Recently, polyclonal and monoclonal antibodies (MAbs) have been raised against the surface components of scale-bearing species, including a scale component of *A. radians* and the putative adhesive(s) that bind siliceous scales/bristles to one another and/or to the plasma membrane in *M. splendens* (Ludwig & Wetherbee, unpublished observations). Preliminary results indicate that a number of algal cell surface components are very antigenic, and that antibodies against these molecules will be valuable tools in determining the composition of the components and their role(s) in the development and deployment of scale layers.

General surface morphology of *M. splendens* and *A. radians*

Mallomonas splendens

As with other synurophytes (for review, see Andersen 1987), *M. splendens* and other species of this genus produce and secrete siliceous scales of distinct morphology that are arranged into a protective scale case with a complex pattern (Figs. 8.1–8.3). In addition to the scale layer, cells produce eight elongate bristles, four at each end of the cell, that attach via an 'adhesive' or fibrillar complex to specialized scales termed base plate scales (Beech *et al.* 1990). The distribution of bristles over the cell surface varies among species of *Mallomonas* (Asmund & Kristiansen 1986). The scale case of *Mallomonas* is composed of spiral rows of overlapping body scales (Figs. 8.1–8.3) (for review, see Siver & Glew 1990; Leadbeater & Barker, this volume) that are presumably anchored to one another by an adhesive boomerang strip similar to those described for species of *Synura* (Leadbeater 1986, 1990). In *M. splendens*, the base plate scales with attached bristles are integrated with the scale layer at the end of each cell (Figs. 8.1, 8.2).

Apedinella radians

In the marine flagellate *A. radians*, cells possess two types of organic body scales as well as six elongate spine-scales which are longer than the cells that produce them (Throndsen 1971; Koutoulis *et al.* 1988). The body

Figures 8.1–8.3. Scanning electron micrographs of *Mallomonas splendens* and *M. adamas* Harris & Bradley (Fig. 8.3), showing the siliceous scales and the organization of the scale case. Note the bristles attached to the base-plate scales in Figs. 8.1 and 8.2. Scale bars represent 2 μm.

scales form an overlapping layer covering the entire cell. Spine-scales are located subapically between the cell's six chloroplasts. These spine-scales exhibit an unusual form of coordinated movement when cells are motile;

that is, the ability to alter their orientation instantaneously. A striated microligament connects each spine-scale to a distinctive plaque that associates with the plasma membrane. Plaques connect on their inner surface with a complex of filamentous bundles that are thought to be responsible for spine-scale movement (Koutoulis *et al.* 1988; see below).

The cytoskeleton during scale/bristle/spine-scale formation and secretion: ultrastructural and immunocytochemical studies

Mallomonas splendens

The role of the cytoskeleton in scale formation in *M. splendens* has recently been reviewed (Wetherbee *et al.* 1989) and will only be summarized here (see Leadbeater & Barker, this volume). The scales and bristles of *Mallomonas* are produced within silica deposition vesicles (SDVs) associated with the scale-forming chloroplast (Schnepf & Deichgräber 1969; Mignot & Brugerolle 1982; Brugerolle & Bricheux 1984; Beech *et al.* 1990). The presence of microtubules overlying the developing SDVs is obvious, though their role is less clear. It has been suggested that they somehow transport the SDVs during their developmental migration along the chloroplast surface and/or have a role in stabilizing the SDV membrane prior to silica deposition (Mignot & Brugerolle 1982; Brugerolle & Bricheux 1984). Actin filaments have been observed associated with the outer membrane of SDVs during the earliest stages of scale/bristle formation, i.e., when the SDVs are still developing in size and shape. The microfilaments form a dense crosslinked pattern that corresponds to the future position of the scale's base plate (Brugerolle & Bricheux 1984). The role of the actin filaments seems purely structural, as they disassemble once silica deposition commences. A motility function has therefore been discounted in favor of a support role that may include strengthening and shaping the membranes of the SDV prior to initial silica deposition.

A description of bristle secretion and deployment has been published recently (Beech *et al.* 1990) and is summarized here in Figs. 8.4 and 8.5. Scales, base plate scales and bristles of *M. splendens* are formed in separate SDVs and secreted independently of one another at different times during the cell cycle. New posterior bristles and their base plate scales are brought together on the surface of the cell following a complex reorientation process that appears to involve a transient, cytoplasmic protuberance (Figs. 8.4, 8.5; Beech *et al.* 1990). Although thin sections suggest that a

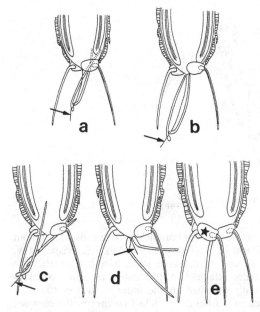

Figure 8.4. The secretion and deployment of two bristles at the posterior end of a *M. splendens* cell. The bristles are secreted base first and are reoriented on the surface, apparently with the aid of a thin protuberance (arrow) that extends from the surface of the cell. Once reoriented, a base plate scale is secreted (star) and the bristle is somehow attached to it. (Redrawn from Beech *et al.* 1990.)

protuberance contains one or more microtubules, the exact role of the cytoskeleton in this activity is not well understood. Likewise, it is not known whether this mechanism, described for a species with only eight bristles, can be extrapolated to species where all scales have an attached bristle.

Apedinella radians

Spine-scales in *A. radians* are longer than the cells they attach to, and the mechanism that produces them accommodates this feature. A spine-scale deposition vesicle forms at the posterior of the cell, just beneath the plasma membrane, on the distal side of the single Golgi body. As the spine-scale elongates during formation, it protrudes outward from the end of the cell and is surrounded by two membranes: the plasma membrane and the spine-scale deposition vesicle membrane (Figs. 8.6–8.8). Microtubules are found in the cytoplasm between the two membranes, as

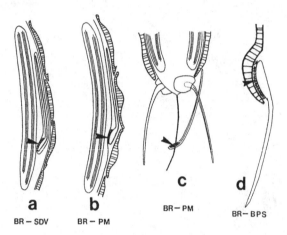

a
BR – SDV

b
BR – PM

c
BR – PM

d
BR– BPS

Figure 8.5. The attachment sites of the fibrillar complex (arrowheads) throughout bristle development and deployment in *Mallomonas splendens*. BR, bristle; SDV, scale deposition vesicle; PM, plasma membrane; BPS, base-plate scale. The fibrillar complex forms and remains attached to its bristle throughout development. The other surface is initially attached to the inner surface of the scale deposition vesicle (a). After secretion, bristles are attached via the fibrillar complex to the plasma membrane (b) and then reoriented with the aid of the protuberance (c). Eventually bristles are retracted back to the cell surface (c) and attached to base-plate scales (c and d), forming essentially a silica–silica attachment.

observed by immunocytochemistry (Fig. 8.9) and electron microscopy, and these appear to be involved in spine-scale morphogenesis (Koutoulis and Wetherbee, unpublished observations).

The cytoskeleton of *Apedinella radians*

The cytoskeleton of *A. radians* is one of the most complex ever described for a unicellular protist (Koutoulis *et al.* 1988), and accounts for a number of functions including the reorientation of the spine-scales during swimming. Details of its microarchitecture have already been published (Koutoulis *et al.* 1988; for review, see Wetherbee *et al.* 1989, 1992), and observations were made possible largely by the use of immunocytochemistry and of antibodies against tubulin, actin and centrin. These initial studies allowed the visualization of the complex cytoskeleton and its interactions (Figs. 8.10–8.12), and thus made the subsequent electron microscopical observations easier to interpret (Koutoulis *et al.* 1988).

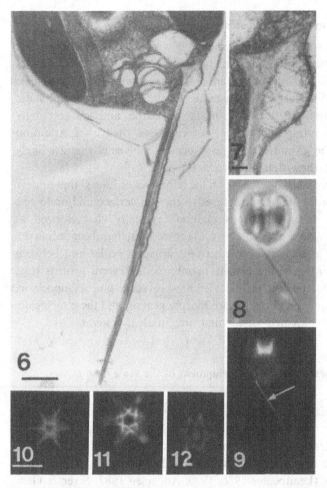

Figures 8.6, 8.7. High-voltage electron micrographs of the development of a spine-scale in *Apedinella radians.* Note how the spine-scale extends out from the posterior end of the cell. Developing spine-scales are surrounded by the plasma membrane and the spine-scale deposition vesicle, with the cytoplasm between containing microtubules. Scale bars represent 1 μm (Fig. 8.6) and 0.125 μm (Fig. 8.7).

Figures 8.8, 8.9. Phase-contrast and fluorescence micrographs of the same cell showing a forming spine-scale of *A. radians.* The cell was stained with an anti-tubulin antibody. Note the microtubules associated with the forming spine-scale (arrow) as well as those associated with the triads at the anterior end of the cell (see Koutoulis *et al.* 1988). Scale bars represent 5 μm.

Figures 8.10–8.12. Fluorescence micrographs showing cells of *A. radians* stained with three antibodies raised against the cytoskeletal proteins tubulin (Fig. 8.10), actin (Fig. 8.11) and centrin (Fig. 8.12). Scale bars represent 5 μm.

172   *Chrysophyte algae*

Scale-associated molecules and structures

Despite the large volume of literature dealing with scales (e.g., Romanovicz 1981), there is little information available on the scale-associated molecules (in some cases, distinct structures) involved in mediating aspects of scale development and surface morphogenesis. Once formed, scales are static wall structures that do not alter in size or shape, and are seemingly incapable of generating any kind of movement on their own. Attention must, therefore, be given to the scale-associated molecules that are likely to be involved in these dynamic processes.

Virtually no work has been done on the adhesives or 'glues' that appear to bind scales and scale-like structures to the cell surface and/or to one another. A thin layer of fibrous material, typically characterized as 'osmiophilic' or 'electron dense' by microscopists, has been observed ultrastructurally at contact points between adjacent scales and between surface components and the plasma membrane in several protists (e.g., Leadbeater 1986). However, little else is known of scale 'glue' composition, distribution, or function in cell surface morphogenesis, and these molecules are now being investigated for the first time in algal systems.

Scale/bristle adhesives and the development of the scale case in
Mallomonas

The scale case of *M. splendens* (Figs. 8.1, 8.2) is believed to be anchored to the cell surface by the adhesion of each scale to the plasma membrane. In addition, scales adhere to one another in the regions of overlap, with scales in a spiral row connected to one another as well as to the scales in adjacent rows (Leadbeater 1986, 1990; Andersen 1987; Siver & Glew 1990). Two conflicting hypotheses have been presented to explain scale case morphogenesis in *Mallomonas* and *Synura* (Leadbeater 1986, 1990; Siver & Glew 1990). The proposals differ as regards both the position of scale/bristle insertion onto the cell surface and the mechanism of their deployment. Unfortunately, little evidence has been presented to date to support either hypothesis. Regardless of the actual mechanism, it would appear that scales must have the ability to detach from one set of scales and re-attach to another, since following cytokinesis the scales on either side of the cleavage furrow must come together to re-establish the intact scale case.

Remembering that development of the scale case occurs on the cell surface, but is presumably controlled by the cell, the composition and role of the adhesives or 'glues' that bind scales to one another and to the plasma membrane are of great interest. In order to characterize these molecules, polyclonal antisera raised to intact scale cases of detergent-treated *M. splendens* cells were used to label isolated, intact scale cases. Using previously described immunofluorescence techniques (Koutoulis *et al.* 1988), labelling was observed at the junctions of body and base-plate scales and also between base-plate scales and attached bristles at the fibrillar complex (Fig. 8.13). Fluorescence was also observed on the base of unattached bristles (Fig. 8.14) (Wetherbee, unpublished observations). Similar labelling patterns were observed when scale cases from detergent-treated *M. splendens* cells were prepared for immunogold electron micro-scopy using techniques essentially as described by Ludwig & Gibbs (1989). Gold particles were observed between all scale types and on bristle bases (Figs. 8.15, 8.16) (unpublished observations). Immunoblots of sodium dodecylsulfate (SDS) polyacrylamide gels of *M. splendens* scale cases using the polyclonal antisera showed several immunoreactive bands.

As a number of surface components of *M. splendens* appeared to be quite antigenic, isolated, intact scale cases were used to generate MAbs (mouse monoclonal antibodies). Recently, we have obtained 15 mono-clonal lines and, on the basis of their immunofluorescence patterns observed during screening, one of these lines was chosen for further characterization. Immunoblot analyses of SDS gels of scale cases indicate that the MAb recognizes a polypeptide epitope. Immunocytochemistry at both the light (Figs. 8.17, 8.18) and electron (Fig. 8.19) microscope levels shows that the MAb labels the area of contact between adjacent scales as well as the fibrillar complex at the base of the bristles (Wetherbee, unpublished observations). This MAb will now be used to characterize the surface antigen further, particularly its origin within the cell and its role in the morphogenesis of the scale case.

Body scale morphogenesis in *Apedinella radians*

Polyclonal and monoclonal antibodies have also been raised to the surface components of *A. radians* and tested in the same manner as described above. Preliminary observations on whole or sectioned cells labelled with one of the MAbs (plus a fluorescein isothiocyanate (FITC)-conjugated or immunogold-conjugated secondary antibody) reveal that the larger body scales and spine-scales, or an adhesive associated with them, is labelled (Figs. 8.20–8.22). Preliminary immunoblot analyses of SDS gels of isolated cell coverings give an immunoreactive banding pattern suggesting that

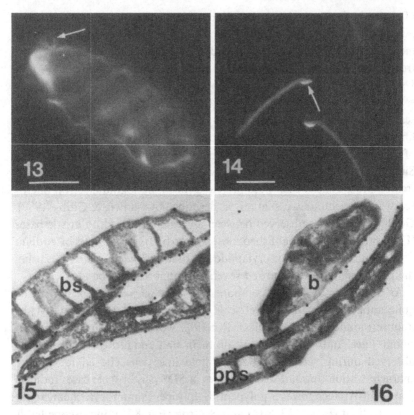

Figures 8.13–8.16. Immunocytochemical labelling of isolated *M. splendens* cell coverings. Figures 8.13, 8.14. Fluorescence micrographs of scale cases labelled with polyclonal antisera directed against the isolated cell coverings followed by an FITC-conjugated secondary antibody. Labelling is observed between body and base-plate scales and between the base of bristles and their associated base-plate scales (Fig. 8.13, arrow). Fluorescence is also seen on the bases of unattached bristles (Fig. 8.14, arrow). Scale bars represent 10 µm. Figures 8.15, 8.16. Immuno-gold electron micrographs of scale cases labelled with the polyclonal antisera followed by a colloidal gold-conjugated secondary antibody. Gold particles are found between body scales (bs) and along the plasma membrane side of the body scales. In cross-section, labelling is also seen between bristles (b) and base-plate scales (bps). Scale bars represent 0.5 µm.

this MAb recognizes a protein of approximately 155 kDa. Further characterization of the MAb and its epitope are under way. If, as suspected, the MAb recognizes a scale protein or glycoprotein, it could be a useful tool in determining the developmental pathway of the scales and/or spine-scales through the cell.

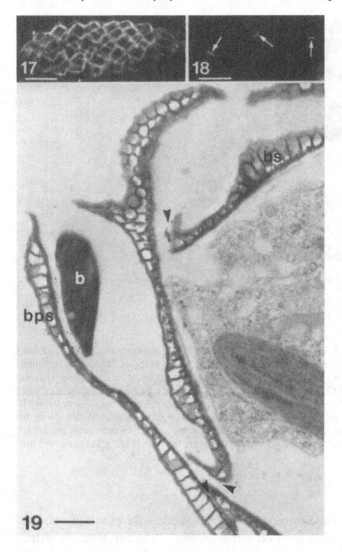

Figures 8.17, 8.18. Fluorescence micrographs of cell surface components of *M. splendens* labelled with a MAb generated against isolated scale cases followed by an FITC-conjugated secondary antibody. Figure 8.17. The MAb labels the junctions of body and base-plate scales. Figure 8.18. Bright fluorescence labelling on the bases of unattached bristles is also demonstrated by the antibody (arrows). Scale bars represent 10 μm.
Figure 8.19. Immunogold electron micrograph of *M. splendens* labelled with the MAb followed by a colloidal gold-conjugated secondary antibody. Gold particles are found on the boomerang strips which attach body scales (bs) to one another (arrowheads). Gold particles are also observed between a bristle (b) and its base-plate scale (bps) at the fibrillar complex. Scale bar represents 0.5 μm.

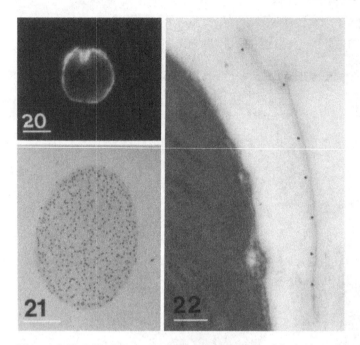

Figures 8.20–8.22. Immunocytochemical labelling of the body scales of *A. radians*.
Figure 8.20. Fluorescence micrograph of the cell surface of *A. radians* labelled
with a monoclonal antibody generated against isolated surface components
followed by an FITC-conjugated secondary antibody. Scale bar represents 5 μm.
Figures 8.21, 8.22. Whole mount and thin section of a body scale(s) stained with
the MAb followed by a colloidal gold-conjugated secondary antibody. Note in
Fig. 8.22 that the gold particles are located on the surface of the scales closest to
the plasma membrane. Scale bars represent 0.5 μm (Fig. 8.21) and 0.2 μm
(Fig. 8.22).

Summary

Ultrastructural studies of scale morphogenesis have provided valuable
information on the role of cellular organelles, the endomembrane system
and, to a lesser degree, the cytoskeleton in this process. Although
microtubules are readily identified in electron micrographs, most other
cytoskeletal components (e.g., actin filaments and centrin) go unnoticed
unless concentrated in distinct bundles. The ability to first visualize
cytoskeletal proteins in whole cells utilizing immunocytochemistry and
fluorescence microscopy greatly reduces the time otherwise required for
reconstructing cells from serial sections.
    Unfortunately, the commercially available antibodies do not recognize
their antigens in all organisms. Furthermore, these antibodies only

recognize very common proteins, and few have been raised specifically against microalgal proteins. To further enhance our knowledge of scale morphogenesis and the formation of scale layers and cases, antibodies should be generated against the cytoskeletal and surface components of the organisms under investigation. Present experience with *M. splendens* and *A. radians*, though limited, suggests that algal cell surface components are very antigenic, and valuable tools can be obtained from the well-documented procedures developed for producing MAbs (e.g., Harlow & Lane 1988). As the above preliminary results clearly demonstrate, these MAbs have considerable potential for answering some of the more difficult questions concerning chrysophyte scale case development.

Acknowledgments

This work was supported by the Australian Research Council. We thank Dr R. McIntosh and the staff of the High Voltage Electron Microscopy Laboratory, University of Colorado, for their assistance and the use of the facility.

References

Andersen, R.A. 1987. Synurophyceae *classis nov.*, a new class of algae. *Am. J. Bot.* **74**: 337–53.
Asmund, B. & Kristiansen, J. 1986. The genus *Mallomonas* (Chrysophyceae). *Opera Bot.* **85**: 1–128.
Beech, P.L., Wetherbee, R. & Pickett-Heaps, J.D. 1990. Secretion and deployment of bristles in *Mallomonas splendens* (Synurophyceae). *J. Phycol.* **26**: 112–22.
Brugerolle, G. & Bricheux, G. 1984. Actin microfilaments are involved in scale formation of the chrysomonad cell. *Synura. Protoplasma* **123**: 203–12.
Harlow, E. & Lane, D. 1988. *Antibodies: A Laboratory Manual*, Cold Spring Harbor Laboratory, New York. 726pp.
Koutoulis, A., McFadden, G.I. & Wetherbee, R. 1988. Spine-scale reorientation in *Apedinella radians* (Pedinellales, Chrysophyceae): the microarchitecture and immunocytochemistry of the associated cytoskeleton. *Protoplasma* **147**: 25–41.
Leadbeater, B.S.C. 1986. Scale case construction in *Synura petersenii* Korsch. (Chrysophyceae). In: Kristiansen, J. & Andersen, R.A. (eds.) *Chrysophytes: Aspects and Problems.* Cambridge University Press, Cambridge. pp. 121–32.
Leadbeater, B.S.C. 1990. Ultrastructure and assembly of the scale case in *Synura* (Synurophyceae Andersen). *Br. Phycol. J.* **25**: 117–32.
Ludwig, M. & Gibbs, S.P. 1989. Localization of phycoerythrin at the luminal surface of the thylakoid membrane in *Rhodomonas lens. J. Cell Biol.* **108**: 875–84.
Mignot, J.-P. & Brugerolle, G. 1982. Scale formation in chrysomonad flagellates. *J. Ultrastruct. Res.* **81**: 13–26.

Romanovicz, D.K. 1981. Scale formation in flagellates. In: Kiermayer, O. (ed.) *Cytomorphogenesis in Plants.* Cell Biology Monographs 8. Springer, New York. pp. 27–62.

Schnepf, E. & Deichgräber, G. 1969. Über die Feinstruktur von *Synura petersenii* unter besonderer Berücksichtigung der Morphogenese ihrer Kieselschuppen. *Protoplasma* **68**: 85–97.

Siver, P.A. & Glew, J.R. 1990. The arrangement of scales and bristles on *Mallomonas* (Chrysophyceae): a proposed mechanism for the formation of the cell covering. *Can. J. Bot.* **68**: 374–80.

Throndsen, J. 1971. *Apedinella gen. nov.* and the fine structure of *A. spinifera* (Throndsen) comb. nov. *Nord. J. Bot.* **18**: 47–62.

Wetherbee, R., Koutoulis, A. & Beech, P.L. 1989. The role of the cytoskeleton during the assembly, secretion and deployment of scales and spines. In: Coleman, A.W., Goff, L.J. & Stein-Taylor, J.R. (eds.) *Algae as Experimental Systems.* A.R. Liss, New York. pp. 93–108.

Wetherbee, R., Koutoulis, A. & Andersen, R.A. 1992. The microarchitecture of the chrysophycean cytoskeleton. In: Menzel, D. (ed.) *The Cytoskeleton of the Algae.* CRC Press, Boca Raton. pp. 1–17.

Part III
Ecology, paleoecology and reproduction

9
Chrysophyte blooms in the plankton and neuston of marine and freshwater systems

KENNETH H. NICHOLLS

Introduction

Excessive accumulations of algae in freshwater lakes and coastal marine environments have been observed for centuries. Homer's *Illiad* mentions discoloration of the sea, and the Bible contains a reference to 'the bloodied waters of the Nile'. Charles Darwin apparently observed a 'red tide' off the coast of Chile, and North American Indians would not eat shellfish from 'shining waters' based on previous experiences with algal-bloom-related shellfish poisonings (Red Tide Newsletter 3(2), April 1990). Among the causes most often cited are enrichment of aquatic systems with nutrients from human activities (Vallentyne 1974; Smayda & White 1990). While there is strong evidence that excessive supply of nitrogen and phosphorus and other nutrients is often the underlying cause of algal blooms, it is also clear that bloom development depends upon the coming together, in appropriate combination, of a number of critical biotic, physical and chemical environmental factors (Paerl 1988).

The implication of the early historical evidence of algal blooms is that the consequences of accelerated human population growth and contemporary urbanization and industrialization activities cannot be the only causative factors for algal blooms. For example, there were especially intense blooms of the dinoflagellate *Alexandrium cantenella* off the Norwegian coast in 1988, which may have been in response to abnormally high water temperatures. Paleo-oceanographic evidence (B. Dale, University of Oslo, unpublished data) suggests that similar blooms developed as far back as 2000 years ago when human-induced influences were undoubtedly negligible.

During the past 5–10 years, however, there is good evidence that the frequency, intensity and geographic extent of algal blooms, in both freshwater (Skulberg *et al.* 1984) and marine environments (Smayda 1990), have increased dramatically in response to human influences related mainly to active transport of bloom-causing organisms (e.g. ship ballast, aquaculture stocking) and stimulation of bloom development by nutrient enrichment. Until recently, the algal bloom phenomenon was generally believed to be monopolized by blue-green algae in lakes and dinoflagellates in the sea, although less significant blooms caused by other algal groups have been recognized (Paerl 1988). The International Conference on Toxic Dinoflagellates (1974, 1978, 1985) changed its name for the June 1989 meetings in Sweden to 'The Fourth International Conference on Toxic Marine Phytoplankton' in order better to reflect the recent catastrophic non-dinoflagellate blooms of the mid to late 1980s. These blooms, which resulted in human death and sickness, severe economic losses and inestimable ecological damage, were caused by the prymnesiophytes *Chrysochromulina polylepis* and *Prymnesium parvum* in Scandinavian coastal waters in 1988 and 1989 (Underdal *et al.* 1989; Kaartvedt *et al.* 1991), the chrysophyte *Aureococcus anophagefferens* in New England in 1985–8 (Nuzzi & Waters 1989; Cosper *et al.* 1990), and the diatom *Nitzschia pungens* f. *multiseries* on Canada's east coast in 1987 (Subba Rao *et al.* 1988). In freshwater, algal blooms of non-blue-green origin with economic and/or nuisance/aesthetic implications, have included the dinoflagellates *Ceratium hirundinella* (Nicholls *et al.* 1980) and *Glenodinium sanguineum* (Dodge *et al.* 1987), the prymnesiophyte *Chrysochromulina breviturrita* (Nicholls *et al.* 1982), and the chrysophytes *Synura petersenii* (Nicholls & Gerrath 1985), *Synura uvella* (Clasen & Bernhardt 1982) and *Uroglena americana* (Kurata 1989). These species are only the tip of the proverbial iceberg however; a recently published guide to the taxonomy of organisms causing 'red tides' in Japan (Fukuyo *et al.* 1990) includes 200 species representing ten algal classes. Interestingly, this comprehensive listing includes only one chrysophyte species (*Uroglena americana*).

It is clear, therefore, that with so many different taxa now implicated in algal blooms, bloom phenomena present special challenges to both the scientist and the user of aquatic resources concerned about understanding and predicting the potential for bloom development and toxin production. This chapter reviews the major developments in recent understanding, as well as the major information deficiencies relating to bloom-causing chrysophytes in both marine and freshwater environments.

Bloom definition

Phytoplankton blooms have been defined in different ways (Paerl 1988; Richardson 1989; Legendre 1990). For the purposes of this chapter I define a chrysophyte bloom as an accumulation of organisms of the classes Chrysophyceae and/or Synurophyceae to a level of intensity which results in one or more of the following:

(1) A visible coloration of the water.
(2) An effect on aesthetic value and/or human use of the water (e.g. taste and odor problems in water supplies).
(3) Major *direct* effects on other aquatic biota (toxin production, physical damage such as clogging of fish gill lamellae).
(4) Major *indirect* effects on other biota (e.g. food web disruption leading to starvation or dissolved oxygen depletion resulting from bloom decomposition).

Marine chrysophyte blooms

The most important species of marine chrysophyte causing blooms (and the only species discussed here) is *Aureococcus anophagefferens*, a pico-planktonic alga first described by Sieburth *et al.* (1988). This species, which may be synonymous with the open ocean taxon *Pelagococcus subviridis* (Cosper *et al.* 1990), developed dense blooms in embayments of Long Island, New York, in 1985–8 and quickly gained notoriety as a 'brown tide' throughout the region including parts of New Jersey and Rhode Island Sound where it was also observed (Nuzzi 1988; Nuzzi & Waters 1989; Smayda & Villareal 1989). After the initial invasion in 1985, the brown tide did not return to Narragansett Bay in 1986, but did to Barnegat Bay and to the Long Island bays during three subsequent years, but at lower densities. Densities during the peak of the 1985 bloom approached 3×10^9 cells l^{-1}.

The effects of the *Aureococcus* blooms were disastrous. Anecdotal evidence suggested that thousands of hectares of eelgrass (*Zostera marine*) beds – plants which are important as settling sites for larval shellfish – were essentially eliminated through light exclusion effects (Nuzzi 1988). However, quantitative data on distribution and density of eelgrass immediately before and after the bloom years apparently do not exist (Dennison *et al.* 1989). Because the life history of the bay scallop (*Argopecten irradians*) through larval, immature and sexually mature adult stages covers less than a 2 year period, the bay scallop fishery was virtually wiped out (Bricelj & Kuenstner 1989) where the *Aureococcus* bloom developed during two consecutive years. The effect was one of physical interference in filter feeding so that the scallops apparently starved to death; no algal toxicity has been implicated (Gallager *et al.* 1989; Ward & Targett 1989). The combined value of the annual bay scallop, blue mussel and cultured oyster (*Crassostrea virginica*) harvest from this region that was lost to the effects of the brown tide was over $2.5 million (Nuzzi 1988). Other damage was not easily quantified. For example, terns and other fish-eating birds left the region, apparently because they could not see their prey (Nuzzi 1988).

The critical questions at the time of the *Aureococcus* bloom were obviously related to the cause. Why did it happen? First of all, it is unfortunate that it seemed to take a catastrophic event such as this to focus attention on a chrysophyte. One major consequence of this bloom was that there followed a flurry of research activity at laboratories in the region. *Aureococcus* was brought into laboratory culture from isolations made in 1986 and effects on growth rates of several environmental factors were determined. Field studies to determine the water quality/productivity relationships in the Long Island Sound area were also initiated (Nuzzi & Waters 1989; Cosper *et al.* 1989). Cosper *et al.* (1987) showed that *Aureococcus* grew much better in enriched coastal bay water than in enriched synthetic ocean media. Dzurica *et al.* (1989) achieved improved growth of *Aureococcus* in media containing organic phosphate compounds such as glycerophosphate and chelating agents such as nitrilotriacetic acid (NTA) and citric acid. This species also demonstrated strong heterotrophic growth capabilities, with rapid uptake of glutamic acid and glucose. Still, the culture studies alone were not able to link these special characteristics of laboratory growth to bloom development: 'How these factors are involved in the dominance of *A. anophagefferens* over other phytoplankton species in nature remains to be further investigated' (Dzurica *et al.* 1989).

One of the most convincing explanations for the cause of the bloom was developed by Cosper *et al.* (1990). I have summarized it here because it appears to represent a particularly good example of the fortuitous interaction of a number of environmental factors with the particular physiological adaptations and requirements of *Aureococcus*. It is an attempt to communicate more widely the Cosper *et al.* hypothesis in the hope that it might serve as a model for explanations of other algal bloom phenomena possibly involving multiple environmental factors.

The first important point to realize is that the brown tide occurrence

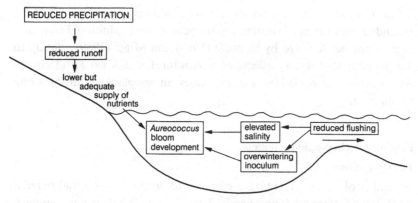

Figure 9.1. A hypothesis relating precipitation and associated physical–chemical changes to development of *Aureococcus* blooms in Long Island embayments. (After Cosper *et al.* 1990.)

in 1985 was over a wide geographic area, and because the bloom did not appear to spread from one coastal bay system to the next, but rather developed more or less simultaneously at all locations, factors initiating the bloom appear to be regional (e.g. weather) rather than localized (e.g. the effects of a single point-source of nutrients). In this regard, the precipitation data are most revealing. Rainfall on Long Island during 1985 was the third lowest since 1949 and was again well below the annual average in 1986 and 1987. This low precipitation resulted in salinity levels that were 20% higher than normal. The laboratory studies had shown a severe reduction in growth rate of *Aureococcus* below 28 ppt, compared with 30 ppt, but growth at the lower salinities could be enhanced by addition of organic rather than inorganic nutrients. This is where the seasonal pattern of annual precipitation is apparently of some importance in bloom development. *Aureococcus* blooms developed during the summers of 1985, 1986 and 1987, but were most intense in 1985 and least intense in 1987. During 1985 and 1986, low winter rainfall was followed by high precipitation in spring. In 1987, spring precipitation was low and the bloom was not nearly as well developed. The implication is that rainfall was important for nutrient supply, but only enhanced bloom development if it followed a period of low rainfall which set up the optimum salinity levels. In addition, the low winter rainfall resulted in generally longer water residence times in the Long Island Sound embayments thereby helping to ensure that an adequate overwintering 'seed' population was present for the summer season (Fig 9.1).

The Cosper *et al.* explanation for the *Aureococcus* bloom might be expanded after further investigation to include some additional facts and suggestions put forward by Sieburth (1989) and Minei (1989) relating to the possible stimulatory effects of agricultural and lawn fertilizers on *Aureococcus* and toxic effects of pesticides on zooplankton in the Long Island area.

Freshwater chrysophyte blooms

Phytoplankton

A number of chrysophyte taxa are known to produce substantial populations in freshwater lakes and ponds. These include *Mallomonas acaroides*, *M. caudata* and *M. crassisquama* (Kristiansen 1971; Thomasson 1970), *Chrysosphaerella longispina* (Pick *et al.* 1984), several *Dinobryon* species (Eloranta 1989), *Uroglena americana* (Kurata 1989) and *Synura* spp. (Nicholls & Gerrath 1985; Clasen & Bernhardt 1982). I will restrict my summary to those genera known for their production of odor and associated problems in domestic water supplies, recreational lakes and reservoirs.

Odors from blooms of *Dinobryon*, *Synura* and *Uroglena* have most often been described as 'fishy', but other descriptors include 'cod liver oil', 'muskmelon' and 'cucumber', especially for the early bloom stages of *Synura* (Whipple *et al.* 1948; Lackey 1950; Palmer 1962; Taft 1965).

Where surface waters provide the raw water source for municipalities, some level of treatment is usually provided before distribution to consumers. This treatment can range from simple disinfection to complex filtration/sedimentation procedures. Excessive densities of algae in the influents to these water treatment plants disrupt their normal functioning and usually require modifications to routine operations, including more frequent filter backwashing, installation of microstrainers, adjustments to chemical flocculation routines, or more expensive activated carbon filtration. At locations where only limited treatment is provided, as for example in regions of chronic economic depression or where the source water supply has historically been of good quality, occasional algal blooms may impart off-flavors which get through the system to the consumer. At such times, complaints from consumers to water treatment operators and waterworks officials are usually described by reference to common naturally occurring odors, such as musty, earthy, fishy, oily, geranium or septic. The general practice of water treatment personnel is at least to record any complaints from consumers and to characterize the type of

Table 9.1. *Common odors in the influents of three drinking water*
utilities expressed as percentage occurrence of all odors

Utility	Odor characterization	Percentage occurrence
Philadelphia Suburban Water Co.	'Sewage'	83
	'Creeky'	20
	'Musty'	10
Philadelphia Water Department	'Decaying vegetation'	62
	'Septic (sewage)'	52
	'Vegetation'	22
	'Earthy'	43
	'Musty'	17
	'Fishy'	13
Lyonnaise Des Eaux	'Muddy'	71
	'Fishy'	71
	'Musty'	38
	'Septic'	29

From Bartels *et al.* (1989).

odor (Table 9.1). There is of course a danger in assuming that all 'fishy' odors in water supplies arise from chrysophyte blooms, because fish-like odors can originate from the decomposition of certain protein complexes. Only rarely have chemical analyses been undertaken to characterize precisely the odor-causing compounds (Mallevialle & Suffet 1987).

A more ideal situation results from microscopic identification of organisms in the raw (untreated) water source at the first sign of an odor problem (e.g. Table 9.2). One difficulty with this scenario is that the odor from an algal population may not arise until near the end of its period of healthy growth. So, by the time it is recognized as a problem and investigated, some other species may have assumed dominance in the phytoplankton. The result is that some 'innocent' species may be mistakenly identified as the problem organism (Fig. 9.2). Clearly, quick reaction times between odor detection and identification of the cause are necessary to avoid erroneous explanations.

At some water treatment plants, the need for timely and accurate diagnosis has led to the establishment of regular phytoplankton monitoring and routine measurements of threshold odor concentrations (APHA 1985; Persson 1983; Persson & Jüttner 1983). The data thus obtained permit

Table 9.2. *Densities of organisms associated with odors in drinking*
water from Lake Lyseren, Norway

Organism	Concentration (cells/ml)	Date
Chlamydobacteria	2000	27 July 1976
Chlorococcales	193	27 July 1976
Asterionella formosa Hass.	60	31 July 1975
Tabellaria fenestrata (Lyngb.) Kütz.	145	31 July 1975
Dinobryon divergens Imh.	560	1 August 1975
D. sociale var. *stipitatum* (Stein) Lemm.	2000	31 July 1975
Chrysomonads	12 000	31 July 1975

From Berglind *et al.* (1983).

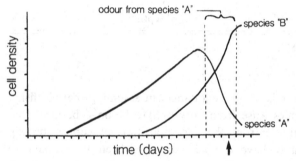

Figure 9.2. Hypothetical population density changes over time for two phytoplankton species, 'A' and 'B', of which, only species 'A' is an odor producer. If
sampling and identification are done at the peak of odor production but after the
population has peaked (arrow), species 'B' may be erroneously implicated as the
cause of the problem.

the judicial application of remedial measures such as activated carbon
filtration in a cost-effective, short-term, proactive manner.

  Chrysophytes tend to dominate the phytoplankton of north-temperate,
oligotrophic lakes (Janus & Duthie 1979; Eloranta 1986, this volume;
Earle *et al.* 1986; Sandgren 1988; Willén *et al.* 1990; Pinel-Alloul
et al. 1990; Nicholls *et al.* 1992). The major human use of such lakes is
often for recreational purposes, rather than for drinking water supplies.
However, as was the case with drinking water supplies, the best source
of information about chrysophyte-related odor problems may originate
from user complaints.

Table 9.3. *Occurrence of* Uroglena *blooms in Ontario lakes during*
1987–1990

Lake	Township	Z_{max} (m)	Total phosphorus (µg P/l)	Alkalinity (mg CaCo$_3$/l)	Secchi disc (m)
Paint	Ridout	21.4	10	5.4	2.1
Pine	Wood/Musk	22.6	8	2.6	3.8
Black	Wood	3.4	12	0.9	1.5
Leech	Oakley	13.7	6	2.2	4.6
Bigwind	Oakley	36.3	6	2.9	5.0
Clear	Oakley	26.8	5	3.0	5.8
Kawagama	McClintock	73.2	3	2.4	9.3
Fletcher	McClintock	30.5	?	3.7	?
Muskoka	Muskoka	66.5	5	4.6	5.1

In Ontario during the past 4 years we have been made aware (through user complaints) of nine episodes of fish-like odor production by *Uroglena* blooms. These nine lakes (Tables 9.3) included a wide range of lake types, from shallow, dystrophic lakes such as Black Lake, to deep, clear-water lakes such as Kawagama Lake. It is clear, therefore, that circumstances other than nutrient supply alone are responsible for such blooms. The species of *Uroglena* causing blooms in Ontario lakes have not been determined. But, in the case of perhaps the best-known *Uroglena* lake in the world, Lake Biwa in Japan, considerable research has been undertaken to determine the factors stimulating growth and causing recurrent blooms of *U. americana* since the early 1970s (Yoshida *et al.* 1983*a, b, c*; Kimura & Ishida 1986; Kimura *et al.* 1986; Kurata 1986, 1989).

Taste and odor problems in Ontario lakes that originate from *Synura* have been more widespread than those induced by *Uroglena*. Some of these episodes were reviewed by Nicholls & Gerrath (1985) and it was concluded that in every case, in both hardwater and softwater lakes, only one of the 20 known species of *Synura* (*S. petersenii*) was the cause.

Characterization of chrysophyte volatile excretory products

There have been only two or three significant studies of the volatile excretory products of chrysophytes. Collins & Kalnins (1965) reproduced typical *Synura* odors in the laboratory in bacteria-free cultures. By applying their solvent extraction and thin layer chromatographic methods

Table 9.4. *Volatile excretory products of three algal species, including the*
chrysophyte Synura uvella

Location and date	Dominant species	Excretory products
Lake Constance 17 October 1978	*Anabaena* sp.	Dimethyldisulfide
Lake Constance 19 July 1979	*Asterionella formosa*	Octadiene
		Octatriene
Wahnbach Reservoir 10 October 1978	*Synura uvella*	Penten-3-one
		Pentanone-3
		Octanone-3
		Octanol-1
		Octanol-3
		Oct-1-en-3-ol
		2-pentylfuran
		Trans,cis-2,4-decadienal
		β-Cyclocitral
		6-Methylhept-5-en-2-one
		β-Ionone

From Jüttner (1981).

to culture filtrate as well as directly to cell concentrates, they showed that
the odor-causing compounds were aldehydes and ketones (mainly *n*-
hapanal) which were excreted by living cells into the growth medium.

Some further work on a *Synura* bloom occurring in the Wahnbach
Reservoir in Germany was reported by Jüttner (1981, 1983), and on the
basis of similar findings from cultures of two *Ochromonas* species, it was
concluded that the Chrysophyceae, like higher terrestrial plants, produce
a wide spectrum of volatile organic compounds including a number of
dienals, alkenols, alkenones, alcohols and ketones (Table 9.4), of which
oct-1-en-3-ol, pentanone-3 and octanone-3 were the most important.
Yano *et al.* (1988) identified the related compounds (E,E)-2,4-heptadienal
and (E,Z)-2,4-heptadienal as the causes of the fishy odor in a *Uroglena*
bloom in the Ninobiki Reservoir in Japan.

Bloom development: ecological considerations

Aside from the explanation offered for the *Aureococcus* bloom, I have not
discussed chrysophyte bloom ecology. The risk in delving into this topic
is that I would be repeating much of the information presented by other
contributors to this volume under the related topics of chrysophyte

distribution, ecology and paleoecology. This is because many of the strategies that permit chrysophytes to survive and flourish in lakes are undoubtedly related to similar factors involved in bloom development (see also review by Sandgren, 1988). Therefore, this section on chrysophyte bloom initiation will include only the essential highlights of the ecology of bloom-forming species, so that more space can be devoted to detailing what I consider to be the major deficiencies in our understanding of the chrysophyte bloom phenomenon. I hope that this approach will provide some direction for future research and management.

An understanding of the ecology of bloom-forming chrysophytes must be based on a solid physiological foundation. A classic example of this is the historical development (spanning a period of nearly half a century) of the understanding of the relative roles of phosphorus and potassium as factors controlling *Dinobryon* populations (see Shapiro 1988: p. 11). Quantitative ecophysiological investigations of the bloom-forming Chrysophyceae probably had their beginnings with the work of Gaidukov (1900) who determined that *Chromophyton rosanoffii* Woronin emend. Couté (Gaidukov's '*Chromulina rosanoffii*') could be grown in a defined mineral medium. Subsequently, Guseva (1935) was able to examine growth rates of *Synura uvella* and *S. petersenii* in cultures based on isolations from the Moscow River Oxbow and from some peat pits in the vicinity of Bolshero. She discovered that iron concentrations were critical for good growth in both species. Optimum iron concentrations of $1.2–1.4$ mg1^{-1} and nitrogen concentrations of $1.0–2.0$ mg $NO_3^- $-$N1^{-1}$ and $0.1–0.2$ mg NH_4^+-$N1^{-1}$ were determined for *S. petersenii*. Today, we know that the relatively high iron requirement of *Synura* species relates to their use of iron-containing cytochrome c as an electron donor in photosystem I (Raven, this issue). The tendency for certain *Synura* species to develop large populations in dystrophic lakes and ponds may depend, in part, on the availability of iron. Dokulil & Skolaut (1991) concluded that the summer populations of *Dinobryon* species in the Mondsee in Austria were dependent on the supply of iron, the availability of which was controlled through chelation by organic substances released by decomposition of the spring diatom populations. Iron and mixed micrometals have been determined to be growth limiting nutrients for chrysophytes in several studies (reviewed in Sandgren 1988).

Light may also be a factor controlling bloom development in dystrophic waters, owing to its rapid attenuation by dissolved organic matter (Jones & Arvola 1984). Little is known, however, of the light quality and quantity preferences and the physiological adaptive responses by phytoplankton as a whole (Falkowski 1984), let alone chrysophytes in particular. In experiments on the role of color and light intensity on the distribution of several phytoplankton species, Wall & Briand (1979) found that chrysophytes represented by *Dinobryon* and *Mallomonas* species tended to be favored more by blue light than were other groups of algae. They suggested that this is consistent with the known vertical distribution of these algae in lakes and that the deep-water occurrence of these species may indicate a competitive advantage over other algae better suited to use the higher-intensity red wavelengths in surface waters.

There is ample evidence of stratified subsurface chrysophyte populations in lakes (Fee 1976; Nygaard 1977; Pick *et al.* 1984; Croome & Tyler 1988) and diurnal migrations of some taxa have been demonstrated (Ilmavirta 1974; Jones 1988). The most obvious attribute that the presence of flagella has provided to each of the genera being discussed here (*Dinobryon*, *Synura*, *Uroglena* and *Mallomonas*) is the ability to 'swim'. Subsurface populations of these organisms are clearly there by choice. The suggestion above was that some relative advantage over other algae relating to light quality and intensity may exist for these motile chrysophytes. The operative word is 'relative' because laboratory studies have shown that species achieving significant populations in the wild at low light and temperature may grow equally well or better at higher light intensities and higher temperatures in the laboratory. For example, Healey (1983) showed that growth rate of *Synura sphagnicola* isolated from a submetalimnetic bloom (ambient light 1% of surface irradiance and temperature 7–8 °C) was a saturable function of light intensity over a temperature range of 5–20 °C in the laboratory (Fig. 9.3).

If light and temperature are not optimal for these organisms at these depths, then they must be actively seeking out these strata for other reasons. Nutrient concentrations are usually higher within and below the metalimnion because of decomposition and mineralization of materials sedimenting from the upper trophogenic zone combined with the upward diffusion of dissolved matter released from the lake sediment and/or from similar processes of decomposition in the hypolimnion. By virtue of its motility, a cell in a homogeneous solution of dissolved nutrients is ensuring a constant concentration gradient across the cell membrane (assuming constant uptake rate and constant swimming speed). While it has been shown that the net energy gain to a motile cell (compared with a non-motile cell) under these conditions is negligible (Sommer 1988), there may be considerable advantage in expending energy to seek out and remain in strata with higher concentrations of essential nutrients. Bloom-

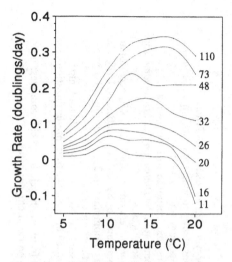

Figure 9.3. The effects of temperature on the growth rate of cultured *Synura sphagnicola* at eight different light intensities ranging from 11 to 110 µEin m^{-2} s^{-1}. (After Healey 1983.)

forming species of *Mallomonas, Dinobryon, Synura* and *Uroglena* are all large and highly motile. One advantage that taxa with the ability to move vertically in the water column may have over those that do not is exposure to 'pockets' with higher nutrient levels; they are thus able to exploit heterogeneities in growth-promoting substances in their environment within the limits of their swimming capabilities and net energy expenditures.

Biotic and anthropogenic factors

In general, chrysophytes tend to comprise a significant proportion of the total phytoplankton of oligotrophic to mesotrophic lakes. The role of nutrients originating from human activities as triggers of chrysophyte blooms is not clear, although it is evident that substantial populations may develop in lakes and ponds with significant inputs of nutrients. In eutrophic lakes, the seasonal distribution of chrysophytes may be restricted mainly to early spring (Kristiansen 1988; Sandgren 1988). Munch (1972) and Roijackers (1985) both reported occasional blooms of *Synura* or *Mallomonas* in productive dimictic lakes. Generally, eutrophied lakes are poorly represented by chrysophytes, but when point-source nutrient loading controls have been initiated, the relative importance of chryso-phytes has increased (Dillon *et al.* 1978; Nicholls *et al.* 1986; Willén 1987).

Table 9.5. *Summary of chrysophyte responses to experimental nutrient additions*

Reference	Lake	Treatment	Response
Langford (1950)	Four lakes in Algonquin Park, Ontario	Agr. fertilizer (12–24–12); 0.05–0.9 mg P/l	All lakes responded 3–4 weeks after treatment with large increases in diatoms and *Dinobryon* (and *Synura* in one lake)
Smith (1969)	Crecy Lake	210 µg N/l; 390 µg P/l; 270 µg K/l	Main response was in *Anabaena* and *Spirogyra* blooms, but *Dinobryon* bloomed in 1961, 3 years after the last fertilization
Schindler *et al.* (1971, 1973), Kling & Holmgren (1972), Findlay & Kling (1975), Findlay (1978, 1981, 1983)	ELA Lake 304	Two years of each of the following treatments: N, P and C; N and C; N and P	*Dinobryon* and *Synura* occasionally among dominants in early summer of some years
See above	ELA Lake 261	Four years of H_3PO_4 additions	*Dinobryon*, *Mallomonas*, and *Synura* consistently among the dominants; *Uroglena* important only in 1975
See above	ELA Lake 227	Inorganic P and N added weekly for several years	Inconsistent response with occasional domination by *Dinobryon* and *Mallomonas* spp. in some years
Findlay & Kasian (1987)	ELA Lake226NE and 226SW	Basin divided by a curtain; 226NE fertilized with N, P and C, and 226SW with N and C	*Dinobryon* less important in 206NE; occasional dominance by *Synura* in both basins; *Uroglena* was a dominant in 226NE during all 3 post-fertilization years
Ramberg (1976)	Lakes Vitalampa and Botjarn, Sweden	Fertilization for 1 year with NH_4NO_4	Chrysophyte portion of total phytoplankton showed little change from pre-fertilization years
Witt (1977)	Vorderer Finstertaler See, Austria	Additions of P as NaH_2PO_4	Shift from dinoflagellates to chlorophytes; no response in chrysophytes

Reference	Location	Treatment	Results
DeNoyelles & O'Brien (1978)	Eight 0.1 ha ponds, Cornell University	N, P and K in Low, medium and high dose rates	Chrysophytes remained dominant in reference ponds, but were replaced by chlorophytes and cyanophytes in treated ponds
Reinertsen (1982)	Lake Langvatn, central Norway	N, P and K supplied as commercial fertilizer in 1975 and 1976	No major response among *Synura*, *Uroglena* or *Dinobryon* (all three present pre-fertilization)
Yan & Lafrance (1984)	Mountaintop and Labelle Lakes, Sudbury, Ontario	P and N added as 20-40-0 commercial fertilizer and as H_3PO_4 and NH_4NO_3	*Dinobryon* and *Mallomonas* spp. replaced by other algae, mainly *Cryptomonas* spp.
Yan & Lafrance (1984)	Middle and Hannah Lakes, Sudbury, Ontario	P added as H_3PO_4	Low-level fertilization resulted in continued domination of chrysophytes; higher fertilization rates led to replacement by other groups
Wilcox & DeCosta (1984)	Cheat Lake, W. Virginia	P added to enclosures as KH_2PO_4	*Dinobryon* replaced by chlorophytes
Holmgren (1984)	Four lakes in the Kuokkel area of N. Sweden	N and P added as NH_4NO_3 and H_3PO_4 in various combinations over several years	Additions of N alone stimulated *Uroglena*; N and P additions to the same lake the following year led to dominance by *Uroglena* and *Dinobryon*, but by cryptophytes in another lake; additions of P alone inhibited *Dinobryon* spp.
Chow-Fraser & Duthie (1987)	An embayment of Lake Matamek, Quebec	N and P added as $NH_4H_2PO_4$	*Dinobryon crenulatum* was one of 17 taxa showing a significant increase
Olofsson et al. (1988)	Lake Hecklan, Central Sweden	Continuous low dosage of N and P in commercial fertilizer	Chrysophytes dominated before and after fertilization with only minor shifts in community composition

Conversely, fertilization of low-productivity lakes has sometimes been followed by blooms of chrysophytes, but the results thus far have not been consistent (Table 9.5). For example, Langford (1950) fertilized four lakes in Ontario with a 12-24-12 formulation of a commercially available fertilizer and achieved epilimnetic concentrations of total phosphorus ranging from 0.05 to 0.9 mgl^{-1} (as P). All lakes, except an unfertilized control lake, responded to treatments 3–4 weeks later with increases in *Tabellaria*, *Asterionella* and *Dinobryon*. One of the lakes produced much higher densities of *Synura*. In contrast, Lake Langvatn in central Norway showed no response among *Synura*, *Uroglena* or *Dinobryon* at fertilizer application rates similar to those used by Langford (Reinertsen 1982). In northern Sweden, a *Uroglena* bloom resulted from fertilization with nitrogen alone in one lake, but no bloom developed in another lake receiving a similar treatment (Holmgren 1984).

The responses of *Dinobryon* to lake fertilization are interesting. The early belief that phosphate was toxic to *Dinobryon*, even at relatively low concentrations, was challenged by Lehman (1976), who showed that when phosphorus was supplied even at high concentrations to *Dinobryon* cultures as K_2HPO_4 or KH_2PO_4, potassium, not phosphorus, was the toxic element. However Lehman's work showed inhibition of growth of *D. sociale* var. *americanum* and *D. cylindricum* at potassium concentrations greater than 500 µgl^{-1} and 7500 µgl^{-1}, respectively, and these results do not entirely explain the inhibition observed by Rodhe (1948) at much lower concentrations of potassium (25 µgl^{-1}). More recently, Wilcox & DeCosta (1984) also observed a rapid disappearance of *Dinobryon* in experimental enclosures fertilized with KH_2PO_4 (100 µg K l^{-1}). Holmgren (1984) found an apparent inhibition of *Dinobryon* spp. in a lake fertilized with phosphate alone (as H_3PO_4), but *Dinobryon* spp. were consistently among the dominants in Lake ELA 261 after fertilization with H_3PO_4 (Table 9.5). Clearly, the role of nutrient supply in chrysophyte bloom development is poorly understood, but it is possible that in combination with other factors, small increases in nutrient supply are stimulatory to some chrysophyte populations.

Factors contributing to chrysophyte blooms probably relate as much to strategies that allow populations to avoid losses as to factors that contribute directly to optimum growth. Another advantage of the capability for vertical movement in lakes may be that it affords protection from zooplankton grazing. In laboratory studies, Sandgren and Walton (this volume) have demonstrated that *Synura* and other large chrysophytes are utilized as food items by large *Daphnia* species. In support of these findings, they assembled data from the literature which showed that the presence of populations of large *Daphnia* in lakes was associated with depressed chrysophyte populations. This finding raises an interesting 'chicken and egg' question: Is the predominance of chrysophytes in softwater lakes a direct physiological response to chemical characteristics of low pH waters, or is it in response to decreased *Daphnia* grazing, since Keller *et al.* (1990) have shown with field data and laboratory bioassays that densities of *Daphnia galeata mendotae* decline in lakes below pH 6.0? Sandgren concluded that there is no refuge from zooplankton grazing by virtue of large size among chrysophytes. There may, however, be protection afforded by the deeper portions of the vertical migratory route (metalimnion or upper hypolimnion) where low water temperatures are associated with low metabolic rates of the grazing animals at these depths.

In summary, large populations of chrysophyte species may develop in softwater lakes because of their ability to grow well at low temperatures and low light intensities (although these may not be optimal), combined with the minimization of zooplankton grazing impacts and adequate nutrient supplies. The latter two factors may be augmented by human influences. For example, removal of large piscivorous fish results in enhanced zooplanktivore populations and increased predation on large zooplankton species. Watershed deforestation, human wastewater seepage and agriculturalization all result in increased nutrient inputs to lakes, and the effects of acid deposition may act as a double-edged sword: (1) by decreasing lakewater alkalinity and thereby putting additional stress on *Daphnia* spp., and (2) by enhancing leaching of growth-promoting substances including trace metals such as cobalt needed for subsequent synthesis of vitamin B_{12}. The combined effects of eutrophication and acidification have been blamed for the recent algal blooms in Scandinavian coastal waters (Sangfors 1988; Granéli *et al.* 1989).

While these factors may explain the occurrence of large chrysophyte populations (Fig. 9.4), they do not necessarily explain some aspects of the bloom phenomenon itself. For example, *Uroglena* blooms are often manifested as sudden mass accumulations at the lake surface, much like the blue-green algal bloom phenomenon. Unlike the blue-green algal bloom, for which an ecophysiological explanation exists (Reynolds & Walsby 1975), no similar explanation for the sudden surface accumulation of *Uroglena* has been suggested. Considerable laboratory study of the now common Lake Biwa *Uroglena* blooms has contributed much to an understanding of phagotrophism, and nutrient, trace metal and vitamin requirements (Yoshida *et al.* 1983*a*, *b*, *c*; Kimura & Ishida 1986; Kimura

198   *Chrysophyte algae*

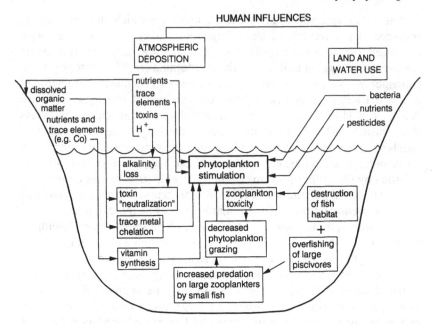

Figure 9.4. The interactions of a number of factors originating from human influences and their stimulatory effect on phytoplankton in a hypothetical softwater lake.

et al. 1986; Kurata 1986, 1989). The apparently now common episodes of *Uroglena* surface blooms in Ontario (Table 9.3), may afford additional opportunities to determine the factors contributing to the surface accumulations.

Neuston

The term 'neuston' was first used by Naumann (1917) to designate the community of organisms associated with the surface microlayer of lakes and ponds. Its use in freshwater and marine contexts has been reviewed by Banse (1975) with reference to subdivisions of the neuston (e.g. epineuston, hyponeuston) and other 'nearby' biotopes in aquatic ecosystems. It is important to realize that organisms inhabiting the neuston are there for reasons related to surface tension, not because of buoyancy adaptations; so, by definition, the neuston excludes surface accumulations of gas vacuolate blue-green algae.

   The neuston communities of freshwaters may include a wide variety of taxa (Frølund 1977; Fuhs 1982*a*, *b*; Pentecost 1984; Estep & Remsen

1984; Timpano & Pfiester 1985). The highly visible oily sheen produced on the water surface by several chrysophytes qualifies this type of growth as a 'bloom' under the definition used here. The most commonly reported neustonic chrysophyte is *Chromophyton rosanoffii* Woronin emend. Couté (Couté 1983), which forms curious epineustic 'pseudocysts'. Some aspects of the life history and seasonal development of this species have been presented by Petersen & Hansen (1958) and Frølund (1977) who found a maximum density of about 2×10^6 cells cm^{-2}. Heynig (1972) has also included *C. rosanoffii* among a listing of other freshwater bloom-forming algae such as *Aphanizomenon, Microcystis, Dinobryon, Ceratium* and *Botryococcus*.

Other Chrysophyceae known to inhabit the neuston (but not necessarily excluded from planktonic, benthic or epiphytic existences in other habitats) include *Chromulina neustophila* Conrad (Conrad 1940), *Paraphysomonas vestita* (Stokes) de Saed. (Frølund 1977), *Epipyxis minuta* (Mack) Hilliard (Petersen & Hansen 1958 [as *Hyalobryon minutum* Mack]), *Hyalocylix stipitata* Pet. & Han. (Petersen & Hansen 1958) and *Kremastochrysis minor* Catalan (Catalan 1987). Two or three other *Chromulina* species are probably synonymous with *Chromophyton rosanoffii* since cell habit and morphology appear identical except for the likely omission of the second short flagellum (Couté 1983). Those neustonic chrysophytes forming visible blooms are probably restricted to *C. rosanoffii, C. neustophila* and *K. minor*.

The surface microlayer may also be enriched with bacteria (Maki & Remsen 1989). Because chrysophytes of the *Ochromonas* type are known facultative phagotrophs (see Raven, this volume; Holen & Boraas, this volume), the generally low light environment of woodland ponds might provide a competitive advantage to such chrysophytes as *Chromophyton* over other algae which must depend only on autotrophic nutrition. Once established in the neuston with some dependence on bacteria for a portion of their energy supplies, other variables might enhance the availability of nutrients and other growth factors in this specialized environment. For example, precipitation can provide nutrients in dissolved, bioavailable form at levels often far exceeding concentrations available to phyto-plankton of oligotrophic systems (Parker *et al.* 1981). Danos *et al.* (1983) found significantly higher concentrations of dissolved inorganic nitrogen, phosphorus, silica and pigments in the surface microlayers of experimental ponds. Also, the surface microlayer of seawater is enriched with surface active substances such as fatty acids and proteinaceous materials (Duce *et al.* 1972; Hardy 1982; Bärlocher *et al.* 1988) which may act as chelators of trace elements required by neustonic species – elements which might not be available to planktonic species. A similar enrichment of the microlayer of freshwater forest pools might be expected given the usually high organic content of their terrestrial surroundings and the associated opportunities for the supply of vitamins and other growth factors.

The surface film algal community may be immune from grazing by micro-crustaceans that are adapted for planktonic filter feeding. This specialized habitat may therefore provide a refuge from one of the important loss mechanisms influencing phytoplankton. However, protists with special adaptations for existence at the air–water interface, such as hypotrichid ciliates (Ricci *et al.* 1991), may exert some influence on phytoneuston communities. Also, because of the specialized habitats of neustonic chrysophytes (mainly physically stable, small, shallow forest pools), these chrysophytes have not in the past created any special problems for human use of these waters. These specialized communities do, however, offer their potential use as model systems for investigation of a number of chrysophyte-related phenomena, including bacterial–flagellate interdependencies and the possible production of volatile compounds which might inhibit the growth of other organisms.

Information needs

A more definite knowledge of the factors involved in the development of blooms in fresh and salt water can certainly be gained from physiological experiments with cultures and simultaneous qualitative and ecological observations. [English translation from German]

Kolkwitz (1914)

This review has, I hope, consolidated some of the known information about the mass occurrences of chrysophytes. However, it has also probably served to point out a number of deficiencies in our understanding of the phenomena. I hope that many of these information gaps will have already become apparent to the reader, but at the risk of stating the obvious, I would like to provide my own thoughts on required future directions for chrysophyte bloom research.

(I)   There is a need to continue development of remote sensing technology, especially as it relates to coastal marine areas. The absorption and fluorescence spectra of *Aureococcus*, the marine 'brown tide' organism, are different from those of other coastal marine phytoplankton species (Yentsch *et al.* 1989); this might be exploited by remote sensing. Recent advances in multispectral scanning technology are leading to real-time assessments of the extent and dynamics of both marine and freshwater algal blooms (Balch *et al.* 1991; Millie *et al.* 1991).

Even if the extent and frequency of marine algal blooms do not increase in the future, the impacts of the marine blooms of the future will undoubtedly be more dramatic as coastal salmon culture intensifies. Net pen farming of both Atlantic and Pacific salmon now accounts for 30% of the world's production of canned salmon. It has been predicted that by the year 2000, farmed salmon will account for 90% of the total (Van Dyk 1990). Penned salmon are vulnerable to algal blooms. On the British Columbia (Canada) coast alone, fish farming losses resulting from diatom blooms have averaged between $2 million and $4 million annually over the period 1986–1990 (Red Tide Newsletter, 3(2): 11, April 1990). The *Prymnesium parvum* bloom on the Norwegian southwest coast in 1989 caused losses of caged salmon valued at $5 million (Kaartvedt *et al.* 1991).

In the path of an advancing algal bloom, the only reactions possible today are: (1) harvest the fish prematurely, (2) tow the net pens out to sea beyond the influence of the bloom, or (3) take a chance on the severity of the effects of the bloom. Early warning systems based in part on remote sensing technology would help in selecting one of these options.

(II)   More biochemical and toxicological work needs to be done on the volatile excretory products of chrysophyte bloomers, especially *Synura* and *Uroglena* species. One striking feature of the big blooms is that they are essentially unialgal. This begs the question as to whether other species are excluded because of toxic excretions. Kamiya *et al.* (1979) have demonstrated the presence of fatty acid ichthyotoxins in *Uroglena volvox*, and toxin production by axenic cultures of *Ochromonas* has also been found (Spiegelstein *et al.* 1969). This is important because there is currently a debate in progress on the role of bacteria in the production of toxins associated with some marine dinoflagellate blooms (Taylor 1990). If bacteria are not implicated in the production of chrysophyte toxins as the scant research data would suggest, then progress in determining factors responsible for toxin production might be relatively rapid because the work would be based on a simpler biotic system.

202 *Chrysophyte algae*

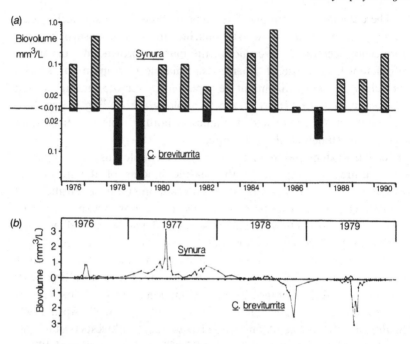

Figure 9.5. (*a*) Annual average biovolume of *Chrysochromulina breviturrita* and *Synura* spp. in the phytoplankton of Dickie Lake, Ontario, 1976–90. (*b*) Seasonal distributions of *C. breviturrita* and *Synura* spp., 1976–79. (Nicholls, unpublished data.)

Some indirect evidence for toxin production by either (or both) *Synura* or *Chrysochromulina breviturrita* may exist in data from Dickie Lake in Ontario. Over a 15 year period of observation, years with high biovolume of *Synura* were also years with low biovolume of *Chrysochromulina breviturrita* and vice versa (Fig. 9.5). The recent *Chrysochromulina polylepis* poisonings in the Baltic (Underdal *et al.* 1989) provided the first evidence for toxin production by members of the prymnesiophyte genus *Chrysochromulina* (but see Nicholls *et al.* 1982). It cannot be determined conclusively from the available Dickie Lake data whether or not *C. breviturrita* and *Synura* have inhibited each other's growth. However, the within-year data suggest that *Synura* may have inhibited *Chrysochromulina* because the *Synura* populations typically developed before the *Chrysochromulina* populations. *Chrysochromulina* did not develop until mid to late summer, and then apparently only in *Synura*'s absence (Fig. 9.5). At this point this is speculative but suggestive, and some experimental work needs to be done on species interactions that include *Synura*, *Uroglena* and other bloom formers. Included under the general topic of species interactions might also be the role of parasitic symbioses and predation by non-crustaceans such as *Bodo* and *Rhizoochromonas* species (Nicholls 1987, 1990) as potential regulators of chrysophyte populations. What are the environmental factors that allow *Bodo crassus* and *Rhizoochromonas endoloricata* to achieve high population densities?

(III) Because their flagellar structure is different from that of phago-trophic chrysophytes with the ochromonad flagellation type, *Mallomonas* and *Synura* are unlikely to utilize bacteria directly as a food source in the same way that several ochromonadalean species do (including *Dinobryon* and *Uroglena*) (Bird & Kalff 1986; Kimura *et al.* 1986; Holen & Boraas, this volume). Nevertheless, the other roles of aquatic bacteria in modifying the growth medium and helping to structure the phytoplankton community (Newhook & Briand 1987) need to be investigated with reference to chrysophyte blooms.

(IV) Although much is now known about the physical and biochemical properties of dissolved organic matter (Gjessing 1976; Steinberg & Muenster 1985), there are still important information gaps relating to the effects of natural organic materials on aquatic organisms (Serrano & Guisande 1990; Steinberg 1990). The role of organic acids and iron in dystrophic systems, both as directly available essential growth factors and as substances regulating the availability of essential trace elements, needs further work with respect to chrysophytes. There is some evidence that the humic contents of Scandinavian lakes is increasing at rates as high as 3% per year (Forsberg & Petersen 1990). Graneli *et al.* (1989) have demonstrated enhanced growth of dinoflagellates in response to experi-mental additions of humic substances in coastal areas of the Baltic influenced by these increased inputs of dissolved humic materials from adjacent watersheds. Similar experiments need to be done in freshwater systems with potential bloom-forming chrysophytes.

(V) Experimental use of 'BIOTRON' type technology (e.g. Ostroff *et al.* 1980; Graham *et al.* 1985) should be encouraged for multifactorial laboratory investigation of light, temperature, nutrients and other variables influencing chrysophyte growth rates, and in particular to answer questions about surface water accumulations of *Uroglena*. Some relevant questions are: (1) Where does the population producing the surface bloom develop? (2) What causes mass movement to the surface and what is the relative importance of active motility and passive buoyancy? (3) Once at the lake surface, are cells viable? (4) What are the time-dependent effects on cell viability of potentially damaging high-intensity solar irradiance at the lake surface? (5) What are the diurnal vertical movements under thermally stratified conditions? (6) What are the relative roles of the water column light, temperature and nutrient gradients as regulators of vertical movements in the population? These questions might be answered through a combination of careful observations of wild populations and laboratory experimentation.

Summary

Although there are many unanswered questions relating to the causes of chrysophyte blooms, an apparent pattern has emerged from the recent studies of algal blooms in coastal marine environments and freshwater lakes. That is that human influences in the form of increased nutrient supply from human waste, drainage from agricultural operations, and contaminated precipitation provide the basic chemical medium for promoting blooms. While this is probably a valid conclusion for algal blooms in general, the relevance of this conclusion to the Chrysophyceae in particular has not been well defined, especially for freshwater populations. Although chrysophyte blooms have resulted from experimental fertilizations, a number of whole-lake and in-lake enclosure fertilization experiments have not produced chrysophyte blooms. The triggering of a chrysophyte bloom probably depends on the achievement of fundamental growth conditions which are species specific and which may be realized only when the 'right' combination of natural and anthropogenically derived variables comes together. These natural environmental variables could include temperature, salinity, hydrologic flushing rate, turbulence, light, grazing, and inter-species competitions and inhibitions. Because these natural influences are multifactorial and highly interactive, their combinations in ways that could trigger algal blooms are not predictable, given our present level of understanding of the ecology of the most important bloom-forming species. A number of experimental approaches to specific questions about the role of volatile excretory substances as inhibitors of other algae, the role of bacteria either directly as an energy source for phagotrophic chrysophytes, or indirectly as facilitators of nutrient availability, and the role of chrysophyte motility (vertical migration), zooplankton grazing and light effects are among those topics suggested here as fruitful lines of future research on the chrysophyte bloom phenomenon.

References

American Public Health Association (APHA) 1985. *Standard Methods for the Examination of Water and Waste water*, 16th edn. American Public Health Association, New York.

Balch, W.M., Holligan, P.M., Ackleson, S.G. & Voss, K.J. 1991. Biological and optical properties of mesoscale coccolithophore blooms in the Gulf of Maine. *Limnol. Oceanogr.* **36**: 629–43.

Banse, K. 1975. Pleuston and neuston: on the categories of organisms in the uppermost pelagial. *Int. Rev. Ges. Hydrobiol.* **4**: 439–47.

Bärlocher, F., Gordon, J. & Ireland, R.J. 1988. Organic composition of seafoam and its digestion by *Corophium volutator* (Pallas). *J. Exp. Mar. Biol. Ecol.* **115**: 179–86.

Bartels, J.H.M., Brady, B.M. & Suffet, I.H. (eds.) 1989. *Taste and Odor in Drinking Water Supplies. Combined Final Report, year 1 and 2 (1984–1986).* American Water Works Association, Denver, Colorado.

Berglind, L., Holtan, H. & Skulberg, O.M. 1983. Case studies on off-flavours in some Norwegian lakes. *Water Sci. Tech.* **15**: 199–209.

Bird, D.F. & Kalff, J. 1986. Bacterial grazing by planktonic lake algae. *Science* **231**: 493–5.

Bricelj, V.M. & Kuenstner, S.H. 1989. The feeding physiology and growth of bay scallops and mussels. In: Cosper, E.M., Bricelj, V.M. & Carpenter, E.J. (eds.) *Novel Phytoplankton Blooms: Causes and Impacts of Recurrent Brown Tides and Other Unusual Blooms.* Springer, Berlin. pp. 491–509.

Brown, O.B., Evans, R.H., Brown, J.W., Gordon, H.R., Smith, R.C. & Baker, K.S. 1985. Phytoplankton blooming off the U.S. East coast: a satellite description. *Science* **229**: 163–7.

Catalan, J. 1987. *Kremastochrysis minor* sp. nov.: a neustonic member of the Chrysophyceae. *Br. Phycol. J.* **22**: 257–60.

Chow-Fraser, P. & Duthie, H.C. 1987. Response of the phytoplankton community to weekly additions of mono-ammonium phosphate in a dystrophic lake. *Arch. Hydrobiol.* **110**: 67–82.

Clasen, J. & Bernhardt, H. 1982. A bloom of the Chrysophycea *Synura uvella* in the Wahnbach reservoir as indicator for the release of phosphates from the sediment. *Arch. Hydrobiol. Beih.* **18**: 61–8.

Collins, R.P. & Kalnins, K. 1965. Volatile constituents of *Synura petersenii*. I. The carbonyl fraction. *Lloydia* **28**: 48–51.

Conrad, W. 1940. Notes Protistologiques 14. Sur une formation neustique. *Bull. Mus. R. Hist. Nat. Belg.* **16**: 1–6.

Cosper, E.M., Dennison, W.C., Carpenter, E.J., Bricelj, V.M., Mitchell, J.G., Kuenstner, S.H., Colflesh, D.C. & Dewey, M. 1987. Recurrent and persistent 'Brown Tide' blooms perturb coastal marine ecosystem. *Estuaries* **10**: 284–90.

Cosper, E.M., Dennison, W., Milligan, A., Carpenter, E.J., Lee, C., Holzapfel, J. & Millanese, L. 1989. An examination of the environmental factors important to initiating and sustaining 'Brown Tide' blooms. In: Cosper, E.M., Bricelj, V.M. & Carpenter, E.J. (eds.) *Novel Phytoplankton Blooms:*

206 Chrysophyte algae

Causes and Impacts of Recurrent Brown Tides and Other Unusual Blooms.
Springer, Berlin. pp. 317–40.
Cosper, E.M., Lee, C. & Carpenter, E.J. 1990. Novel 'brown tide' blooms in
Long Island embayments: a search for the causes. In: Granéli, E.,
Sundström, B., Edler, L. & Anderson, D.M. (eds.) Toxic Marine
Phytoplankton. Proceedings of the Fourth International Conference on
Toxic Marine Phytoplankton, Lund, Sweden. Elsevier, New York.
pp. 17–28.
Couté, A. 1983. Ultrastructure de Chromophyton rosanoffii Woronin emend.
Couté et Chr. vischeri (Bourrel.) nov. comb. (Chrysophyceae,
Ochromonadales, Ochromonadaceae). Protistologica 19: 393–416.
Croome, R.L. & Tyler, P.A. 1988. Phytoflagellates and their ecology in
Tasmanian polyhumic lakes. Hydrobiologia 161: 245–53.
Danos, S.C., Maki, J.S. & Remsen, C.C. 1983. Stratification of microorganisms
and nutrients in the surface microlayer of freshwater ponds. Hydrobiologia
98: 193–202.
Dennison, W.C., Marshall, G.J. & Wigand, C. 1989. Effect of 'brown tide'
shading on eelgrass (Zostera marine L.) distributions. In: Cosper, E.M.,
Bricelj, V.M. & Carpenter, E.J. (eds.) Novel Phytoplankton Blooms: Causes
and Impacts of Recurrent Brown Tides and Other Unusual Blooms. Springer,
Berlin. pp. 675–92.
DeNoyelles, F., Jr & O'Brien, W.J. 1978. Phytoplankton succession in nutrient
enriched experimental ponds as related to changing carbon, nitrogen and
phosphorus conditions. Arch. Hydrobiol. 81: 137–65.
Dillon, P.J., Nicholls, K.H. & Robinson, G.W. 1978. Phosphorus removal at
Gravenhurst Bay, Ontario: an 8 year study on water quality changes.
Verh. Int. Ver. Limnol. 20: 263–71.
Dodge, J.D., Mariani, P., Paganelli, A. & Trevisan, R. 1987. Fine structure of
the red-bloom dinoflagellate Glenodinium sanguineum, from Lake Tovel (N.
Italy). Arch. Hydrobiol. (Suppl.) 78: 125–38.
Dokulil, M.T. & Skolaut, C. 1991. Aspects of phytoplankton seasonal
succession in Mondsee, Austria, with particular reference to the ecology of
Dinobryon Ehrenb. Verh. Int. Ver. Limnol. 24: 968–73.
Duce, R.A., Quinn, J.G., Olney, C.E., Piotrowicz, S.R., Ray, B.J. & Wade, T.L.
1972. Enrichment of heavy metals and organic compounds in the surface
microlayer of Narragansett Bay, Rhode Island. Science 176: 161–3.
Dzurica, S., Lee, C., Cosper, E.M. & Carpenter, E.J. 1989. Role of
environmental variables, specifically organic compounds and
micronutrients, in the growth of the chrysophyte Aureococcus
anophagefferens. In: Cosper, E.M., Bricelj, V.M. & Carpenter, E.J. (eds.)
Novel Phytoplankton Blooms: Causes and Impacts of Recurrent Brown Tides
and Other Unusual Blooms. Springer, Berlin. pp. 229–52.
Earle, J.C., Duthie, H.C. & Scruton, D.A. 1986. Analysis of the phytoplankton
composition of 95 Laborador lakes, with special reference to natural and
anthropogenic acidification. Can. J. Fish. Aquat. Sci. 43: 1804–11.
Eloranta, P. 1986. Phytoplankton structure in different lake types in central
Finland. Holarctic Ecol. 9: 214–24.
Eloranta, P. 1989. On the ecology of the genus Dinobryon in Finnish Lakes.
Beih. Nova Hedwigia 95: 99–109.
Estep, K.W. & Remsen, C.C. 1984. The relationship of individual algal species
to the surface microlayer of a small freshwater pond. J. Plankton Res. 6:
123–35.

Chrysophyte blooms in the plankton and neuston

Falkowski, P.G. 1984. Physiological responses of phytoplankton to natural light regimes. *J. Plankton Res.* **6**: 295–307.

Fee, E.J. 1976. The vertical and seasonal distribution of chlorophyll in lakes of the Experimental Lakes Area, northwestern Ontario: implications for primary production estimates. *Limnol. Oceanogr.* **21**: 767–83.

Findlay, D.L. 1978. Seasonal successions in phytoplankton in seven lake basins in the Experimental Lakes Area, northwestern Ontario, following artificial eutrophication. Data from 1974 to 1976. *Can. Fish Mar. Serv. MS Rep.* **1466**: 41pp.

Findlay, D.L. 1981. Seasonal successions in phytoplankton in seven lake basins in the Experimental Lakes Area, northwestern Ontario, following artificial eutrophication. Data from 1977 to 1979. *Can. MS Rep. Fish. Aquat. Sci.* **1627**: 40pp.

Findlay, D.L. 1983. Seasonal successions of phytoplankton in seven lake basins in the Experimental Lakes Area, northwestern Ontario, following artificial eutrophication. Data from 1980 to 1982. *Can. MS Rep. Fish. Aquat. Sci.* **1710**: 30pp.

Findlay, D.L. & Kasian, S.E.M. 1987. Phytoplankton community responses to nutrient additions in Lake 226, Experimental Lakes Area, northwestern Ontario. *Can. J. Fish. Aquat. Sci.* **44** (Suppl. 1): 35–46.

Findlay, D.L. & Kling, H.J. 1975. Seasonal succession of phytoplankton in seven lake basins in the Experimental Lakes Area, northwestern Ontario, following artificial eutrophication. *Fish. Mar. Serv. Res. Dev. Tech. Rep.* **513**: 53pp.

Forsberg, C. & Petersen, R.C. Jr 1990. A darkening of Swedish lakes due to increased humus inputs during the last 15 years. *Verh. Int. Ver. Limnol.* **24**: 289–92.

Frølund, A. 1977. The seasonal variation of the neuston of a small pond. *Bot. Tidsskr.* **72**: 45–56.

Fuhs, G.W. 1982a. Microbiota in surface films: an historical perspective. *J. Great Lakes Res.* **8**: 312–15.

Fuhs, G.W. 1982b. Overview of microbiota in surface films. *J. Great Lakes Res.* **8**: 310–11.

Fukuyo, Y., Takano, H., Chihara, M. & Matsuoka, K. 1990. *Red Tide Organisms in Japan.* Uchida Rokakuho, Tokyo. 430pp.

Gaidukov, N. 1900. Über die Ernährung der *Chromulina rosanoffii*. *Nova Hedwigia* **39** (Suppl. 4): 115–45.

Gallager, S.M., Stoecker, D.K. & Bricel, V.M. 1989. Effects of the brown tide alga on growth, feeding physiology and locomotory behavior of scallop larvae (*Argopecten irradians*). In: Cosper, E.M., Bricelj, V.M. & Carpenter, E.J. (eds.) *Novel Phytoplankton Blooms: Causes and Impacts of Recurrent Brown Tides and Other Unusual Blooms.* Springer, Berlin. pp. 511–41.

Gjessing, E.T. 1976. *Physical and Chemical Characteristics of Aquatic Humus.* Ann Arbor Science Publishers, Ann Arbor. 120pp.

Graham, J.M., Kranzfelder, J.A. & Auer, M.T. 1985. Light and temperature as factors regulating seasonal growth and distribution of *Ulothrix zonata* (Ulvophyceae). *J. Phycol.* **21**: 228–34.

Granéli, E., Carlsson, P., Olsson, P., Sundtrom, B., Granéli, W. & Lindahl, O. 1989. From anoxia to fish poisoning: the last ten years of phytoplankton blooms in Swedish marine waters. In: Cosper, E.M., Bricelj, V.M. & Carpenter, E.J. (eds.) *Novel Phytoplankton Blooms: Causes and Impacts of*

Recurrent Brown Tides and Other Unusual Blooms. Springer, Berlin.
pp. 407–27.
Guseva, K.A. 1935. The conditions of mass development and the physiology of
nutrition of Synura. Mikrobiologiya 4: 24–43. (Translated from the Russian;
Technical Translation 1091, National Research Council of Canada, 1963.)
Hardy, J.T. 1982. The sea-surface microlayer: biology, chemistry and
anthropogenic enrichment. Prog. Oceanogr. 11: 307–28.
Healey, F.P. 1983. Effect of temperature and light intensity on the growth rate
of Synura sphagnicola. J. Plankton Res. 5: 767–74.
Heynig, H. 1972. Algenmassenentwicklungen: ein Zeichen der
Gewässereutrophierung: I. Mikrokosmos 11: 325–30.
Holmgren, S.K. 1984. Exerimental lake fertilization in the Kuokkel area,
northern Sweden: phytoplankton biomass and algal composition in natural
and fertilized subarctic lakes. Int. Rev. Ges. Hydrobiol. 69: 781–817.
Ilmavirta, V. 1974. Diel periodicity in the phytoplankton community of the
oligotrophic Lake Pääjärvi, southern Finland. I. Phytoplanktonic primary
production and related factors. Ann. Bot. Fenn. 11: 136–77.
Ilmavirta, V. 1988. Phytoflagellates and their ecology in Finnish brown-water
lakes. Hydrobiologia 161: 255–70.
Janus, L.L. & Duthie, H.C. 1979. Phytoplankton and primary production of
lakes in the Matamek watershed, Quebec. Int. Rev. Ges. Hydrobiol. 64: 89–98.
Jones, R.I. 1988. Vertical distribution and diel migration of flagellated
phytoplankton in a small humic lake. Hydrobiologia 161: 75–87.
Jones, R.I. & Arvola, L. 1984. Light penetration and some related
characteristics in small forest lakes in southern Finland. Verh. Int. Ver.
Limnol. 22: 811–16.
Jüttner, F. 1981. Biologically active compounds released during algal blooms.
Verh. Int. Ver. Limnol. 21: 227–30.
Jüttner, F. 1983. Volatile odorous excretion products of algae and their occurrence
in the natural aquatic environment. Water Sci. Technol. 15: 247–57.
Kaartvedt, S., Johnsen, T.M., Aksnes, D.L., Lie, U. & Svendsen, H. 1991.
Occurrence of the toxic phytoflagellate Prymnesium parvum and associated
fish mortality in Norwegian fjord system. Can. J. Fish. Aquat. Sci. 48:
2316–23.
Kamiya, H., Naka, K. & Hashimoto, K. 1979. Ichthyotoxicity of a flagellate
Uroglena volvox. Bull. Jpn. Soc. Sci. Fish. 45(1): 129.
Keller, W., Yan, N.D., Holtze, K.E. & Pitblado, J.R. 1990. Inferred effects of
lake acidification on Daphnia galeata mendotae. Environ. Sci. Technol. 24:
1259–61.
Kimura, B. & Ishida, Y. 1986. Possible phagotrophic feeding of bacteria in a
freshwater red tide Chrysophyceae Uroglena americana. Bull. Jpn. Soc. Sci.
Fish. 52: 697–701.
Kimura, B., Ishida, Y. & Kadota, H. 1986. Effect of naturally collected bacteria
on growth of Uroglena americana, a freshwater red tide Chrysophyceae.
Bull. Jpn. Soc. Sci. Fish. 52: 691–6.
Kling, H.J. & Holmgren, S.K. 1972. Species composition and seasonal
distribution of phytoplankton in the Experimental Lakes Area,
northwestern Ontario. Fisheries Research Board of Canada, Technical
Report 337. 56pp.
Kolkwitz, R. 1914. Über Wasserblüten. Bot. Jahrb. 50: 355.
Kristiansen, J. 1971. A Mallomonas bloom in a Bulgarian mountain lake. Nova
Hedwigia 21: 877–82.

Kristiansen, J. 1988. Seasonal occurrence of silica-scaled chrysophytes under eutrophic conditions. *Hydrobiologia* **61**: 171–84.

Kurata, A. 1986. Blooms of *Uroglena americana* in relation to concentrations of B group vitamins in *Lake Biwa*. In: Kristiansen, J. & Andersen, R.A. (eds.) *Chrysophytes: Aspects and Problems*. Cambridge University Press, Cambridge. pp. 185–96.

Kurata, A. 1989. The relationship between metal concentrations and *Uroglena americana* blooms in Lake Biwa, Japan. *Beih. Nova Hedwigia* **95**: 119–29.

Lackey, J.B. 1950. Aquatic biology and the water works engineer. *Public Works* **81**: 39–41.

Langford, R.R. 1950. Fertilization of Lakes in Algonquin Park, Ontario. *Trans. Am. Fish. Soc.* **78**: 133–44.

Legendre, L. 1990. The significance of microalgal blooms for fisheries and for the export of particulate organic carbon in oceans. *J. Plankton Res.* **12**: 681–99.

Lehman, J.T. 1976. Ecological and nutritional studies on *Dinobryon* Ehrenb.: seasonal periodicity and the phosphate toxicity problem. *Limnol. Oceanogr.* **21**: 646–58.

Maki, J.S. & Remsen, C.C. 1989. Examination of a freshwater surface microlayer for diel changes in the bacterial neuston. *Hydrobiologia* **182**: 25–34.

Mallevialle, J. & Suffet, I.H. (eds.) 1987. *Identification and Treatment of Tastes and Odors in Drinking Water*. American Water Works Association, Denver, Colorado.

Millie, D., Baker, M., Tucker, C. & Dionigi, C. 1991. High-resolution remote-sensing of bloom-forming phytoplankton. *J. Phycol.* (Suppl.) **27**(3): 50.

Minei, V.A. 1989. The possible role of lawn fertilizers and pesticide use in the occurrence of the brown tide. In: Cosper, E.M., Bricelj, V.M. & Carpenter, E.J. (eds.) *Novel Phytoplankton Blooms: Causes and Impacts of Recurrent Brown Tides and Other Unusual Blooms*. Springer, Berlin. p. 785.

Munch, C.S. 1972. An ecological study of the planktonic chrysophytes of Hall Lake, Washington. PhD thesis, University of Washington, Seattle.

Naumann, E. 1917. Beiträge zur Kenntnis des Teichnannoplanktons. II. Über das Neuston des Süsswassers. *Biol. Zentralbl.* **37**: 98–106.

Newhook, R. & Briand, F. 1987. Bacteria as structuring agents in lakes: field manipulations with bacterioplankton. *Arch. Hydrobiol.* **109**: 121–38.

Nicholls, K.H. 1987. Predation on *Synura* spp. (Chrysophyceae) by *Bodo crassus* (Bodonaceae). *Trans. Am. Microsc. Soc.* **106**: 359–63.

Nicholls, K.H. 1990. Life history and taxonomy of *Rhizoochromonas endoloricata* gen. et sp. nov., a new freshwater chrysophyte inhabiting *Dinobryon* loricae. *J. Phycol.* **26**: 558–63.

Nicholls, K.H. & Gerrath, J.F. 1985. The taxonomy of *Synura* (Chrysophyceae) in Ontario with special reference to taste and odor in water supplies. *Can. J. Bot.* **63**: 1482–93.

Nicholls, K.H., Kennedy, W. & Hammett, C. 1980. A fish-kill in Heart Lake, Ontario, associated with the collapse of a massive population of *Ceratium hirundinella* (Dinophyceae). *Freshwater Biol.* **10**: 553–61.

Nicholls, K.H., Beaver, J.L. & Estabrook, R.H. 1982. Lakewide odors in Ontario and New Hampshire caused by *Chrysochromulina breviturrita* Nich.(Prymnesiophyceae). *Hydrobiologia* **96**: 91–5.

Nicholls, K.H., Heintsch, L., Carney, E., Beaver, J. & Middleton, D. 1986. Some

effects of phosphorus loading reductions on phytoplankton in the Bay of
Quinte, Lake Ontario. In: Minns, C.K., Hurley, D.A. & Nicholls, K.H.
(eds.) Project Quinte: point source phosphorus control and ecosystem
response in the Bay of Quinte, Lake Ontario. *Can. Spec. Publ. Fish. Aquat.
Sci.* **86**. pp. 145–58.

Nicholls, K.H., Nakamoto, L. & Keller, W. 1992. The phytoplankton of
Sudbury area lakes (Ontario) and relationships with acidification status.
Can. J. Fish. Aquat. Sci., **49** (Suppl. 1): 40–51.

Nuzzi, R. 1988. New York's brown tide. *Conservationist* [Sept.–Oct.]: 30–5.

Nuzzi, R. & Waters, R.M. 1989. The spatial and temporal distribution of 'Brown
Tide' in eastern Long Island. In: Cosper, E.M., Bricelj, V.M. & Carpenter,
E.J. (eds.) *Novel Phytoplankton Blooms: Causes and Impacts of Recurrent
Brown Tides and Other Unusual Blooms*. Springer, Berlin. pp. 117–37.

Nygaard, G. 1977. Vertical and seasonal distribution of some motile freshwater
plankton algae in relation to some environmental factors. *Arch. Hydrobiol.*
(Suppl.) **51**: 67–76.

Olofsson, H., Blomqvist, P., Olsson, H. & Broberg, O. 1988. Restoration of the
pelagic food web in acidified and limed lakes by gentle fertilization.
Limnologica **19**: 27–35.

Ostroff, C.R., Karlander, E.P. & Van Valkenburg, S.D. 1980. Growth rates of
Pseudopedinella pyriforme (Chrysophyceae) in response to 75 combinations
of light, temperature and salinity. *J. Phycol.* **16**: 421–3.

Paerl, H.W. 1988. Nuisance phytoplankton blooms in coastal, estuarine and
inland waters. *Limnol. Oceanogr.* **33**: 823–47.

Palmer, C.M. 1962. *Algae in Water Supplies*. U.S. Department of Health
Education and Welfare, Public Health Service, Washington. 88pp.

Parker, J.I., Tisue, G.T., Kennedy, C.W. & Seils, C.A. 1981. Effects of
atmospheric precipitation additions on phytoplankton photosynthesis in
Lake Michigan water samples. *J. Great Lakes Res.* **7**: 21–8.

Pentecost, A. 1984. Observations on a bloom of the neuston alga, *Nautococcus
pyriformis*, from southern England with an explanation of the flotation
mechanism. *Br. Phycol. J.* **19**: 227–32.

Persson, P. 1983. Off-flavours in aquatic ecosystems: an introduction. *Water
Sci. Tech.* **15**: 1–11.

Persson, P. & Jüttner, F. 1983. Threshold odor concentrations of odorous algal
metabolites occurring in Lake water. *Aquat. Fenn.* **13**: 3–7.

Petersen, J.B. & Hansen, J.B. 1958. On some neuston organisms: I. *Bot.
Tidsskr.* **54**: 93–110.

Pick, F.R., Lean, D.R.S. & Nalewajko, C. 1984. Nutrient status of metalimnetic
phytoplankton peaks. *Limnol. Oceanogr.* **29**: 960–71.

Pinel-Alloul, B., Méthot, G., Verrault, G. & Vigeault, Y. 1990. Phytoplankton
in Quebec Lakes: variation with lake morphometry, and with natural and
anthropogenic acidification. *Can. J. Fish. Aquat. Sci.* **47**: 1047–57.

Ramberg, L. 1976. *Relations between Phytoplankton and Environment in Two
Swedish Forest Lakes*. Klotenprojektet Report 7. Scripta Limnologica
Upsaliensia 426. 97pp.

Reinertsen, H. 1982. The effect of nutrient addition on the phytoplankton
community of an oligotrophic lake. *Holarctic Ecol.* **5**: 225–52.

Reynolds, C.S. & Walsby, A.E. 1975. Water-blooms. *Biol. Rev.* **50**: 437–81.

Ricci, N., Erra, F., Russo, A. & Banchetti, R. 1991. The air–water interface: a
microhabitat for hypotrichous settlers (Protista, Ciliata). *Limnol. Oceanogr.*
36: 1178–88.

Richardson, K. 1989. Algal blooms in the North Sea: the good, the bad and the ugly. *Dana* **8**: 83–93.

Rodhe, W. 1948. Environmental requirements of freshwater plankton algae. *Symb. Bot. Ups.* **10**: 1–149.

Roijackers, R.M.M. 1985. Phytoplankton studies on a nymphaeid-dominated system. PhD, thesis, Agricultural University, Wageningen, The Netherlands. 172pp.

Sandgren, C.D. 1988. The ecology of chrysophyte flagellates: their growth and perennation strategies as freshwater phytoplankton. In: Sandgren, C.D. (ed.) *Growth and Reproductive Strategies of Freshwater Phytoplankton.* Cambridge University Press, Cambridge. pp. 9–104.

Sangfors, O. 1988. Are synergistic effects of acidification and eutrophication causing excessive algal growth in Scandinavian coastal waters? *Ambio* **17**: 296.

Schindler, D.W., Armstrong, F.A.J., Holmgren, S.K. & Brunskill, D.J. 1971. Eutrophication of Lake 227, Experimental Lakes Area, northwestern Ontario, by addition of phosphate and nitrate. *J. Fish. Res. Bd. Can.* **28**: 1763–82.

Schindler, D.W., Kling, H., Schmidt, R.V., Prokopowich, J., Frost, V.E., Reid, R.A. & Capel, M. 1973. Eutrophication of Lake 227 by addition of phosphate and nitrate: the second, third, and fourth years of enrichment, 1970, 1971, and 1972. *J. Fish. Res. Bd. Can.* **30**: 1415–40.

Serrano, L. & Guisande, C. 1990. Effects of polyphenolic compounds on phytoplankton. *Verh. Int. Ver. Limnol.* **24**: 282–8.

Shapiro, J. 1988. Introductory lecture at the international symposium 'Phosphorus in Freshwater Ecosystems', Uppsala, Sweden, October 1985. *Hydrobiologia* **170**: 9–17.

Sieburth, J. McN. 1989. Epilogue to the Second Brown Tide Conference: Are *Aureococcus* and other nuisance algal blooms selectively enriched by the runoff of turf chemicals? In: Cosper, E.M., Bricelj, V.M. & Carpenter, E.J. (ed.) *Novel Phytoplankton Blooms: Causes and Impacts of Recurrent Brown Tides and Other Unusual Blooms.* Springer, Berlin. pp. 779–84.

Sieburth, J. McN., Johnson, P.W. & Hargraves, P.E. 1988. Ultrastructure and ecology of *Aureococcus anophagefferens* gen. et sp. nov. (Chrysophyceae): the dominant picoplankter during a bloom in Narragansett Bay, Rhode Island, summer 1985. *J. Phycol.* **24**: 416–25.

Skulberg, O.M., Codd, G.A. & Carmichael, W.W. 1984. Toxic blue-green algal blooms in Europe: a growing problem. *Ambio* **13**: 244–7.

Smayda, T.J. 1990. Novel and nuisance phytoplankton blooms in the sea: evidence for a global epidemic. In: Granéli, E., Sundström, B., Edler, L. & Anderson, D.M. (eds.) *Toxic Marine Phytoplankton.* Proceedings of the Fourth International Conference on Toxic Marine Phytoplankton, Lund, Sweden. Elsevier, New York. pp. 29–40.

Smayda, T.J. & Villareal, T.A. 1989. The 1985 'Brown Tide' and the open phytoplankton niche in Narragansett Bay during summer. In: Cosper, E.M., Bricelj, V.M. & Carpenter, E.J. (eds.) *Novel Phytoplankton Blooms: Causes and Impacts of Recurrent Brown Tides and Other Unusual Blooms.* Springer, Berlin. pp. 159–87.

Smayda, T.J. & White, A.W. 1990. Has there been a global expansion of algal blooms? If so, is there a connection with human activities? In: Granéli, E., Sundström, B., Edler, L. & Anderson, D.M. (eds.) *Toxic Marine Phytoplankton.* Proceedings of the Fourth International Conference on Toxic Marine Phyplankton, Lund, Sweden. Elsevier, New York. pp. 516–17.

Smith, M.W. 1969. Changes in environment and biota of a natural lake after fertilization. J. Fish. Res. Bd. Can. 26: 3101–32.
Sommer, U. 1988. Some relationships in phytoflagellate motility. Hydriobiologia 161: 125–31.
Spiegelstein, M., Reich, K. & Bergman, F. 1969. The toxic principles of Ochromonas and related Chrysomonadina. Verh. Int. Ver. Limnol. 17: 778–83.
Steinberg, C.E.W. 1990. Alteration of organic substances during eutrophication and effects of the modified organic substances on trophic interactions. In: Perdue, E.M. & Gjessing, E.T. (eds.) Organic Acids in Aquatic Ecosystems. Wiley, New York. pp. 189–208.
Steinberg, C. & Muenster, U. 1985. Geochemistry and ecological role of humic substances in lakewater. In: Aiken, G.R., McNight, D.M., Wershaw, R.L. & MacCarthy, P. (eds.) Humic Substances in Soil, Sediment, and Water: Geochemistry, Isolation, and Characterization. Wiley, New York. pp. 105–45.
Subba Rao, D.V., Quilliam, M.A. & Pocklington, R. 1988. Domoic acid: a neurotoxic amino acid produced by the marine diatom Nitzschia pungens in culture. Can. J. Fish. Aquat. Sci. 45: 2076–9.
Taft, C.E. 1965. Water and Algae: World Problems. Educational Publications, Chicago. 236pp.
Taylor, F.J.R. 1990. Red tides, brown tides and other harmful algal blooms: the view into the 1990s. In: Granéli, E., Sundström, B., Edler, L. & Anderson, D.M. (eds.) Toxic Marine Phytoplankton. Proceedings of the Fourth International Conference on Toxic Marine Phytoplankton, Lund, Sweden. Elsevier, New York. pp. 527–33.
Thomasson, K. 1970. A Mallomonas population. Sven. Bot. Tidskr. 64: 303–11.
Timpano, P. & Pfiester, L.A. 1985. Colonization of the epineuston by Cystodinium bataviense (Dinophyceae): behaviour of the zoospore. J. Phycol. 21: 56–62.
Underdal, B.O., Skulberg, E.D. & Aune, T. 1989. Disastrous bloom of Chrysochomulina polylepis (Prymnesiophyceae) in Norwegian coastal waters 1988: mortality in marine biota. Ambio 18: 265–70.
Vallentyne, J.R. 1974. The Algal Bowl: Lakes and Man. Miscellaneous Special Publication 22. Department of the Environment (Canada), Fisheries and Marine Service, Ottawa. 185pp.
Van Dyk, J. 1990. Long journey of the Pacific salmon. Nat. Geographic 178: 3–37.
Wall, D. & Briand, F. 1979. Response of lake phytoplankton communities to in situ manipulations of light intensity and colour. J. Plankton Res. 1: 103–12.
Ward, J.E. & Targett, N.M. 1989. Are metabolites from the brown tide alga, Aureococcus anophagefferens, deleterious to mussel feeding behavior? In: Cosper, E.M., Bricelj, V.M. & Carpenter, E.J. (eds.) Novel Phytoplankton Blooms: Causes and Impacts of Recurrent Brown Tides and Other Unusual Blooms. Springer, Berlin. pp. 543–56.
Whipple, G.C., Fair, G.M. & Whipple, M.C. 1948. The Microscopy of Drinking Water, 4th edn. Wiley, New York.
Wilcox, G.R. & DeCosta, J. 1984. Bag experiments on the effect of phosphorus and base additions on the algal biomass and species composition of an acid lake. Int. Rev. Ges. Hydrobiol. 69: 173–99.
Willén, E. 1987. Phytoplankton and reversed eutrophication in Lake Mälaren, Central Sweden, 1965–1983. Br. Phycol. J. 22: 193–208.
Willén, E., Haydu, S. & Pejler, Y. 1990. Summer phytoplankton in 73 y

nutrient-poor Swedish lakes: classification, ordination and choice of long-term monitoring objects. *Limnologica* **20**: 217–27.

Witt, V.U. 1977. Effects of artificial fertilization of a high-mountain lake. *Arch. Hydriobol.* **81**: 211–32.

Yan, N.D. & Lafrance, C. 1984. Responses of acidic and neutralized lakes near Sudbury, Ontario, to nutrient enrichment. In: Nriagu, J. (ed.) *Environmental Impacts of Smelters*. Wiley, New York. pp. 457–521.

Yano, H., Nakahara, M. & Ito, H. 1988. Water blooms of *Uroglena americana* and the identification of odorous compounds. *Water Sci. Technol.* **20**: 75–80.

Yentsch, C.S., Phinney, D.A. & Shapiro, L.P. 1989. Absorption and fluorescent characteristics of the Brown Tide chrysophyte. In: Cosper, E.M., Bricelj, V.M. & Carpenter, E.J. (eds.) *Novel Phytoplankton Blooms: Causes and Impacts of Recurrent Brown Tides and Other Unusual Blooms*. Springer, Berlin. pp. 77–83.

Yoshida, Y., Mitamura, O., Tanaka, N. & Kadota, H. 1983a. Studies on a freshwater red tide in Lake Biwa. I. Changes in the distribution of phytoplankton and nutrients. *Jpn. J. Limnol.* **44**: 21–7.

Yoshida, Y., Matsumoto, T. & Kadota, H. 1983b. Studies on a freshwater red tide in Lake Biwa. II. Relation between occurrence of red tide and environmental factors. *Jpn. J. Limnol.* **44**: 28–35.

Yoshida, Y., Kawaguchi, K. & Kadota, H. 1983c. Studies on a freshwater red tide in Lake Biwa. III. Patterns of horizontal distribution of *Uroglena americana*. *Jpn. J. Limnol.* **44**: 293–7.

10
Biogeography of chrysophytes in Finnish lakes

PERTTI ELORANTA

Introduction

Chrysophytes are known to be characteristic of slightly acid, soft waters with low alkalinity and conductivity, and with moderate or low productivity (Sandgren 1988; Siver & Hamer 1989). These characteristics are typical for Finnish lakes. In addition to the diatoms and cryptophytes, the chrysophytes are important phytoplankton groups in Finnish lakes (e.g. Heinonen 1980; Eloranta 1986a). Most of the studies concerning chrysophytes in Finnish lakes are general phytoplankton surveys, but there are also some floristic studies that concern only chrysophytes (Kristiansen 1964; Eloranta 1985, 1989a, Asmund & Kristiansen 1986; Hällfors & Hällfors 1988). Eloranta (1989b) also studied the ecology of the genus *Dinobryon* in Finnish lakes. Some chrysophytes have siliceous scales which remain in lake sediments, and have been used since the late 1960s for paleolimnological purposes (Fott 1966; Munch 1980; Smol 1980; Smol, this volume). Some records of scales in sediments of Finnish lakes have also been reported (Battarbee *et al.* 1980; Tolonen *et al.* 1986; Christie *et al.* 1988).

This study investigates chrysophyte ecology and seasonality in different areas of Finland, and explores some of the relationships of chrysophyte distribution with environmental factors.

Materials and methods

Phytoplankton collections from 329 lakes located in different parts of Finland are included in this survey. Information for a total of 55 lakes, mostly from coastal areas and northern Finland, was obtained from literature records (Lepistö *et al.* 1981). In addition, results from 32 large

214

lakes in different parts of Finland were taken from the biological data bank of the National Board of Waters and Environment. A total of 34 waterbodies from the subarctic North Lapland region were studied in 1980 (Eloranta 1986c) and collections from lakes in western Lapland were sampled in 1988 (Eloranta, unpublished data). Material from lakes in central and southern Finland (93 lakes) collected since 1970 was partially published in an earlier communication (Eloranta 1986a). During the summers of 1987 and 1988, 103 national park lakes in southern and central Finland were studied (Eloranta 1988, 1989a).

Samples for the seasonal analyses were collected from oligotrophic to mesotrophic lakes in the Finnish Lake District (Lake Keurusselkä: Eloranta 1974a; Eloranta & Eloranta 1978; Lake Tuomiojärvi: Eloranta, unpublished results). Lake Keurusselkä was sampled monthly over a year at 16 stations; the lake has spatial differences in the humic matter from polyhumic to oligohumic areas. Although some sections of Lake Keurusselkä are eutrophied by sewage, the main portion of the lake is oligotrophic (Eloranta 1974a, b).

All other samples were taken in July and August, preserved with Lugol's solution and counted with an inverted microscope using phase contrast optics. Counting was done using the Utermöhl technique, most often from 50 ml chambers.

The size of the waterbodies ranged from 0.004 to 1100 km². For practical purposes, Finland was divided into six areas (Fig. 10.1). The

Table 10.1. *Median water quality in the statistically selected lakes (987 lakes)*

	S subregion	N subregion	All
Lake surface area (km²)	0.07	0.20	
pH	6.1	6.8	6.3
Alkalinity (μeq l^{-1}) [Gran]	69	90	75
Conductance (mS m^{-1})	3.4	2.2	3.1
Color (mg Pt l^{-1})	120	40	100
Total N (μg l^{-1})	510	270	450
Total P (μg l^{-1})	18	8	15
TOC (mg l^{-1})	14	6.4	12
COD$_{Mn}$ (mg O$_2$ l^{-1})	17	6.7	15
SiO$_2$ (mg l^{-1})	4.6	2.5	4.2

Adapted from Forsius *et al.* (1990).
TOC, total organic carbon; COD$_{Mn}$, chemical oxygen demand.

coastal area is characterized mostly by rivers and only a few lakes. Many
of the lakes in this area are eutrophic due to clay soil and agricultural
activities. The Finnish Lake District was divided into western and eastern
sections. The northern part of the country was divided into three areas:
Middle Finland, South Lapland and North Lapland. The average water
quality in large lakes of these three areas differs in some degree, but in
this study many small lakes were included with widely varying water
quality. The lakes in the eastern and nothern parts of Finland are typically
oligotrophic with rather low humic content. Differences in water quality
in statistically selected lakes from the northern and southern part of
Finland are obvious (Table 10.1; Forsius *et al.* 1990). The boundary
between these two parts runs through the Middle Finland region of this
study.

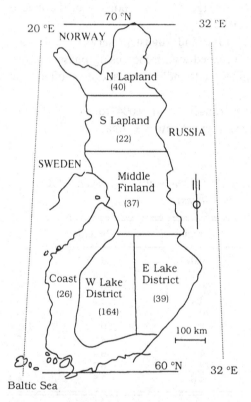

Figure 10.1. Areas of Finland used in the regional analyses (number of lakes
studied in each area is indicated in parentheses).

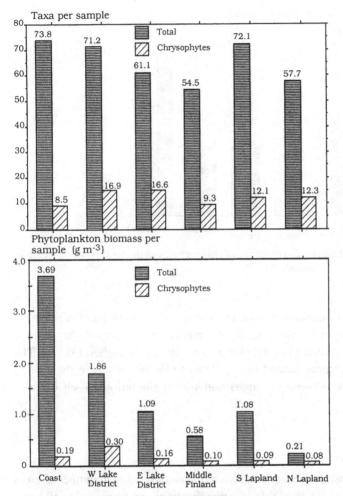

Figure 10.2. Average total number of phytoplankton and chrysophyte taxa, and average phytoplankton and chrysophyte biomass, in the different areas of Finland.

Results

Regional analyses

The average total phytoplankton biomass was the highest in lakes along the coastal region (Fig. 10.2), probably because of high nutrient concentrations from rich soils, a high degree of agriculture, dense population and small lake size with small drainage basins. In these eutrophic lakes, the relative contribution of chrysophytes is clearly lower than in all the other areas, both for biomass and the number of taxa (Fig. 10.3). Although

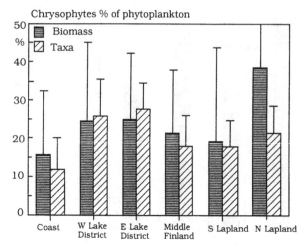

Chrysophytes % of phytoplankton

Figure 10.3. Relative contributions of chrysophytes to the total phytoplankton in different areas of Finland.

the total phytoplankton biomass was lower in subarctic lakes of northern Lapland than in the other areas, the relative proportion of chrysophyte biomass to the total phytoplankton biomass was the highest (38.8%). The regional differences seemed to be related to the differences in the general water quality between the areas and not to the latitude itself (cf. also Fig. 10.4).

Seasonal results

Although seasonal studies are depicted from only a small number of lakes from the Finnish Lake District, generalizations can be made for all areas (Fig. 10.5). Phytoplankton density was low during the winter under the ice and the total number of taxa was around 20, from which the number of chrysophytes was three to five. The most common chrysophyte taxa in winter were *Bicosoeca* spp., *Ochromonas* spp. *Stenocalyx inconstans* Schmid and occasionally a few cells of *Dinobryon* spp. Winter taxa seem to be mostly heterotrophic forms or taxa capable of heterotrophic growth. The number of chrysophyte taxa and chrysophyte biomass reach their maxima during the warmest period of the year, from July to September (Fig. 10.5). The average water temperature in July, August and September was 19 °C, 17 °C and 11 °C, respectively. The lakes were oligotrophic or mesotrophic, and each had a clear summer biomass maximum. The

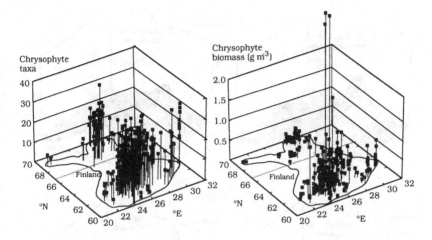

Figure 10.4. Number of chrysophyte taxa and chrysophyte biomass in the lakes studied.

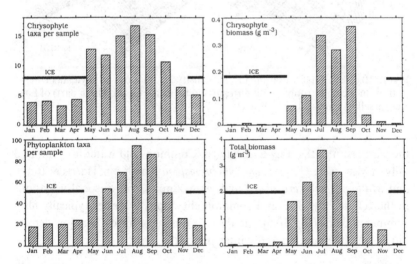

Figure 10.5. Average monthly biomass and number of taxa in seasonally studied lakes.

summer succession of chrysophytes starts in late spring or early summer with a dominance of *Uroglena americana* Calcins together with *Synura* spp., followed in the summer by *Dinobryon* spp. and *Mallomonas* spp. Many of the common species of *Mallomonas* have their maxima during

220   *Chrysophyte algae*

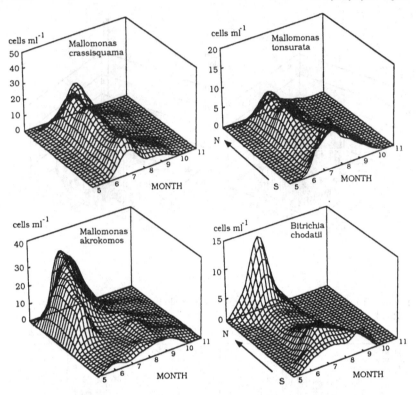

Figure 10.6. Seasonal variation in cell density of four chrysophyte taxa at 11
stations in the oligotrophic (southern part) to mesotrophic (northern part) of Lake
Keurusselkä, Finnish Lake District.

the warmest months, July and August. Common cold water forms, found
only in May, were *Chrysosphaerella brevispina* Korsh. em. Harris & Bradl.
and *Mallomonas heterospina* Lund. As examples, the seasonal variations
in the cell density of several common chrysophyte species typically have
summer maxima (Fig. 10.6), and they are absent or have a very low density
during the late autumn and winter periods.

Lake surface area

The surface area of the lake was not significantly correlated with the
number of phytoplankton taxa per sample ($r = 0.004$), nor with the
number of chrysophyte taxa per sample ($r = -0.117$). The highest values,
however, were found in moderately small (0.1–1.0 km^2) to medium-sized

Biogeography of chrysophytes in Finnish lakes 221

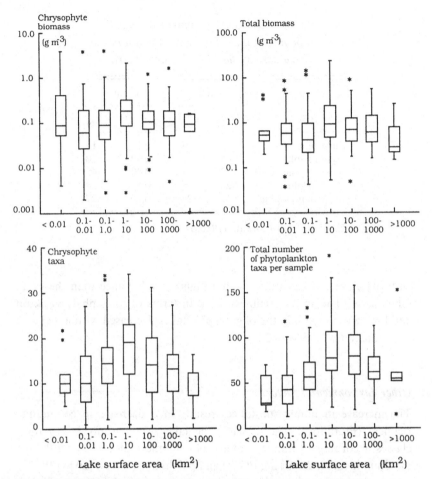

Figure 10.7. Box plot of numbers of taxa and biomass versus the logarithmic size class of lakes. Center line, median; the edges of central box, 25th and 75th percentiles; outer lines, 5th and 95th percentiles; stars, values outside the 5th and 95th percentiles.

(10–100 km^2) lakes, with the maximum being in the small lake group (1–10 km^2; median 19 chrysophyte taxa per sample; Fig. 10.7). Chryso-phyte biomass was also not correlated with lake size. Both maximum median total biomass and median chrysophyte biomass were found in lakes with a surface area of 1–10 km^2, and the median values decreased with either a decrease or an increase in lake size. This is explained by the fact that the majority of the smaller lakes are unproductive forest and peatland lakes, while the largest lakes are typically oligotrophic, often

Table 10.2. *Relative contributions of*
chrysophytes to the total phytoplankton in the
western part of the Finnish Lake District

Lake type	Taxa		Biomass	
	No.	%	g m^{-3}	%
Eutrophic	15	13	0.74	18
Dyseutrophic	18	23	0.69	20
Mesotrophic	22	23	0.49	40
Oligotrophic	23	25	0.15	28
Acid oligotrophic	12	28	0.10	27

Adapted from Eloranta (1986*a*).

with pH close to neutral and with a higher conductance than the small
forest lakes. The species richness, both in terms of total phytoplankton
and for chrysophytes, in the oligotrophic lakes decreased with a decrease
in productivity (Table 10.2).

Other environmental factors

The increase in humic substances results in a decrease in community
richness (Eloranta 1974*a*, 1986*a*). The lakes were grouped into three
classes according to humic content: oligohumic (color ≤ 20 mg Pt l^{-1}),
mesohumic (color 25–80 mg Pt l^{-1}) and polyhumic (color > 80 mg Pt l^{-1}).
Lakes were also divided into three classes on the basis of their
trophic state as evidenced by total phytoplankton biomass: oligotrophic
(biomass < 1 g m^{-3}), mesotrophic (biomass 1.0–2.5 g m^{-3}) and eutrophic
(biomass > 2.5 g m^{-3}).

The lowest median numbers of chrysophyte taxa were found in
polyhumic, oligotrophic peatland lakes, and in oligohumic, oligotrophic
lakes (Fig. 10.8). This group contains many subarctic lakes and several
acid, clearwater lakes. The highest median number of chrysophyte taxa
was found in mesohumic mesotrophic lakes. The highest median chrysophyte biomass values were in mesotrophic and eutrophic lakes with
mesohumic or polyhumic water.

Chrysophyte biomass increased with an increase in the total phytoplankton biomass for oligotrophic and mesotrophic lakes. High chryso-

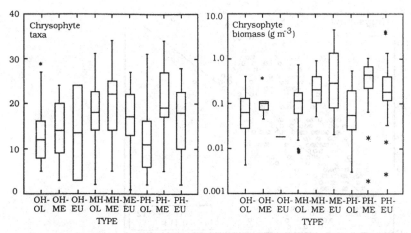

Figure 10.8. Box plot of the number of chrysophyte taxa and chrysophyte biomass in lake groups with different trophic states (OL, oligotrophy; ME, mesotrophy; EU, eutrophy) and content of humic matter (OH, oligohumic; MH, mesohumic; PH, polyhumic).

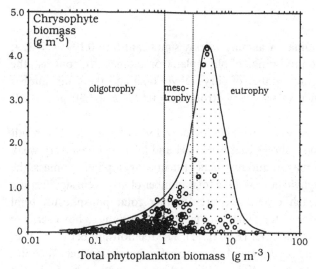

Figure 10.9. Relationship between chrysophyte biomass and total phytoplankton biomass in Finnish lakes (July–August results only).

phyte values were found in some lakes with a moderate level of eutrophy (<5 g m^{-3}), but chrysophytes decreased dramatically in waterbodies with a high degree (>5 g m^{-3}) of eutrophy (Fig. 10.9). Due to this non-linear relation, the linear correlation between chrysophyte biomass and total

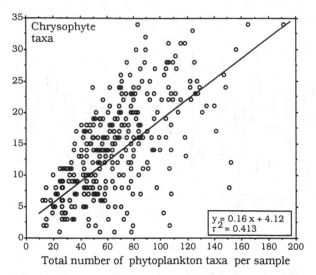

Figure 10.10. Relationship between the number of chrysophyte taxa and the total number of phytoplankton taxa in samples in Finnish lakes (July–August data only).

phytoplankton biomass was only weakly significant ($r = 0.199$; $n = 329$; $p < 0.05$) over the total range of phytoplankton biomass. In contrast, the relation between the number of chrysophyte taxa and the total number of phytoplankton taxa was very strong ($r = 0.643$; $n = 329$; $p < 0.001$) (Fig. 10.10).

Nutrient analyses were available for only 130 of the lakes. The result of the factor analysis shows that water pH and lake area have very weak loading to the biomasses and numbers of taxa. Chrysophyte biomass, the number of chrysophyte taxa and the total number of taxa were significantly correlated with each other, as were summer total phosphorus, total nitrogen and water color. The two groups of variables, however, are uncorrelated with each other (Fig. 10.11). The relationship between mean annual phosphorus loading and chrysophyte abundance (see Nicholls, this volume; Sandgren & Walton, this volume) could not be evaluated from these data.

Studies of the vertical distribution of phytoplankton are rare in Finnish lakes; in most studies the samples are taken as integrated samples from the epilimnion (0–2 m or 0–4 m). In clearwater lakes rather high phytoplankton biomass can be found below the epilimnion, and the community structure in the epilimnion differs greatly from that at greater depths

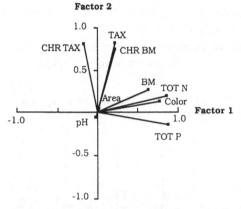

Figure 10.11. Plot of the results from the factor analysis on chrysophytes and some environmental factors. CHR TAX, chrysophyte taxa; TAX, taxa; CHR BM, chrysophyte biomass; BM, biomass; TOT N, total nitrogen; TOT P, total phosphorus.

(Sandgren 1988). In the oligotrophic and meromictic Lake Valkiajärvi, a large biomass maximum was found at the thermocline and in the upper part of metalimnion (Eloranta 1986*b*). This community was dominated (70–80%) by chrysophytes (*Mallomonas akrokomos, M. caudata, M. crassisquama, M. lychenensis, M. hamata, Dinobryon acuminatum, D. suecicum, D. borgei* and some *Kephyrion* spp.). Chrysophytes were almost absent in the epilimnion. This type of vertical distribution may cause an underestimation of chrysophytes in oligotrophic clearwater lakes if only the epilimnion is sampled.

Floristic notes

Järnefelt (1952) developed an indicator quotient system based on the frequencies of different species from lakes with a known trophic level. He counted the occurrences of each species in eutrophic, dyseutrophic, oligohumic oligotrophic and in meso- to polyhumic lakes. In this list he named ten chrysophyte taxa (Table 10.3), eight of which he considered to be indicators of oligotrophy. Järnefelt named *Mallomonas producta* Iwanoff as an indicator of eutrophy, but the taxon has not been verified by electron microscopic examination, and so its taxonomic identity remains in question. The scarcity of chrysophyte taxa in Järnefelt's material is mostly explained by the use of old microscope techniques (only the light microscope was used) and the use of formalin as a preservative.

In the late summer material collected from the larger lakes monitored

Table 10.3. *Chrysophytes listed by Järnefelt (1952) in his quotient
system*

	Frequency (number of lakes)				
Taxon	EU	DEu	OL-OH	MH-PH	E/O ratio
Mallomonas producta Iwanoff	4	0	0	0	–
Synura uvella Ehr. (coll.)	14	33	7	5	3.9
Kephyrion spirale (Lack.) Conr.	6	16	7	3	2.2
Uroglenopsis americana Calkins	6	11	8	2	1.7
Kephyrion ovale (Lackey)	6	11	8	3	1.5
Mallomonas akrokomos Ruttner	6	16	10	7	1.3
M. reginae Teiling	11	33	15	22	1.2
Dinobryon borgei Lemm.	3	13	5	10	1.1
D. suecicum Lemm.	11	36	32	23	0.85
Mallomonas caudata Iwanoff	11	40	38	23	0.84

	Frequency (% of lakes in group)			
Indicators for oligotrophy	EU	DEu	OL-OH	MH-PH
Mallomonas allorgei (Defl.) Conr.	0	2	13	8
Dinobryon cylindricum Imhof	0	2	6	8
Kephyrion spp. (without those above)	3	14	23	18
Diceras spp. (= *Bitrichia* spp.)	3	5	8	11
Ochromonas sp.	3	5	16	2
Dinobryon bavaricum Imhof	30	38	60	46
D. divergens Imhof	42	47	82	53
Stichogloea olivacea Chod.	15	17	32	20

EU, eutrophic; DEu, dyseutrophic; OL-OH, oligotrophic oligohumic; MH-PH,
mesohumic to polyhumic; E/O ratio, ratio between the frequency in the eutrophic
lakes (EU + DEu) and the oligotrophic lakes (OL-OH + MH-PH).

by the National Board of Waters, Finland (Heinonen 1980), nine chrysophyte taxa are listed in the 50 most frequent phytoplankton taxa (Table
10.4). These taxa were identified only with the light microscope from water
samples during routine phytoplankton counting. *Synura uvella* was used
as a collective name for all *Synura* species lumped together, and *Mallomonas tonsurata* most obviously also includes *M. alpina*. The other
taxa listed in Table 10.4 can be identified and distinguished from each
other using the light microscope.

Hällfors & Hällfors (1988) reported a total of 73 taxa of scaled
chrysophytes in 264 samples from 141 lakes. These were mostly large lakes

Table 10.4. *The chrysophyte taxa included in the 50 most frequent phytoplankton in larger Finnish lakes in late summer (Heinonen 1980)*

	Frequency (%) of samples
Mallomonas caudata Iwanoff	69
M. reginae Teiling (= *M. punctifera*)	66
Dinobryon bavaricum Imhof	65
D. divergens Imhof	64
Mallomonas akrokomos Ruttner	61
M. tonsurata Teiling	50
Diceras chodatii Rev. (= *Bitrichia ch.*)	46
Salpingoeca frequentissima (Zach.) Lemm.	44
Dinobryon acuminatum Ruttner	41
Synura uvella (collective category)	< 38

Table 10.5. *Genera of the most common scaled chrysophytes found in Finnish lakes by Hällfors & Hällfors (1988; 141 lakes) and Eloranta (1989a; 64 lakes)*

	No. of species	
	Hällfors & Hällfors (1988)	Eloranta (1989a)
Mallomonas	40	18
Synura	9	6
Chrysosphaerella	3	2
Spiniferomonas	8	6
Paraphysomonas	13	2

($100-1000$ km^2) but also included some small forest lakes. They studied those samples using dry preparations for both light microscopy and transmission electron microscopy (TEM). Eloranta (1989a) found, using scanning electron microscopy (SEM), a total of 34 scaled chrysophyte taxa in 64 small national park lakes in the southern part of Finland. Those taxa belonged to five genera, as shown in Table 10.5. Many of the frequently occurring taxa reported in these two studies on scaled chrysophytes were missing from the above mentioned list of the 50 most common taxa (Table 10.6). This is mostly explained by the different methods used (light microscopy and water samples versus TEM or SEM techniques). The dominant genera in the phytoplankton (*Dinobryon, Mallomonas,*

Chrysophyte algae

Table 10.6. *Relative frequencies of the most common scaled chrysophytes in Finnish lakes according to Hällfors & Hällfors (1988; 141 lakes) and Eloranta (1989a; 59 lakes)*

	Frequency (%)	
	Hällfors & Hällfors (1988)	Eloranta (1989a)
Chrysosphaerella coronacircumspina Wujek & Krist.	21	33
Mallomonas akrokomos Ruttn. in Pascher	42	53
M. allorgei (Defl.) Conrad	0	62
M. canina Krist.	11	41
M. caudata Iwanoff em. Krieger	68	72
M. crassisquama (Asmund) Fott	78	66
M. hamata Asmund	22	21
M. punctifera Korsh.	47	25
M. tonsurata Teiling em. Krieger	52	26
Paraphysomonas vestita (Stokes) De Saed.	13	10
Spiniferomonas bilacunosa Takah.	12	18
S. bourrellyi Takah.	34	36
S. serrata Nicholls	10	26
S. trioralis Takah.	36	44
Synura echinulata Korsh.	13	15
S. petersenii Korsh.	41	33
S. sphagnicola (Korsh.) Korsh.	17	18
S. spinosa Korsh.	34	5

Synura and *Uroglena*) were typically observed in larger lakes. Besides the taxa mentioned earlier, *Bicosoeca* spp., *Chromulina* spp., *Chrysidiastrum catenatum* Lauterb., *Chrysococcus* spp., *Chrysolykos* spp. (esp. *C. planctonicus* Mack), *Chrysostephanosphaera globulifera* Scherffel and *Epipyxis* spp. are fairly common in plankton samples. The author has recorded *Eusphaerella turfosa* Skuja in live plankton samples from Lake Konnevesi (Lake District), but never in the preserved samples. The colonies of this species obviously break apart into single cells when preserved, causing the taxon to be overlooked or misidentified.

Recommendations for future work

Although there are many phytoplankton records from Finland, more detailed work using different techniques is needed. Also, results gathered from standard phytoplankton counts are generally not sufficient for floristic or taxonomic purposes. Studies using live samples with video recording and photography are needed. More seasonal monitoring is also required, especially during the early spring phases, to find chrysophytes characteristic of cold water. In clearwater lakes, metalimnetic sampling should be done to reach forms avoiding high illumination and to illustrate the vertical distribution of biomass and differences in community structure in different layers.

There are only a few reports on scaled chrysophytes based on SEM and TEM from Finnish waters. This work should be extended, as should the use of chrysophyte scales and statospores (stomatocysts) in sediments for paleolimnological purposes.

Summary

Chrysophytes are, along with diatoms and cryptophytes, one of the most characteristic and diverse phytoplankton groups in Finnish lakes during the open water period. In small forest lakes with dark brown water, in eutrophic lakes and in acid clearwater lakes, chrysophytes are, in general, less abundant. Differences in the abundance or species richness of chrysophytes seem to be related more to water quality than to latitude. The abundance and species richness of chrysophytes increase with increasing trophic state until the level of mesotrophy or slight eutrophy, but thereafter decrease dramatically in more eutrophic waters. Although many chrysophyte taxa thrive only during cold water periods, both chrysophyte biomass and chrysophyte species richness have their maxima during the three summer months (from June to August or early September) in Finnish lakes.

Acknowledgments

I am grateful to Professor John P. Smol and Dr. Peter A. Siver for numerous valuable comments on the manuscript and for checking and correcting my English.

References

Asmund, B. & Kristiansen, J. 1986. The genus *Mallomonas* (Chrysophyceae). *Opera Bot.* **85**: 1–128.
Battarbee, R.W., Cronberg, G. & Lowry, S. 1980. Observations on the occurrence of scales and bristles of *Mallomonas* spp. (Chrysophyceae) in

the micro-laminated sediments of a small lake in Finnish North Karelia. *Hydrobiologia* **71**: 225–32.

Christie, C.E., Smol, J.P., Huttunen, P. & Meriläinen, J. 1988. Chrysophyte scales recorded in lake sediments from eastern Finland. *Hydrobiologia* **161**: 237–43.

Eloranta, P. 1974*a*. Studies on the phytoplankton in Lake Keurusselkä, Finnish Lake District. *Ann. Bot. Fenn.* **11**: 13–24.

Eloranta, P. 1974*b*. Lake Keurusselkä: physical and chemical properties of water, phytoplankton, zooplankton and fishes. *Aqua Fenn.* **1973**: 18–43.

Eloranta, P. 1985. Notes on the scaled chrysophytes (Synuraceae, Chrysophyceae) in small lakes in and near Salamajärvi National Park, western Finland. *Mem. Soc. Fauna Flora Fenn.* **61**: 77–83.

Eloranta, P. 1986*a*. Phytoplankton structure in different lake types in central Finland. *Holarctic Ecol.* **9**: 214–24.

Eloranta, P. 1986*b*. Vertical distribution of phytoplankton and bacterioplankton in the meromictic Lake Valkiajärvi, central Finland. *Acta Acad. Aboensis* **47**: 117–24.

Eloranta, P. 1986*c*. The phytoplankton of some subarctic subalpine lakes in Finnish Lapland. *Mem. Soc. Fauna Flora Fenn.* **62**: 41–57.

Eloranta, P. 1988. Etelä- ja Keski-Suomen kansallispuistojen järvien kasviplanktonista heinäkuussa 1987. (Summary: Phytoplankton in national park lakes of South and Central Finland in 1987.) *Jyväskylän yliopiston biol. laitoksen tied.antoja* **51**: 1–39.

Eloranta, P. 1989*a*. Scaled chrysophytes (Chrysophyceae and Synurophyceae) from national park lakes in southern and central Finland. *Nord. J. Bot.* **8**: 671–81.

Eloranta, P. 1989*b*. On the ecology of the genus *Dinobryon* in Finnish lakes. *Beih. Nova Hedwigia* **95**: 99–109.

Eloranta, P. & Eloranta, A. 1978. Tutkimus kalaston rakenteesta ja kalojen kasvusta Kuusvedessä, Ahvenisessa ja Leivonvedessä (Laukaa). *Jyväskylän yliopiston biol. laitoksen tied.antoja* **10**: 1–46.

Eloranta, P. & Palomäki, A. 1986. Phytoplankton in Lake Konnevesi with special reference to eutrophication of the lake by fish farming. *Aqua Fenn.* **16**: 37–45.

Forsius, M., Kämäri, J., Kortelainen, P., Mannio, J. Verta, M. & Kinnunen, K. 1990. Statistical lake survey in Finland: regional estimates of lake acidification. In: Kauppi, P., Anttila, P. & Kenttämies, K. (eds.) *Acidification in Finland*, Springer, Berlin. pp. 759–80.

Fott, B. 1966. Elektronenmikroskopischer Nachweis von *Mallomonas*-Schuppen in Seeablagerungen. *Int. Rev. Ges. Hydrobiol.* **51**: 787–90.

Hällfors, G. & Hällfors, S. 1988. Records of chrysophytes with siliceous scales (Mallomonadaceae and Paraphysomonadaceae) from Finnish inland waters. *Hydrobiologia* **161**: 1–29.

Heinonen, P. 1980. Quantity and composition of phytoplankton in Finnish inland waters. *Publ. Water Res. Inst., Nat. Board of Waters, Finland* **37**: 1–91.

Järnefelt, H. 1952. Plankton als Indikator der Trophiengruppen der Seen. *Ann. Acad. Sci. Fenn. Ser. A IV* **18**: 1–29.

Kristiansen, J. 1964. Flagellates from the Finnish Lapland. *Bot. Tidsskr.* **59**: 315–33.

Lepistö, L., Kokkonen, P. & Puumala, R. 1981. Kasviplanktonin määristä ja koostumuksesta Suomen vesistöissä 1971. *National Board of Waters, Finland Rep.* **207**: 1–146.

Munch, S. 1980. Fossil diatoms and scales of Chrysophyceae in the recent history of Hall Lake, Washington. *Freshwater Biol.* **10**: 61–6.

Sandgren, C.D. 1988. The ecology of chrysophyte flagellates: their growth and perennation strategies as freshwater phytoplankton. In: Sandgren, C.D. (ed.) *Growth and Reproductive Strategies of Freshwater Phytoplankton,* Cambridge University Press, Cambridge. pp. 9–104.

Siver, P.A. & Hamer, J.S. 1989. Multivariate statistics analysis of factors controlling the distribution of scaled chrysophytes. *Limnol. Oceanogr.* **34**: 368–81.

Smol, J.P. 1980. Fossil synuracean (Chrysophyceae) scales in lake sediments: a new group of paleoindicators. *Can. J. Bot.* **58**: 458–65.

Tolonen, K., Liukkonen, M., Harjula, R. & Pätilä, A. 1986. Acidification of small lakes in Finland documented by sedimentary diatom and chrysophycean remains. In: Smol, J.P., Battarbee, R.W., Davis, R.B. & Meriläinen, J. (eds.) *Diatoms and Lake Acidity. Hydrobiologia* **156**: 169–99.

11
The distribution of chrysophytes along environmental gradients: their use as biological indicators

PETER A. SIVER

Introduction

It is difficult to generalize about the ecological tolerances and distributions along environmental gradients of a group of organisms as large and diverse as the chrysophytes. Many excellent floristic papers lack quantitative ecological data, while others provide ecological data but lack critical microscopical detail. There are many more detailed studies pertaining to the scale-bearing forms than for non-scaled taxa. Thus, we know more about the forms with scales and have much to learn about the non-scaled species. In addition, many methodologies have been used by the different researchers to assemble their data. Despite the apparent obstacles, many generalizations have been made for the chrysophytes as a group as well as for specific and subspecific taxa.

Although there is much scattered information concerning the ecology of the chrysophytes (Kristiansen 1986; Siver & Hamer 1989; Siver 1991), especially for individual taxa, there are few comprehensive studies (Kristiansen 1986; Hartmann & Steinberg 1989; Wee & Gabel 1989). Sandgren (1988) provided an excellent review of the ecology of the group as a whole. Siver (1991) and Eloranta (1989b) summarized ecological tolerances for many taxa of *Mallomonas* and *Dinobryon*, respectively; however, similar works are lacking for most other common genera. Kristiansen (1986) and Smol (1986, 1990, this volume) provide excellent starting points for the use of the chrysophytes as biological indicators and in paleolimnological work, respectively.

What are the primary habitat requirements for chrysophytes?

Until recently the chrysophytes had long been considered as a group of organisms primarily restricted to cold, oligotrophic conditions (Kristiansen

232

& Takahashi 1982), most often occurring in the early spring (Kristiansen 1975). However, most recent work supports the idea that even though for individual waterbodies or taxa such conditions may hold true, the generalization cannot be used to describe the group as a whole (Kristiansen 1981, 1986; Dürrschmidt & Croome 1985; Cronberg 1989; Eloranta, this volume). Two basic difficulties emerge from such a broad generalization. First, the statement could adequately describe biomass characteristics for a suite of lakes, but not species diversity, or vice versa. Second, the taxonomic scale must be considered. The statement may be true for the group as a whole, but not for individual species or subspecies.

Rich chrysophyte floras have been described from warmer and/or more tropical regions such as Greece (Kristiansen 1980), India (Philipose 1953; Saha & Wujek 1990; Wujek & Saha, this volume), Bangladesh (Takahashi & Hayakawa 1979), Malaysia (Prowse 1962; Dürrschmidt & Croome 1985), Central America (Wujek 1984*a*, 1986), Australia (Dürrschmidt & Croome 1985), Florida (Wujek 1984*a*; Wujek & Gardiner 1985; Wujek & Bland 1991) and Africa (Wujek & Asmund 1979; Cronberg 1989). Cronberg (1989) has recently summarized much of our knowledge concerning the distribution and importance of scaled chrysophytes in tropical regions.

Although there is ample evidence that the chrysophytes form a greater percentage of the phytoplanktonic biomass in oligotrophic lakes (see below), large numbers of species can also be found in eutrophic waterbodies (Kristiansen 1985, 1986, 1988; Hickel & Maass 1989; Hartmann & Steinberg 1989). Siver & Hamer (1989) observed no significant difference in the number of scaled chrysophytes found in a given collection made along a total phosphorus gradient. Likewise, although the chrysophytes are important components of the spring floras in many waterbodies (Kristiansen 1986), they can also become dominant during the winter, summer and autumn (Dürrschmidt 1980; Kristiansen 1986; Sandgren 1988; Siver 1991; Siver & Hamer 1992; Nicholls, this volume).

It is clear that as a group, as well as for individual species, chrysophytes have different tolerances along various environmental gradients. Based on current knowledge of the group, localities supporting the richest floras of chrysophytes are ones that are neutral to slightly acidic (Siver & Hamer 1989; Siver 1992), low in specific conductance, alkalinity and nutrient content (Dürrschmidt 1980, 1982; Roijackers & Kessels 1986; Cronberg & Kristiansen 1980; Sandgren 1988; Siver & Hamer 1989; Siver 1991) and with low to moderate amounts of dissolved humic substances (Cronberg & Kristiansen 1980; Dürrschmidt 1980, 1982; Kristiansen

1981; Croome & Tyler 1985, 1988; Wee & Gabel 1989; Eloranta, this volume). In addition, chrysophytes are more abundant in smaller water-bodies (Dürrschmidt 1980; Kristiansen 1981; Cronberg 1989; Eloranta, this volume).

An important aspect of the ecology of any group of organisms is an understanding of the relative importance of each environmental variable in controlling the occurrence and distribution of the group. For example, is a slightly acid condition more important than the level of a given nutrient? Application of indirect (e.g. correspondence analysis) and direct (e.g. canonical correspondence analysis) gradient analyses to existing and future sets of data will be instrumental in identifying the relative impor-tance of variables controlling the distribution of the chrysophytes in different regions. There are only a few studies of chrysophytes that have utilized such ordination methods in their analyses. Except for the work by Roijackers & Kessels (1986) and Siver & Hamer (1989), most multi-variate analyses on chrysophyte assemblages have been paleolimno-logical in nature (e.g. Smol *et al.* 1984; Dixit *et al.* 1988*b*, 1989*b*, 1990; Cumming *et al.* 1991, 1992; Smol, this volume).

Distribution along a pH gradient

Waterbodies harboring the most diverse chrysophyte floras are generally neutral to slightly acidic in nature (Cronberg & Kristiansen 1980; Dürrschmidt 1980, 1982; Sandgren 1988; Siver & Hamer 1989). Siver & Hamer (1989) reported that pH had a significant effect on the number of scaled chrysophyte taxa found per collection for lakes in Connecticut. A mean of 8.5 taxa per collection was found between a pH of 5.5 and 6.5; significantly fewer taxa per collection were recorded above and below this range. Fewer than four and two taxa were observed below and above a pH of 5 and 8, respectively. A similar pattern for the number of taxa found per collection along a pH gradient was observed using literature records (Siver & Smol 1993), indicating that the effect of pH on chrysophyte diversity is similar between different regions of the world. Roijackers (1981) also reported that most species of scaled chrysophytes in lakes from The Netherlands were found between pH 6 and 7. Even though a pH of 5 appears to represent a point below which many species of chrysophytes disappear (Siver 1989*a*, 1991; Hartmann & Steinberg 1989), a number of taxa have their maximum occurrence below this level (see below). Only rarely have high numbers of species been reported from alkaline localities (Hartmann & Steinberg 1989) with a high pH (Kristiansen 1985, 1988).

The importance of pH as a primary factor controlling the distribution of chrysophytes, especially scale-bearing members, has been reported in studies from widely separated geographic regions. Takahashi (1967) concluded that pH was more important than water temperature in determining the distribution of species in Japan, and he later (1978) arranged 33 species into the pH categories described by Hustedt (1939). Kristiansen (1975) and Hartmann & Steinberg (1989) noted the importance of pH in studies from Denmark and central Europe, respectively. None of these studies incorporated statistical methods to determine the relative importance of pH as compared with other parameters in controlling the distribution of the chrysophytes.

A common result in all of the ordination studies incorporating scaled chrysophytes has been the identification of pH as an important variable controlling the occurrence of the group. Smol *et al.* (1984) used correspondence analysis to show that assemblages of scaled chrysophytes in surface sediments of Adirondack lakes were closely correlated with pH and pH-related parameters. Similarly, Dixit *et al.* (1989*b*, 1990) and Cumming *et al.* (1991, 1992) used canonical correspondence analysis to demonstrate the importance of pH in controlling the distribution of scaled chrysophytes in Ontario, northern New England, Norway, and New York respectively. In a study of 332 collections from 62 waterbodies from Connecticut and the Adirondacks, Siver & Hamer (1989) found pH and specific conductance as important factors controlling the distribution of scaled chrysophytes; 80% and 83% of the total variance within the pH and specific conductance variables were explained by the first principal component. Similarly, in a study of 50 lakes in The Netherlands, Roijackers & Kessels (1986) found that pH and water temperature dominated the first and second principal component axes, respectively. Siver & Hamer (1989) and Roijackers & Kessels (1986) concluded that temperature was more important in controlling the time period and extent to which a population developed in a given lake. It is clear that similar ordination studies will be invaluable in further unravelling the parameters controlling distribution of the chrysophytes.

Many species and subspecific taxa of scaled chrysophytes have definitive and well-documented distributions along a continuum of pH, and most taxa can easily be placed into one of the pH categories described by Hustedt in 1939 (Takahashi 1978; Smol *et al.* 1984; Siver 1989*a*, 1991). Siver (1989*a*) found that the distributions and weighted mean pH values for many taxa studied in waterbodies from Connecticut and the Adirondacks were similar to ones reconstructed from literature records; he concluded that many taxa have similar distributional patterns along a pH gradient in widely separated geographic regions.

Siver & Smol (1993) identified four groups of species along a pH gradient. The low acid group included *Mallomonas canina*, *M. hindonii*, *M. paludosa*, *M. pugio*, *M. acaroides* v. *muskokana*, *M. hamata*, *Synura sphagnicola* and *S. echinulata*, all taxa found predominantly in acidic waters. The maximum frequency of occurrence and weighted mean pH for species in the low acid group are consistently reported below pH 6 (Smol *et al.* 1984; Roijackers & Kessels 1986; Charles & Smol 1988; Dixit *et al.* 1988*a*, 1989*b*; Dixit & Dixit 1989; Hartman & Steinberg 1989; Siver 1989*a*; Siver & Smol 1993). A second group of species, referred to as the mid-acid group, consisted of species reported primarily from acidic localities above a pH of 5. Taxa in the mid-acid group, including *Mallomonas galeiformis*, *M. heterospina*, *M. punctifera*, *M. duerrschmidtiae*, *M. transsylvanica*, *M. dickii*, *M. doignonii*, *M. torquata*, *Synura spinosa* and *Chrysosphaerella longispina*, have often been reported to have weighted mean pH values between a pH of 5.5 and 6.5 and are commonly absent below pH 5 (Smol *et al.* 1984; Charles & Smol 1988; Dixit *et al.* 1989*b*, 1990; Siver 1988*b*, 1989*a*, 1991; Siver & Hamer 1990; Siver *et al.* 1990).

A third group of taxa, including *Mallomonas akrokomos*, *M. crassisquama*, *M. caudata*, *Synura petersenii*, *S. uvella*, *Spiniferomonas trioralis*, *S. bilacunosa*, *S. coronacircumspina*, *S. cornuta* and *Chrysosphaerella brevispina*, have their center of distribution and weighted mean pH around pH 7 (Siver 1989*a*). Many of these species can also be found in very acidic or alkaline waters. The fourth group identified by Siver & Smol (1993), referred to as the high pH group, consists of organisms distributed primarily along the alkaline end of the pH gradient with weighted mean values above pH 7. Organisms commonly placed in this group include *Mallomonas acaroides* v. *acaroides*, *M. corymbosa*, *M. alpina*, *M. pseudocoronata*, *M. tonsurata*, *M. elongata* and *M. portae-ferreae*. Because the distributions of many species along a pH continuum are rather narrow, it is not surprising that fossil (e.g. Charles & Smol 1988; Dixit *et al.* 1989*b*) and extant (Siver & Hamer 1990) assemblages of scaled chrysophytes have been successfully utilized in the preparation of pH inference models.

Several mechanisms regarding the effect of pH on individual taxa have been proposed. Changes in the relative amounts and form of inorganic carbon are directly influenced by pH. As the pH drops to 5 and lower, the point where many taxa disappear, the bicarbonate buffering system is lost (Yan 1979) and the total amount of inorganic carbon decreases (Galloway *et al.* 1976; Dixit & Smol 1989). It is possible that for some taxa the photosynthetic process may become carbon dioxide limited over short time periods at low pH (e.g. a day; Dixit & Smol 1989); other species may be limited by the available source of carbon (Roijackers & Kessels 1986). As the pH rises above 7, bicarbonate becomes the primary source of inorganic carbon for most algae. Sandgren (1988) suggested that the chrysophytes may not be as capable of utilizing bicarbonate as are other algal groups, giving them a competitive disadvantage at high pH. Cassin (1974) and Nalewajko & Paul (1985) believed that high H^+ levels may adversely affect membrane uptake mechanisms in some species.

The predominance of the chrysophytes in acidic habitats may be related to their ability to produce acid, but not alkaline phosphatases (Jansson 1981; Olsson 1983; Sandgren 1988). Olsson (1983) suggested that since small chrysophytes produced acid phosphatases, they may be favored in acidic environments low in phosphorus (see below). Further physiological work needs to be done in order to determine how widespread the ability or inability to manufacture acid or alkaline phosphatases is within the chrysophytes.

It is not uncommon for the epilimnetic pH of a lake to change by 1 or 2 units over the course of a day, especially when primary productivity rates are high (Wetzel 1983). It is also common for the pH to drop in metalimnetic and hypolimnetic waters, especially in well stratified and deep lakes. Under these conditions species unable to use bicarbonate or to produce alkaline phosphatases at high pH may be able to compete by taking up inorganic carbon or phosphorus in the early morning hours when the pH is lower, or by actively migrating (Ilmavirta 1974; Pick & Lean 1984; Arvola 1983) to deeper, more acidic waters.

There is ample evidence that changes in lakewater pH can significantly affect the concentrations of metals in solution (e.g. Galloway & Likens 1979; Stokes 1983; Schofield *et al.* 1985). In particular, aluminum levels are often elevated in acidified lakes (Cronan & Schofield 1979; Hultberg & Johansson 1981; Schofield *et al.* 1985). The phosphorus levels in dilute acidic lakes may, in turn, be partially regulated by aluminum concentrations (Siver 1992). The ability of aluminum to reduce phosphorus levels via flocculation and precipitation is greatest between pH 5 and 6 (Dickson 1978). Although it is possible that phosphorus concentrations will increase below pH 5 because of a reduction in the precipitation with aluminum, the picture becomes complicated by the fact that the leaching of aluminum from the watershed increases as the pH of the runoff drops below 5.

If phosphorus levels are indeed decreased by elevated aluminum levels, then taxa capable of producing acid phosphatases could have a competitive advantage. Jansson (1981) found that as the pH decreased, aluminum levels increased and total phosphorus concentrations decreased, which triggered a rise in acid phosphatase activity. Sandgren (1988) suggested that species capable of producing acid phosphatases in response to phosphorus limitation would be competitively favored.

There is preliminary evidence that scaled chrysophytes may be sensitive to elevated concentrations of metals besides aluminum (Gibson *et al.* 1987; Dixit *et al.* 1989*a*). Decreases in the concentrations of *Mallomonas acaroides* v. *muskokana* (Dixit *et al.* 1988*b*, 1989*b*), *M. hamata* (Dixit *et al.* 1988*b*) and *M. crassisquama* (Dixit *et al.* 1989*a*) were correlated with increased concentrations of metals in paleolimnological studies from the Sudbury, Ontario region. The decline in abundance of scales at high concentrations of metals prompted Dixit *et al.* (1989*a*) to suggest that chrysophytes may be less sensitive indicators of pH in lakes with high metal concentrations. On the other hand, *Mallomonas hindonii* was able to withstand high aluminum levels (Gibson *et al.* 1987).

Distribution along a specific conductance gradient

Except for the work of Siver & Hamer (1989, 1992) and Siver (1991), few studies have resulted in the quantification of taxa along a specific conductance gradient. Most evidence indicates that the chrysophytes are more tolerant of localities low in specific conductance (Sandgren 1988; Siver & Hamer 1989). Using principal component analysis, Siver & Hamer (1989) identified water conductivity as being as important a variable as pH in controlling the occurrence of scaled chrysophytes. Siver & Hamer (1989) further observed that the specific conductance had a significant effect on the number of taxa found per collection. A mean of seven taxa per collection was found in localities with values less than 40 μS. Significantly fewer species per collection were recorded as the conductance increased above 40 μS; less than two taxa per collection were found in localities with values above 160 μS.

Despite finding fewer taxa at any given instant in time, moderate numbers of species may still be recorded throughout the year from localities high in salt content. For example, Gutowski (1989) observed 23 taxa of scaled chrysophytes over the course of a year in a waterbody with a specific conductance ranging from 345 to 800 μS. It is of interest to note that many of the species described by Gutowski (1989) were restricted in

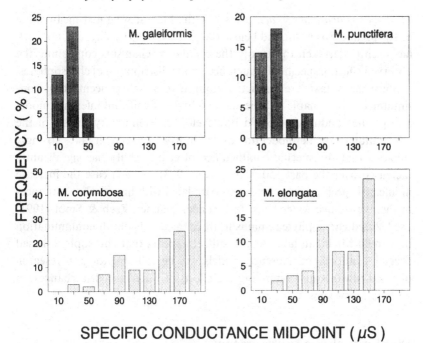

SPECIFIC CONDUCTANCE MIDPOINT (μS)

Figure 11.1. The frequencies of occurrence of four species of *Mallomonas* along a specific conductance gradient. *Mallomonas galeiformis* and *M. punctifera* are distributed along the low end of the gradient, while *M. corymbosa* and *M. elongata* are more common under higher salt levels. (Data adapted from Siver (1991).)

Connecticut to localities high in specific conductance; the flora was dramatically different from that described from the Adirondacks, where specific conductance values are rarely above 60 μS (Siver 1988*a*).

Many species of scaled chrysophytes are restricted to narrow portions of a specific conductance gradient (Fig. 11.1). Siver (1991) identified a low specific conductance group of *Mallomonas* taxa that each had a weighted mean value below 40 μS and was not reported above 90 μS. The low specific conductance group consisted of *Mallomonas acaroides* v. *muskokana*, *M. canina*, *M. hindonii*, *M. dickii*, *M. doignonii*, *M. duerrschmidtiae*, *M. galeiformis*, *M. paludosa*, *M. punctifera*, *M. pugio*, *M. torquata* and *M. transsylvanica*. *Synura sphagnicola* is also classified in the low specific conductance group (Siver & Hamer 1989). Common species with weighted mean values above 80 μS included *Mallomonas annulata*, *M. elongata*, *M. heterospina*, *M. portae-ferreae*, *M. pseudocoronata*, *M. tonsurata*, *M. corymbosa* and *Synura curtispina*. In Connecticut localities,

Mallomonas acaroides v. *acaroides* had the highest weighted mean value of 167 μS and was not found below 100 μS (Siver 1991). Virtually no work has been undertaken to identify the cellular mechanisms controlling the observed effect of specific conductance on the distribution of chrysophytes.

Since many taxa are restricted along a total salt concentration continuum, it is reasonable to believe that a highly significant inference model for specific conductance could be developed. Preliminary results of the application of such a model down-core in five Connecticut lakes strongly suggests that the specific conductance of each locality has significantly increased over the past 150 years (Siver 1993). In each case the increase in inferred specific conductance was correlated with historical disturbances in their respective watersheds. In a similar manner, Zeeb & Smol (1991) used scaled chrysophyte remains to trace the effects of salt contamination in a small Michigan lake. As a result, it appears that chrysophyte-based specific conductance inference models will be valuable tools in research directed towards understanding the effects of watershed disturbances on aquatic ecosystems.

Distribution along a trophic gradient

One difficulty in reviewing the literature describing the distribution of chrysophytes along a trophic gradient is the working definition of an oligotrophic versus a eutrophic condition. A small, shallow pond with abundant macrophyte growth is considered eutrophic by some researchers, yet the open water may be low in nutrient content and unproductive, characteristic of oligotrophy (Wetzel 1983). I believe, as stated by Roijackers (1981), that assemblages should be related to and characterized by the physicochemical state at the time and not the type of waterbody investigated. Taxa found in water samples devoid of nutrients during the time of collection should be noted as such, despite the overall classification of the waterbody as 'eutrophic'. When reviewing the literature one should not be surprised to find a typical oligotrophic taxon in a eutrophic setting. Sandgren (1988) also hints at the idea that operational definitions for oligotrophy and eutrophy differ among researchers. It is possible that more definitive gradients will become apparent for individual taxa if conditions at the time of collection are considered.

Although it is now known that high numbers of species can be found in eutrophic habitats (Kristiansen 1986, 1988; Hickel & Maass 1989; Gutowski 1989), the chrysophytes are clearly more important and account for a greater percentage of the total phytoplankton biomass in oligotrophic waterbodies. Chrysophytes have been shown to dominate oligotrophic conditions in many parts of the world, including Australia (Croome & Tyler 1985, 1988), Greenland (Jacobsen 1985), Scandanavia (Rosén 1981; Arvola 1986; Eloranta 1986), the arctic (Kalff 1967; Kalff *et al.* 1975; Sheath 1986), Africa (Hecky & Kling 1981) and temperate regions of North America (Schindler & Holmgren 1971; Kling & Holmgren 1972; Ostrofsky & Duthie 1975; Munawar & Munawar 1986; Siver & Chock 1986). It is not uncommon for chrysophytes to account for between 10% and 75% of the phytoplankton biomass of an oligotrophic lake, despite differences in size, geographic location or the annual mixing pattern of the waterbody (Sandgren 1988). Planktonic chrysophytes become less important in mesotrophic and eutrophic lakes, often comprising less than 20% of the total biomass. Under very eutrophic conditions the chrysophytes may only account for 5% or less of the annual biomass (see Sandgren, 1988, for a thorough review). Generally, the presence of chrysophytes indicates that the locality is not heavily polluted (Kristiansen 1986). Kristiansen (1988) and Hickel & Maass (1989) both reported over 30 species of scaled chrysophytes from highly eutrophic localities. However, their contribution to the total phytoplankton biomass was low.

On the basis of a thorough search of the literature, Sandgren (1988) summarized that the importance of chrysophytes declined as the concentrations of nitrogen and phosphorus increased (also see Nicholls, this volume). Using nutrient enrichment experiments, DeNoyelles & O'Brien (1978) showed that the chrysophytes disappeared when nitrogen and phosphorus were increased and concluded that the group was better able to compete at low nutrient levels. Schindler & Holmgren (1971) reported that the chrysophytes accounted for over 50% of the phytoplanktonic biomass throughout the year in most nutrient-poor Experimental Lakes Area (ELA) waterbodies. A major shift in phytoplankton composition and biomass from the Chrysophyceae to the Chlorophyceae and/or Cyanophyceae was documented in oligotrophic lakes from the ELA region following enrichment with nitrogen and phosphorus (Schindler *et al.* 1973; Findlay 1978). In each instance there was a shift back to chrysophyte dominance once enrichment was stopped. In a detailed study of the restoration of Lake Trummen (Sweden), Cronberg (Cronberg *et al.* 1975; Cronberg 1982) found an increase in the importance of the chrysophytes following dredging when the lake became more oligotrophic. In a study of 58 waterbodies from central Finland, Eloranta (1986) found the chrysophytes to be most important in oligotrophic and mesotrophic lakes low in nutrients; the chrysophytes accounted for 27–40% of the planktonic biomass. Chrysophytes also dominated the phytoplankton flora, in terms of both diversity and biomass, in a suite of subarctic and subalpine lakes in Finland characterized as oligotrophic and ultraoligotrophic in nature (Eloranta 1986; also see Eloranta, this volume).

Many nutrient-poor waters are believed to be growth limited by phosphorus or less often nitrogen. However, Sandgren (1988) reported that chelated iron compounds may be as important a limiting factor as phosphorus for chrysophyte growth. He further observed that the addition of EDTA often overcame the chelated iron limitation and concluded that chrysophytes may be inferior to other algal classes in sequestering iron in clearwater lakes. On a species level many scaled chrysophytes are also often reported to 'avoid larger lakes' or to 'avoid clear lakes'. Typically, the larger a waterbody is, the less humus-stained it is. The increase in abundance of chrysophytes in slightly humus-stained waters and their avoidance of larger clearwater lakes are probably related to the degree of iron chelation (see Eloranta, this volume). More work also needs to be done relating ratios of phosphorus, nitrogen and silicon to chrysophyte abundance.

Species and subspecific taxa are differentially distributed along a trophic gradient. Despite the importance of the group in oligotrophic waters, a fair number of species and subspecific taxa are distributed primarily in eutrophic localities. Siver (1991) summarized the literature regarding taxa of *Mallomonas* along a total phosphorus gradient as well as a trophic continuum. A group of taxa, including *Mallomonas duerrschmidtiae, M. acaroides v. muskokana, M. galeiformis, M. asmundiae, M. pugio, M. paludosa, M. torquata* and *M. hamata*, were observed primarily in oligotrophic or mesotrophic habitats characterized by low total phosphorus levels. At the other end of the spectrum, species found mostly from mesotrophic and eutrophic conditions included *Mallomonas tonsurata, M. corymbosa, M. portae-ferreae, M. acaroides v. acaroides* and *M. lychenensis*. Some taxa, including *Mallomonas elongata* and *M. pseudocoronata*, were most often reported from more eutrophic habitats yet were also collected from oligotrophic sites. Other species, including *Mallomonas crassisquama, M. caudata, M. akrokomos* and *M. annulata* were equally observed in oligotrophic and eutrophic water bodies. *Mallomonas heterospina* is one taxon that has repeatedly been found in dung-contaminated localities (Harris & Bradley 1957; Kristiansen 1986) and Siver (1991) postulated that *M. matvienkoae* was capable of tolerating severely culturally eutrophied conditions.

Species of other genera of chrysophytes are differentially distributed along a trophic gradient. The number of species of the heterotrophic flagellate genus *Paraphysomonas* often increases with increasing eutrophy (Hickel & Maass 1989). Hickel & Maass (1989) reported approximately half of the known species of *Paraphysomonas* from a suite of eutrophic localities from the Baltic lowlands of northern Germany; *P. vestita*, *P. takahashi* and *P. imperforata* were among the six most common scaled chrysophytes. Eloranta (1989*b*) reported that most species of *Dinobryon*, except *D. pediforme*, occurred in oligotrophic as well as eutrophic habitats, although the genus is more typical and abundant in oligotrophic waters. Ito & Takahashi (1982) and Eloranta (1989*a*) suggested that more species of *Spiniferomonas* could be collected from oligotrophic localities. Species of *Synura* are also differentially distributed along a trophic gradient. Some species, such as *S. sphagnicola* (Kristiansen 1986) and *S. spinosa* f. *longispina* (Dürrschmidt 1982), are more typical of oligotrophic habitats, while others, such as *S. curtispina* appear more common under eutrophic conditions (Gutowski 1989). Other taxa of *Synura*, such as *S. petersenii*, *S. echinulata* and *S. uvella*, can be found in oligotrophic as well as eutrophic water, but tend to be lacking under heavily eutrophied conditions (Kristiansen 1975, 1981; Hartmann & Steinberg 1989).

Distribution along a temperature gradient

Water temperature is another important variable controlling the distribution of chrysophytes. Roijackers & Kessels (1986) and Siver & Hamer (1989) concluded that although pH and related parameters were more important than water temperature in determining whether a given waterbody would harbor a particular taxon, water temperature played a role in determining when, and to what degree, the taxon grew. If this is true, then the importance of temperature may not become evident in large-scale surveys involving a few samples from many lakes, especially if the samples are taken from the same time of year. In addition, since the temperature variable is not included in paleolimnological inference studies, where surface sediment populations are correlated with lakewater characteristics, the importance of temperature is not considered (e.g. Smol *et al.* 1984; Dixit *et al.* 1988*b*, 1989*b*). Thus, studies on individual lakes involving close interval sampling, as well as culture work, will be instrumental in determining the role temperature plays in the ecology of the chrysophytes (see Eloranta, this volume).

  Although seasonal studies of a year or more in duration are few, they illustrate the importance of water temperature. Roijackers (1986) concluded that temperature was an important variable controlling the occurrence of scaled chrysophytes in two shallow lakes in The Netherlands. Gutowski (1989) reported that temperature was the most important factor controlling the occurrence of species in a small lake in Germany. In a 5 year study of a small lake in New England harboring over 50 taxa of scaled chrysophytes, Siver & Hamer (1992) reported that water temperature, as indicated using canonical correspondence analysis, played a key role in the periodic succession of species and subspecific taxa. Siver & Hamer (1992) also found that the chrysophyte community could be used to model and significantly track seasonal changes in water temperature over the 5 year period. Thus, water temperature appears to play a role in the occurrence and succession of species and subspecific taxa. However, further study will be needed in order to determine whether the role of temperature is of a direct or indirect nature (e.g. related to nutrient or zooplankton concentrations).

As a group, the chrysophytes have often been reported to be most important at low temperatures (Hutchinson 1967; Wetzel 1983; see Sandgren 1988); such an observation correlates well with the broadly based seasonal succession models presented by Sandgren (1988) for the group as a whole (see below). There is additional support for the low-temperature hypothesis. On the basis of the analysis of 141 water-bodies, Sandgren (1988) reported a maximum abundance of chrysophytes between 10 °C and 20 °C and a general decline at higher temperatures. Also, the maximum biomass of chrysophytes is often observed in the spring at temperatures less than 12 °C (Kristiansen 1975, 1986; Dürrschmidt 1980; Kies & Berndt 1984; Siver & Chock 1986; Arvola 1986; Siver 1991).

Despite the idea that chrysophyte biomass tends to be higher at lower temperatures, other factors may be of equal, or perhaps greater, importance. For example, in many temperate lakes the spring peak in chrysophyte biomass occurs at temperatures only a few degrees higher than just prior to the loss of an ice cover, inviting the idea that increased turbulence, light and/or nutrients are also important factors regulating the occurrence of chrysophytes at lower temperatures. Perhaps as a group chrysophytes are better adapted to surviving conditions under an ice cover, including low temperature, giving them the advantage (e.g. a seed stock of cells) once ice-out occurs. Even more important than the actual water temperature may be the rate at which the variable changes. In temperate dimictic lakes the greatest increase and decrease in water temperature occurs in the spring and autumn, respectively (Likens 1985); these periods often

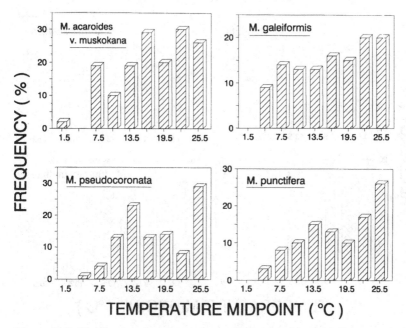

TEMPERATURE MIDPOINT (°C)

Figure 11.2. The frequencies of occurrence of four species of *Mallomonas* along the high end of a temperature gradient. *Mallomonas acaroides* v. *muskokana* is classified as a warm-water taxon. *M. galeiformis*, *M. punctifera*, and *M. pseudocoronata* are classified as warm-cool water organisms. (Data adapted from Siver (1991).)

correlate with the appearance and disappearance of individual taxa of chrysophytes (see below). Lastly, Sandgren (1988; Sandgren & Walton, this volume) has proposed that the biomass of chrysophytes may be maximal at lower temperatures because of reduced grazing.

The role of temperature in regulating species diversity among the chrysophytes is less apparent. Siver & Hamer (1989) observed no significant difference in the number of taxa of scaled chrysophytes found per collection taken between 1 °C and 27 °C. Siver (1991) also reported no difference in the number of *Mallomonas* taxa per collection over the same temperature gradient. Siver & Hamer (1989) did, however, observe a significant reduction in the number of species of *Synura* and *Spinifero-monas* above 21 °C and below 6 °C, respectively.

As was described for trophic, pH, conductance and seasonal gradients (see below), individual species and subspecific taxa of chrysophytes are also commonly restricted along a temperature gradient (Figs. 11.2, 11.3). Despite the role temperature may or may not play in controlling the

Chrysophyte algae

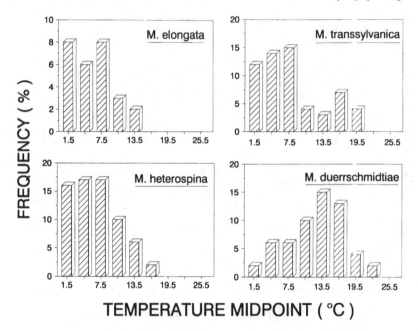

Figure 11.3. The frequencies of occurrence of four species of *Mallomonas* along the low end of a temperature gradient. *Mallomonas elongata*, *M. heterospina* and *M. transsylvanica* are classified as cold-water organisms, and *M. duerrschmidtiae* as a cool-water species. (Data adapted from Siver (1991).)

biomass or diversity of the chrysophytes as a group, clear patterns have been documented at the specific and subspecific levels. Siver (1991) defined five temperature groups for *Mallomonas* based on the distributions of taxa above and below the 12–15 °C interval. At one end of the temperature gradient is the warm-water group, comprising taxa rarely encountered below 12 °C and with weighted mean temperatures above 15 °C. At the other end of the temperature gradient is the cold-water group, comprising taxa with a weighted mean temperature below 12 °C and rarely found above 15 °C. Taxa in the cold-cool group also have weighted mean temperatures below 12 °C, but can be found more often in waters with a temperature above 15 °C than can taxa in the cold-water group (see *Mallomonas transsylvanica*, Fig. 11.3). Organisms in the cool-water group, such as *Mallomonas duerrschmidtiae*, have a maximum distribution between 12 °C and 15 °C and become less abundant above and below this interval. Lastly, many of the more common species of *Mallomonas*, such as *M. caudata* and *M. crassisquama*, are temperature indifferent.

Species from other genera, such as *Synura* and *Spiniferomonas*, are also differentially distributed along temperature gradients. For example, *Synura lapponica* (Siver & Hamer 1992), *S. spinosa* (Kristiansen 1975; Cronberg & Kristiansen 1980; Kies & Berndt 1984; Roijackers & Kessels 1986) and *S. echinulata* (Asmund 1968; Takahashi 1978; Roijackers & Kessels 1986; Siver & Hamer 1989) are primarily found in colder waters, while *S. sphagnicola* (Asmund 1968; Kristiansen 1975; Siver & Hamer 1992) is a warm-water taxon. *Spiniferomonas abei* (Kristiansen 1986) and *S. bourrellyi* (Ito & Takahashi 1982; Kristiansen 1988) are often reported at high and low temperatures, respectively. In contrast, Eloranta (1989*b*) reported many species of *Dinobryon* to be similarly distributed between approximately 5 °C and 22 °C.

The development of asexual or sexual cysts in many phytoplankton groups is believed to be exogenous in nature, that is, directly related to physical and chemical conditions such as temperature changes (Sandgren 1986, 1988; Sandgren & Flanagin 1986). This type of encystment was termed extrinsic by Sandgren & Flanagin (1986). In contrast, Sandgren & Flanagin (1986) suggested that sexual cyst formation in the chrysophytes was dependent on vegetative cell density and not external factors such as temperature; this pattern of encystment was termed intrinsic. Sandgren (1988) does point out, however, that temperature and other extrinsic factors, because of their effect on vegetative growth rates, may regulate formation of cysts within the chrysophytes. Sandgren (1988) further suggested that temperature may not be a controlling factor regulating the germination of cysts. The role that temperature may or may not play in cyst formation and germination needs to be further investigated, especially since it has a direct and obvious effect on succession.

Distribution along a seasonal gradient

In many localities the chrysophytes, especially the scale-bearing species, are often generalized as being an important component of the flora during the spring period (e.g. Kristiansen 1975, 1988; Siver & Chock 1986; Arvola 1986; Hickel & Maass 1989; Gutowski 1989). However, other studies also point to the importance of the chrysophytes during the winter (Takahashi 1978; Dürrschmidt 1980; Siver 1991; Siver & Hamer 1992), autumn (Siver 1991; Siver & Hamer 1992) and summer (Eloranta 1986, this volume). In a comprehensive survey, Siver (1991) reported that the month of the year did not affect the number of taxa of *Mallomonas* found in a given collection. In the same data set slightly more taxa of *Synura* per collection were found during the colder months from October through March; *Spiniferomonas* exhibited a slight tendency towards occurring in the warmer summer months. In a review, Kristiansen (1986) stated that 'most species show maximum development during spring, or spring and autumn, but many of them also occur throughout the year'. Small chrysomonad species are often found to be abundant throughout the year in arctic, subarctic and alpine lakes (Kalff 1967; Reinertsen 1982; Sheath 1986) as well as larger (Munawar & Munawar 1986) and smaller (Schindler & Holmgren 1971) north temperate water bodies. Thus, although in many instances the number of species and the overall biomass of the chrysophytes may peak during the vernal mixing period, the underlying factors controlling seasonal distribution patterns appear to be much more complicated.

Even though many observations exist in the literature concerning the seasonality of the chrysophytes as a group, Sandgren (1988) concluded that patterns were, at best, poorly understood. In a thorough survey and synthesis of the literature, Sandgren (1988) was able to characterize patterns of seasonal periodicity for many different lake types covering a wide geographical area. He concluded that the three factors underlying the seasonal patterns of planktonic chrysophytes were the trophic status and annual mixing cycle of the lake and the morphological size of the taxa. Sandgren's findings can be summarized as follows:

(1) Small unicellular chrysomonads often dominate the phytoplankton biomass throughout the year in: (a) small, oligotrophic, cold monomictic ponds in the arctic; (b) large, deep, oligotrophic, cool monomictic lakes of the north temperate zone; and (c) large, deep, oligotrophic monomictic lakes in the tropics. In all instances net forms (greater than 20–30 µm maximum linear dimension) were not significant.

(2) In cold monomictic lakes of moderate or high productivity, small chrysomonads still dominate the planktonic biomass (20–50%) throughout the year. In addition, larger-sized netplankton forms, such as species of *Mallomonas* and colonial forms of *Synura* and *Dinobryon*, often form short monospecific peaks.

(3) Chrysophytes account for a much lower percentage of the biomass (0–20%) in monomictic lakes located in warmer climates that are mesotrophic to eutrophic in nature. Nanoplanktonic chrysomonads are of minimal importance and are sporadically present throughout the year. Larger forms develop brief monospecific peaks, primarily in the cooler periods of the year. This scenario holds for both cool and warm monomictic lakes.

(4) In oligotrophic lakes with a dimictic mixing pattern the chrysophytes often comprise from 25 to 75% of the total phytoplankton biomass throughout the year. Unlike the monomictic lakes, the larger netplankton forms are generally co-dominant with the smaller chrysomonad nano-plankton species. This lake group includes smaller north temperate water bodies, larger temperate lakes and smaller alpine or subarctic localities.

(5) In dimictic lakes with a mesotrophic to eutrophic condition, the chrysophytes account for only 5–20% of the total annual phytoplankton biomass. The smaller flagellate forms are uncommon, while the larger single-celled and colonial taxa generally form maxima during vernal mixing, early summer or under the ice.

(6) The chrysophytes are virtually absent, accounting for less than 5% of the total biomass, in eutrophic and hypereutrophic lakes with a dimictic mixing pattern.

In studies of eutrophic localities Gutowski (1989), Kristiansen (1988) and Hickel & Maass (1989) all reported a maximum number of species during the spring; however, in each case they accounted for a small amount of the total biomass.

Although factors such as the temperature, pH, water color and chemistry do not appear to regulate seasonal distribution patterns for planktonic chrysophytes as a group (Sandgren 1988), I believe these parameters are instrumental in controlling the seasonal occurrences of individual species and subspecific taxa. For example, if a species is indeed a warm- or cold-water organism, its seasonal occurrence could be molded in a number of ways. First, the upper temperature limit of a species will play a role in determining whether the taxon will be present during the warmest period of the summer. If two taxa, species A and B, exhibited temperature ranges of 12–20 °C and 12–25 °C, respectively, they may be expected to be found together in the spring and autumn in a typical small, temperate dimictic lake. However, species A would disappear during mid-summer as temperatures rose above 20 °C, while species B could continue to survive, yielding different seasonal strategies for the two taxa. In waterbodies where the mid-summer temperature was below 20 °C (e.g. further north in latitude), both taxa may exhibit similar seasonal strategies. It is also possible that species A could survive during the warmer portion of summer by taking refuge in the metalimnetic or submetalimnetic zones (Pick & Lean 1984; Pick *et al.* 1984; Pick & Cuhel 1986). Second, the minimum temperature limit of a taxon will directly influence whether it could be present below 4 °C under an ice cover. The difference between taxa possessing a pattern III or IV seasonality (see below) may simply be that species with pattern IV have a lower temperature limit enabling them to persist under the ice. Thus, species possessing pattern IV have a greater chance of persisting and dominating in the spring after ice-out.

Since studies of individual lakes utilizing close interval sampling and spanning a year or longer are few (Gutowski 1989; Siver & Hamer 1992), the seasonal occurrences of many species must be pieced together using many observations from discrete collections. Siver & Hamer (1992) postulated that if the distributions of scaled chrysophyte taxa along a seasonal gradient could be determined for a given geographic region, then sediment assemblages could be divided into warm- and cold-water floras. Once warm- and cold-water floras were reconstructed, they could be used in paleolimnological work for seasonal inferences.

Siver & Hamer (1992) found that many of the seasonal distribution patterns of 51 taxa of scaled chrysophytes observed over a 5 year period in a small oligotrophic dimictic lake, Bigelow Pond, fit into one of five categories (Fig. 11.4). The first two patterns represented taxa restricted to either warm summer months (pattern I) or the cold winter months (pattern II). Taxa exhibiting pattern I generally begin growth in the spring, have a maximum occurrence in the summer and disappear by mid-autumn. Cold-water pattern II taxa begin growth in late autumn or early winter, persist under the ice, and disappear soon after ice-out in the spring. Pattern III represented taxa that began growth in the summer, persisted through the autumn and disappeared in the winter; species fitting this pattern were also commonly recorded in the spring, but were lacking under the ice. Pattern IV was an extension of pattern III where the species were initially observed in late summer, persisted through the autumn and winter, and disappeared after ice-out in the spring. Pattern IV is also similar to pattern II, except that species with pattern IV begin growth in the latter part of the summer or early autumn. Pattern V depicted species that occurred throughout the year, often sporadically. A sixth pattern, not exhibited by any taxa in the study of Bigelow Pond, describes species found primarily during the spring and early summer after ice-out and again in the autumn; species fitting this pattern are lacking in the warmer summer months and under the ice.

It should be noted that these seasonal patterns have been developed using data from temperate dimictic lakes that typically reach a maximum summer temperature between 23 °C and 27 °C. Further work would be needed in order to document how each of the seasonal patterns would become modified in a monomictic lake, or a dimictic lake with a lower

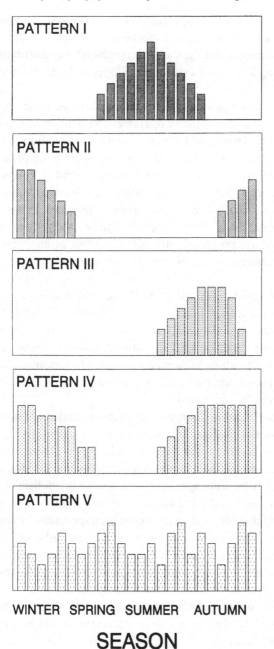

Figure 11.4. Five generalized seasonal distributional patterns used by Siver & Hamer (1992) to describe the seasonality of species in a small dimictic lake. The patterns are discussed in detail in the text.

or higher summer maximum temperature. It should also be remembered that organisms may consistently exhibit the same seasonal pattern in a suite of lakes; however, when the data are combined, the distribution of a given taxon along a seasonal gradient may 'appear' wider than for any one lake.

Many species of *Mallomonas* (Siver 1991), as well as other common species of scaled chrysophytes (Siver & Hamer 1992), have been fitted to one of the five seasonal patterns. Taxa consistently fitting pattern I include *Mallomonas acaroides* v. *muskokana*, *M. dickii*, *M. galeiformis*, *M. pseudocoronata*, *M. punctifera* and *M. tonsurata*. Species found in Connecticut and the Adirondacks that exhibited pattern II include *Mallomonas elongata*, *M. heterospina*, *M. doignonii*, *Synura lapponica* and *Chrysosphaerella brevispina* (Siver 1991; Siver & Hamer 1992). Taxa with a slightly wider seasonal range, but still best described as having pattern II, include *Mallomonas annulata*, *M. transsylvanica* and *M. torquata* (Siver 1991).

In the Bigelow Pond study (Siver & Hamer 1992) *Chrysosphaerella longispina* and *Synura sphagnicola* both exhibited pattern III, while *Synura spinosa*, *S. echinulata*, *S. uvella* and *S. petersenii* consistently fitted pattern IV over the 5 year period. In other localities *Mallomonas portae-ferreae* also fitted pattern III, while *Mallomonas striata*, *M. paludosa*, and to a lesser degree *M. duerrschmidtiae* and *M. acaroides* v. *acaroides*, are best described as pattern VI organisms.

Many of the most commonly reported species of scaled chrysophytes, including *Mallomonas caudata*, *M. crassisquama*, *M. akrokomos* and *M. hamata*, can be found throughout the year and are best placed in pattern V. However, using records from many collections, Siver (1991) found that the occurrence of these four species over a seasonal gradient differed. *Mallomonas akrokomos* and *M. caudata* were found all year, but were more abundant in the colder and warmer months, respectively. *Mallomonas crassisquama* was found more often near vernal and autumnal mixing periods, whereas *M. hamata* exhibited a sporadic seasonal occurrence. Siver (1991) compared the seasonal patterns of taxa from Connecticut and Adirondack waterbodies with literature records.

Many of the common species of scaled chrysophytes represent netplankton as defined by Sandgren (1988). As reviewed above, Sandgren (1988) pointed out that netplankton species are most often found during discrete times of the year and rarely persist throughout the year. Thus, it is no surprise that many of the taxa show distinct distributions along a seasonal gradient. However, the roles that variables such as temperature, pH, specific conductance, nutrient levels and grazing play in molding the seasonal patterns for individual taxa are largely unknown. In the 5 year study of Bigelow Pond, Siver & Hamer (1992) found not only temperature, but also pH and specific conductance to vary consistently along a seasonal gradient; all three variables were also highly correlated with the primary canonical correspondence analysis axis. Thus, even though the abundances of species were successfully used to infer water temperature in Bigelow Pond, the model may work simply because temperature is covarying with salt content and/or pH. Clearly, much more work needs to be done in order to unravel the relative importance of different factors in governing the seasonal periodicity patterns of individual species.

Chrysophytes as biological indicators

Long-term monitoring programs aimed at detecting shifts in lakewater conditions have and will continue to be invaluable tools in the protection of our aquatic resources. Changes in lakewater conditions beyond natural variability, whether unidirectional or oscillating, whether subtle or episodic, and whether long or short lived, could go undetected in a routine chemical monitoring program (Siver & Smol 1993). However, such changes would leave their 'mark' on the living organisms. Thus, the incorporation of a biological component into primarily chemical-based monitoring programs will inevitably enhance the ability of the programs to detect changes in the quality of the waterbody that could otherwise be missed with seasonal or yearly chemical measurements. The concept is that the composition of organisms within a lake is constantly molded by daily and longer-term fluctuations in lakewater conditions. Thus, if changes in the condition of a waterbody could be quantified using compositional shifts in the abundances of organisms, it would provide an additional tool for estimating alterations in the quality of the waterbody.

Kristiansen (1986) listed three prerequisites for an organism to be useful as a bioindicator. First, it must be taxonomically well defined. Second, it must be relatively easy to identify. Third, it must be distributed within a 'narrow ecological spectrum'. Many specific and subspecific taxa of chrysophytes fit all three criteria, rendering their use as indicators invaluable (Siver 1991; Siver & Smol 1993). On the basis of the above discussion it is also clear that since the chrysophytes respond to slight changes in many environmental variables, they would be valuable bio-indicators (Siver & Smol 1993). In addition, since scaled chrysophytes leave behind species-specific components, they are also excellent indicators of historical environments (see Smol 1990, this volume and references therein).

Fossil (see Smol, this volume) and extant (e.g. Siver & Hamer 1990) populations of scaled chrysophytes have been successfully utilized in pH inference work. In a similar manner scaled chrysophytes will become valuable tools in the inference of specific conductance (Siver 1993) and trophic conditions, as well as in global climate studies (see Smol, this volume for further discussion).

Even without the development and use of sophisticated inference models, the presence of individual taxa of chrysophytes can indicate much about a given body of water. For example, the presence of *Mallomonas galeiformis* and *M. acaroides* v. *muskokana* would be indicative of a collection made during the warmer months from a waterbody that was acidic, but above pH 5.5, low in specific conductance and nutrient levels and possibly slightly humic in nature (see Siver 1988*b*, 1989*b*). *Mallomonas transsylvanica* would also be indicative of such a locality, except for colder months (Siver 1991). The disappearance of either *M. galeiformis* or *M. transsylvanica* could signal a drop in pH below 5 or a rise in nutrient levels; the disappearance of only one taxon may indicate a seasonally related change. On the other hand, *Mallomonas acaroides* v. *muskokana* would be able to tolerate a higher nutrient level or a drop in pH below 5.

In a similar fashion, the occurrence of species such as *Mallomonas acaroides* v. *acaroides*, *M. alpina*, *M. tonsurata*, *M. elongata* and *M. pseudocoronata* would clearly indicate alkaline waters, probably high in specific conductance. The former three species would be more tolerant of eutrophic conditions. *Mallomonas elongata* and *M. pseudocoronata* would be more common in mesotrophic or oligo-mesotrophic habitats, during the winter or summer months, respectively. The presence of other alkaline taxa such as *Mallomonas heterospina* and *M. matvienkoae* would be indicative of polluted habitats. It seems reasonable that with further work other species will become valuable indicators.

Comparison of the chrysophyte floras from Connecticut, the Adirondacks and the literature

A question often asked is whether the distribution of individual species along various gradients is the same in different geographic regions. Although this question has been addressed for pH to a limited degree (e.g. Siver 1989*a*, 1991), much work needs to be done for other environmental variables. Siver (1991) presented a comparison of *Mallomonas* floras from localities in Connecticut, the Adirondacks and a review of the literature. A summary of that work is warranted here in order to illustrate the wealth of information that can be gleaned by comparing floras from different regions.

The distribution of collections made along pH, specific conductance and total phosphorus gradients can be used to illustrate some of the basic differences in the water chemistry between Connecticut and the Adirondacks localities. In general, samples from Connecticut represent localities higher in pH, specific conductance and total phosphorus levels than those from the Adirondacks. Most collections from Connecticut had a pH between 5.5 and 7.5; 5% and 15% were below and above this range, respectively. In contrast, 44% of the samples from the Adirondacks had a pH below 5.5 and none were above 7.5, illustrating the more acidic nature of the waterbodies. A total of 78% of the collections from Connecticut had specific conductance values between 20 µS and 120 µS; 6% and 16% of the samples were above and below this range, respectively. All of the collections from the Adirondacks were below 60 µS. There is a greater degree of development of the watersheds surrounding lakes in Connecticut, resulting in more mesotrophic and eutrophic localities when compared with the Adirondacks. A total of 50% and 76% of the collections from Connecticut and the Adirondacks had a total phosphorus concentration below 15 µg P l^{-1}, respectively. Similar distributions could not be assembled for literature records due to the lack of numerical data. However, on the basis of the available data and written descriptions of sites, it is concluded that the literature records represent waterbodies that are on average more eutrophic and higher in pH and specific conductance than those in Connecticut.

The distributions of most common scaled chrysophytes in the three data sets were similar along pH, specific conductance, temperature and total phosphorus gradients. There was a close match in the weighted mean pH values for most common taxa between the three geographic regions (Fig. 11.5). Siver & Hamer (1990) were able to develop a highly significant inference model for pH using literature records, and concluded that such models may have applicability over large geographic areas; this adds support to the idea that taxa have similar distributions along a pH gradient over wide geographic regions.

There are, however, significant differences in the relative occurrences of species between the Connecticut, Adirondack and literature data sets (Figs. 11.6, 11.7). Figure 11.6 depicts the relative abundances for taxa of *Mallomonas*, while Fig. 11.7 displays results for common species of *Synura*,

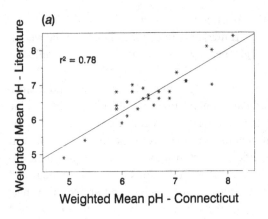

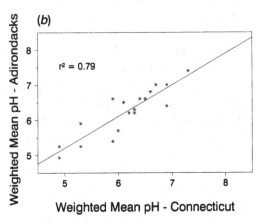

Figure 11.5. A comparison of weighted mean pH values for species of chrysophytes. Comparisons are made between Connecticut and literature records (*a*) or Adirondack localities (*b*). Regression lines are depicted.

Chrysosphaerella and *Spiniferomonas*. For the most part the differences correspond to differences in the chemical nature of the waterbodies between regions. Species such as *Mallomonas duerrschmidtiae*, *M. galeiformis*, *M. acaroides* v. *muskokana*, *Synura sphagnicola* and *S. echinulata* that tolerate more acidic localities low in specific conductance and total phosphorus, were more common in the Adirondacks. On the other hand, species that are more common in localities higher in pH, specific conductance and total phosphorus, such as *Mallomonas tonsurata*, *M. akrokomos*, *M. acaroides* v. *acaroides*, *M. annulata*, *M. striata* and *M. heterospina*, were of greater importance in Connecticut and/or the literature

Species

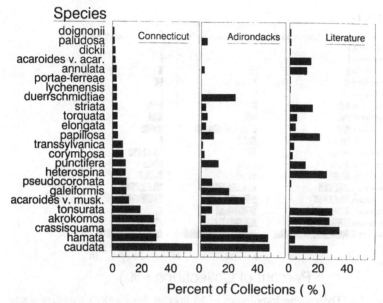

Figure 11.6. The relative importance of 24 taxa of *Mallomonas* in waterbodies from Connecticut, the Adirondacks and in a survey of the literature. Taxa are arranged according to their increasing importance in the Connecticut lakes. (Reprinted with permission from Siver (1991).)

records. Taxa known to tolerate highly eutrophic localities, high pH or high salt levels, such as *Mallomonas heterospina*, *M. annulata* and *M. acaroides* v. *acaroides*, were most common in the literature, supporting the idea that the literature records represent collections from more eutrophic habitats with high salt and pH conditions. Interestingly, *Synura petersenii*, one of the most common chrysophyte species, was present in about 55% of the collections from all three data sets.

In conclusion, comparative surveys of floras separated by relatively short distances (e.g. Connecticut and the Adirondacks) as well as within (e.g. the tropics; Cronberg 1989) and between (e.g. southeast Asia and North America) widely separated zones will aid in our understanding of chrysophyte biogeography.

The use of studies on extant populations in paleolimnological work

Over the past 10 years, scaled chrysophyte fossils have become integral components of paleolimnological efforts focused on the reconstruction of historical lakewater conditions (see Smol, this volume). The emphasis of

Taxon

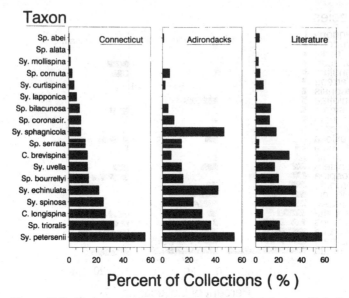

Percent of Collections (%)

Figure 11.7. The relative importance of 18 taxa of *Synura* (Sy.), *Spiniferomonas* (So.) or *Chrysosphaerella* in waterbodies from Connecticut, the Adirondacks and in a survey of the literature. Taxa are arranged according to their increasing importance in the Connecticut lakes.

the work has centered on the use of the group in reconstructing past pH conditions in order to establish trends in acidification patterns (e.g. Charles & Smol 1988; Dixit *et al.* 1988*b*, 1989*b*, 1990; Cumming *et al.* 1991). In a similar fashion scaled chrysophytes are being incorporated into the reconstruction of other environmental variables, such as specific conductance and trophic levels, as well as in the study of global climate change.

Much valuable ecological data, especially at the species and subspecific levels, has been and will continue to be learned through paleolimnological investigations. For example, the distributions along a pH gradient for many taxa from widely separated geographical regions have been characterized through the correlations of surface sediment remains with current environmental data. The descriptions of pH tolerance limits and weighted mean pH values obtained through paleolimnological studies are similar to those derived from the study of extant populations (Siver 1988*a*, 1989*a*). Thus, it is reasonable to conclude that distributions of taxa along other environmental gradients of concern (e.g. phosphorus) can be learned using the paleolimnological approach.

Some of the current limitations to the use of scaled chrysophytes in paleolimnological work have and will be overcome through careful analysis of extant populations. For example, although *Mallomonas acaroides* was a valuable indicator species in acid deposition studies of the Adirondack region (Smol *et al.* 1984), there were discrepancies in the literature concerning the true distribution of the taxon along a pH continuum (Siver 1989*b*). Smol *et al.* (1984) found *M. acaroides* as an acidobiont, however, Roijackers & Kessels (1986) and Kristiansen (1986) reported it from much higher pH values. Soon thereafter, Nicholls (1987) described a new variety, *M. acaroides* v. *muskokana*, from dilute softwater habitats in Ontario. By separating populations into v. *muskokana* and v. *acaroides*, Siver (1989*b*) found that the two varieties were clearly partitioned along a pH gradient. Variety *muskokana* was acidobiontic with a maximum occurrence and a weighted mean pH below 5.5. However, v. *acaroides* was observed to be alkalibiontic in nature with a maximum distribution and a weighted mean pH above 8; a similar distributional pattern was described for v. *acaroides* using literature records (Siver 1989*b*). Thus, critical taxonomic detail (e.g. Nicholls 1988) coupled with careful eco-logical data on extant populations has enhanced the use of *M. acaroides* in paleolimnological research.

Another example where the study of extant populations along environ-mental gradients has resulted in a fine-tuning of paleolimnological inference models involves the species *Mallomonas duerrschmidtiae* (Siver *et al.* 1990). From a taxonomic viewpoint *Mallomonas duerrschmidtiae* is closely related to *M. pseudocoronata* and even more so to *M. crassisquama*. From an ecological standpoint the three taxa are differentially distributed along pH, temperature, specific conductance, total phosphorus and seasonal gradients. All three taxa are commonly found in the same geographic region (e.g. the Adirondacks), making taxonomic separation mandatory if the species are to be used in inference work. For example, *M. crassisquama* has a higher weighted mean pH than *M. duerrschmidtiae* (Siver *et al.* 1990). If the species were not separated, the combined calculated weighted mean pH would be higher than for *M. duerrschmidtiae* alone. Thus, if the combined pH data were applied to sediment layers composed of only *M. duerrschmidtiae* or *M. crassisquama*, an erroneously high or low pH, respectively, would result. It is clear that correct identification of these species will provide more appropriate paleolim-nological reconstructions.

Another limitation with paleolimnological analyses is that they do not include seasonal attributes of the species utilized to form the inference models. In most paleolimnological studies inference models are based on environmental data corresponding to a given time of the year (e.g. spring pH). However, the fossil chrysophyte remains used in the analysis represent organisms that have different distributions along a seasonal gradient. Thus, one potential source of variation in this example is that spring pH is the appropriate measurement to use only for taxa that grew primarily during the spring. The incorporation of a 'seasonal element', derived from the study of extant populations, into paleolimnological studies could potentially result in more appropriate inference work.

Siver & Hamer (1992) proposed that seasonal data gleaned from extant populations could be utilized to fine-tune inference models in at least two ways. First, they showed that the surface sediment assemblage from Bigelow Pond was most similar to the flora that occurred in the lake during the late autumn/early winter period. As such, they concluded that the surface sediment flora correlated most closely with chemical data from the late autumn/early winter.

The second way in which Siver & Hamer (1992) suggested that seasonal data could be applied to paleolimnological work was through the reconstruction of seasonal fossil floras. With an understanding of the seasonal occurrences of species it would be possible to reconstruct seasonal floras from a microfossil assemblage. The seasonal floras could then be used to reconstruct seasonal lakewater conditions. The fact that scaled chrysophyte floras were successfully used to infer water temperature in Bigelow Pond for a 5 year period strongly supports the idea that separation of a fossil flora into warm- and cold-water components would enhance paleolimnological reconstructions. It is clear that ecological and taxonomic work on extant floras will continue to support and complement paleolimnological investigations.

Summary

Despite the recent advances in our understanding of the ecology of the chrysophytes, we have essentially just begun to identify distributional patterns along environmental gradients. We have learned more about taxa with siliceous scales than those without. The relative importance of different variables in controlling the occurrence of the group is just beginning to be elucidated; for most variables it remains unknown whether the measured response of a taxon along a given gradient is of a direct or indirect nature. Is the distribution of a species in the acid end of a pH gradient related to the proper functioning of the cell membrane, the available carbon source, the activity of an enzyme or a combination of factors? Is a cold-water taxon better able to grow at low temperature, or is it simply not grazed at this end of the gradient? The chrysophytes have been successfully utilized as bioindicators and in the development of paleolimnological inference models. The potential use of inference models and the degree of detail yielded with such models will be greatly enhanced once the mechanisms underlying the distribution of the group are understood.

The recent development of the constrained ordination technique canonical correspondence analysis (ter Braak 1986, 1989; Jongman *et al.* 1987), and its application to large sets of data, will be instrumental in identifying the relative importance of different environmental factors in controlling the distributions of the chrysophytes as a whole, as well as for individual species. However, when applying techniques such as ordination analysis to data sets it is important to remember that even though many taxa are differentially distributed along various environmental gradients, the correlations do not necessarily indicate the most appropriate conditions for maximum growth. Nor does a highly correlative relationship between a given variable and the occurrence of a species necessarily indicate why a species is present and perhaps dominating in a given habitat. In many instances an organism may be clearly defined along a given environmental gradient; however, it may be responding to another unmeasured variable. In this case the relationship between the organism and the measured variable will be maintained only if the two variables continue to covary.

As is often observed with other algal classes, a clonal culture of a chrysophyte species may respond differently to environmental variables from expectations based on observations of wild populations. For example, Sandgren (1988) reported a linear increase in growth between about 5 °C to 10 °C and 20 °C for seven clones (four species). However, the clones did not exhibit positive growth below 5 °C, which is in conflict with field observations. Another important consideration, expressed by Kristiansen (1986), is that species with wider ecological tolerances may actually represent different ecomorphs with similar morphologies.

A wealth of ecological data, especially for scaled chrysophytes, was collected during the 1980s. Even though I believe that many additional field data are needed, especially for selected regions, it is to be hoped that the 1990s will be the decade when we begin to incorporate much-needed culture/field experiments. Only then will we be able to verify many of the relationships we now observe in nature, and begin to identify the underlying cellular mechanisms. In summary, I believe the following types of studies should be incorporated into future research initiatives.

(1) Continued data collection, especially from unexplored regions, for use in delineating distributions along environmental gradients. Studies may range from larger-scale regional surveys including biological, chemical and physical data to smaller-scale floristic surveys. Every attempt should be made to include chemical/physical data for the collecting sites. At a minimum, descriptions of the waterbodies should be included.

(2) Application of multivariate statistical analyses to existing and new data sets in order better to understand the role that each factor plays in controlling the distribution of taxa.

(3) Investigations aimed at comparing data sets from different regions.

(4) Seasonal studies utilizing close interval sampling from a range of different types of localities.

(5) Laboratory culture experiments in order to verify observed patterns from field studies.

(6) Increased efforts towards understanding the physiological/biochemical mechanisms controlling species distributions.

(7) The assembling of ecological data for non-scale-bearing forms in a similar fashion to that done for scale-bearing taxa during the 1980s.

My last comment concerns the value of and need for good taxonomy. The taxonomic framework, the result of the painstaking efforts of many scientists, represents one of the primary bases by which ecologists catalog and analyze their data. Such a framework has formed the basis on which most of the data collected by many investigators and summarized here were initially cataloged. Thus, an updating of the taxonomic framework for an organismal group will result in a more correct catalog for ecological data (see Moestrup, this volume; Preisig, this volume). Although the interdependence of the two disciplines is perhaps obvious, it is rarely mentioned.

References

Arvola, L. 1983. Primary production and phytoplankton in two small, polyhumic forest lakes in southern Finland. *Hydrobiologia* **101**: 105–10.

Arvola, L. 1986. Spring phytoplankton of 54 small lakes in southern Finland. *Hydrobiologia* **137**: 125–34.

Asmund, B. 1968. Studies on Chrysophyceae from some ponds and lakes in Alaska. VI. Occurrence of *Synura* species. *Hydrobiologica* **31**: 497–515.

Cassin, P.E. 1974. Isolation, growth and physiology of acidophilic chlamydomonads. *J. Phycol.* **10**: 439–46.

Charles, D.F. & Smol, J.P. 1988. New methods for using diatoms and chrysophytes to infer past pH of low-alkalinity lakes. *Limnol. Oceanogr.* **33**: 1451–62.

Cronan, C.S. & Schofield, C.L. 1979. Aluminum leaching response to acid precipitation: effects on high elevation watersheds in the northeast. *Science* **204**: 304–6.

Cronberg, G. 1982. Phytoplankton changes in Lake Trummen induced by restoration. *Folia Limnol. Scand.* **18**: 1–119, appendices.

Cronberg, G. 1989. Scaled chrysophytes from the tropics. *Beih. Nova Hedwigia* **95**: 191–232.

Cronberg, G. & Kristiansen, J. 1980. Synuraceae and other Chrysophyceae from central Smaland, Sweden. *Bot. Not.* **133**: 595–618.

Cronberg, G., Gelin, C. & Larsson, K. 1975. Lake Trummen restoration project. II. Bacteria, phytoplankton and phytoplankton productivity. *Verh. Int. Ver. Limnol.* **19**: 1907–96.

Croome, R.L. & Tyler, P.A. 1985. Distribution of silica-scaled Chrysophyceae (Paraphysomonadaceae and Mallomonadaceae) in Australian inland waters. *Aust. J. Mar. Freshwater Res.* **36**: 839–53.

Croome, R.L. & Tyler, P.A. 1988. Phytoflagellates and their ecology in Tasmanian polyhumic lakes. *Hydrobiologia* **161**: 245–53.

Cumming, B.F., Smol, J.P. & Birks, H.J.B. 1991. The relationship between sedimentary chrysophyte scales (Chrysophyceae and Synurophyceae) and limnological characters in 25 Norwegian lakes. *Nord. J. Bot.* **11**: 231–42.

Cumming, B.F., Smol, J.P. & Birks, H.J.B. 1992. Scaled chrysophytes (Chrysophyceae and Synurophyceae) from Adirondack drainage lakes and their relationship to environmental variables. *J. Phycol.* **28**: 162–78.

DeNoyelles, F. & O'Brien, W.J. 1978. Phytoplankton succession in nutrient enriched experimental ponds as related to changing carbon, nitrogen and phosphorus concentrations. *Arch. Hydrobiol.* **81**: 137–65.

Dickson, W. 1978. Some effects of the acidification of Swedish lakes. *Verh. Int. Ver. Limnol.* **20**: 851–6.

Dixit, A.S. & Dixit, S.S. 1989. Surface-sediment chrysophytes from 35 Quebec lakes and their usefulness in reconstructing lakewater pH. *Can. J. Bot.* **67**: 2071–6.

Dixit, S.S. & Smol, J.P. 1989. Algal assemblages in acid-stressed lakes with particular emphasis on diatoms and chrysophytes. In: Rao, S.S. (ed.) *Acid Stress and Aquatic Microbial Interactions.* CRC Press, Boca Raton. pp. 91–114.

Dixit, S.S., Dixit, A.S. & Evans, R.D. 1988*a*. Chrysophyte scales in lake sediments provide evidence of recent acidification in two Quebec (Canada) lakes. *Water Air Soil Pollut.* **38**: 97–104.

Dixit, S.S., Dixit, A.S. & Evans, R.D. 1988*b*. Scaled chrysophytes (Chrysophyceae) as indicators of pH in Sudbury, Ontario lakes. *Can. J. Fish. Aquat. Sci.* **45**: 1411–21.

Dixit, S.S., Dixit, A.S. & Evans, R.D. 1989*a*. Paleolimnological evidence for trace metal sensitivity in scaled chrysophytes. *Environ. Sci. Technol.* **23**: 110–15.

Dixit, S.S., Dixit, A.S. & Smol, J.P. 1989*b*. Relationship between chrysophyte assemblages and environmental variables in seventy-two Sudbury lakes as examined by canonical correspondence analysis (CCA). *Can. J. Fish. Aquat. Sci.* **46**: 1667–76.

Dixit, S.S., Smol, J.P., Anderson, D.S. & Davis, R.B. 1990. Utility of scaled chrysophytes for inferring lakewater pH in northern New England lakes. *J. Paleolimnol.* **3**: 269–87.

Dürrschmidt, M. 1980. Studies on the Chrysophyceae from Rio Cruces, Prov. Valdivia, south Chile by scanning and transmission microscopy. *Nova Hedwigia* **33**: 353–88.

Dürrschmidt, M. 1982. Studies on the Chrysophyceae from southern Chilean inland waters by means of scanning and transmission electron microscopy, II. *Arch. Hydrobiol.* (Suppl.) **63**: 121–63.

Dürrschmidt, M. & Croome, R. 1985. Mallomonadaceae (Chrysophyceae) from Malaysia and Australia. *Nord. J. Bot.* **5**: 285–98.

Eloranta, P. 1986. Phytoplankton structure in different lake types in central Finland. *Holarctic Ecol.* **9**: 214–24.

Eloranta, P. 1989a. Scaled chrysophytes (Chrysophyceae and Synurophyceae) from national park lakes in southern and central Finland. *Nord. J. Bot.* **8**: 671–81.

Eloranta, P. 1989b. On the ecology of the genus *Dinobryon* in Finnish lakes. *Beih. Nova Hedwigia* **95**: 99–109.

Findlay, D.L. 1978. Seasonal succession of phytoplankton in seven lake basins in the Experimental Lakes Area, northwestern Ontario, following artificial eutrophication. Data from 1974–1976. *Can. Fish. Mar. Serv. M.S. Rep.* **466**: 41pp.

Galloway, J.N. & Likens, G.E. 1979. Atmospheric enhancement of metal deposition in Adirondack lake sediments. *Limnol. Oceanogr.* **24**: 427–33.

Galloway, J.N., Likens, G.E. & Edgerton, E.S. 1976. Acid precipitation in the northeastern United States: pH and acidity. *Science* **194**: 722–4.

Gibson, K.N., Smol, J.P. & Ford, J. 1987. Chrysophycean microfossils provide new insight into the recent history of a naturally acidic lake (Cond Pond, New Hampshire). *Can. J. Fish. Aquat. Sci.* **44**: 1584–8.

Gutowski, A. 1989. Seasonal succession of scaled chrysophytes in a small lake in Berlin. *Beih. Nova Hedwigia* **95**: 159–77.

Harris, K. & Bradley, D.E. 1957. An examination of the scales and bristles of *Mallomonas* in the electron microscope using carbon replicas. *Proc. R. Microsc. Soc.* **76**: 37–46.

Hartmann, H. & Steinberg, C. 1989. The occurrence of silica-scaled chrysophytes in some central European lakes and their relation to pH. *Beih. Nova Hedwigia* **95**: 131–58.

Hecky, R.E. & Kling, H.J. 1981. The phytoplankton and protozooplankton of the euphotic zone of Lake Tanganyika: species composition, biomass, chlorophyll content and spatio-temporal distribution. *Limnol. Oceanogr.* **26**: 548–64.

Hickel, B. & Maass, I. 1989. Scaled chrysophytes, including heterotrophic nanoflagellates from the lake district in Holstein, northern Germany. *Beih. Nova Hedwigia* **95**: 233–57.

Hultberg, H. & Johansson, S. 1981. Acid groundwater. *Nordic Hydrol.* **12**: 51–64.

Hustedt, F. 1939. Systematische und ökologische Untersuchungen über die Diatomeen-Flora von Java, Bali und Sumatra nach dem Material der Deutschen Limnologischen Sunda-Expedition. III. Die ökologischen Faktoren und ihr Einfluss auf die Diatomeen-flora. *Arch. Hydrobiol.* (Suppl.) **16**: 1–394.

Distribution of chrysophytes along environmental gradients 265

Hutchinson, G.E. 1967. *A Treatise on Limnology*, vol. II, *Introduction to Lake Biology and the Limnoplankton*. Wiley, New York. 1115pp.
Ilmavirta, V. 1974. Diel periodicity in the phytoplankton community of the oligotrophic lake Pääjärvi, southern Finland. I. Phytoplankton primary production and related factors. *Ann. Bot. Fenn.* **11**: 136–77.
Ito, H. & Takahashi, E. 1982. Seasonal fluctuation of *Spiniferomonas* (Chrysophyceae, Synuraceae) in ponds on Mt. Rokko, Japan. *Jpn. J. Phycol.* **30**: 272–8.
Jacobsen, B.A. 1985. Scale-bearing Chrysophyceae (Mallomonadaceae and Paraphysomonadaceae) from west Greenland. *Nord. J. Bot.* **5**: 381–98.
Jansson, M. 1981. Induction of high phosphatase activity by aluminum in acid lakes. *Arch. Hydrobiol.* **93**: 32.
Jongman, R.H.G., ter Braak, C.J.F. & van Tongeran, O.F.R. 1987. *Data Analysis in Community and Landscape Ecology*. Pudoc, Wageningen. 299pp.
Kalff, J. 1967. Phytoplankton dynamics in an arctic lake. *J. Fish. Res. Bd. Can.* **24**: 1861–71.
Kalff, J., Kling, H.J., Holmgren, S.H. & Welch, H.E. 1975. Phytoplankton, phytoplankton growth and biomass cycles in an unpolluted and in a polluted polar lake. *Verh. Int. Ver. Limnol.* **19**: 487–95.
Kies, L. & Berndt, M. 1984. Die *Synura*-arten (Chrysophyceae) Hamburgs und seiner nordöstlichen Umgebund. *Mitt. Inst. Allg. Bot. Hamburg* **19**: 99–122.
Kling, H. & Holmgren, S. 1972. Species composition and seasonal distribution of phytoplankton in the Experimental Lakes Area, northwestern Ontario. Canadian Fishery Marine Service Technical Report No. 337.
Kristiansen, J. 1975. On the occurrence of the species of *Synura* (Chrysophyceae). *Verh. Int. Ver. Limnol.* **19**: 2709–15.
Kristiansen, J. 1980. Chrysophyceae from some Greek lakes. *Nova Hedwigia* **33**: 167–94.
Kristiansen, J. 1981. Distribution problems in the Synuraceae (Chrysophyceae). *Verh. Int. Ver. Limnol.* **21**: 1444–8.
Kristiansen, J. 1985. Occurrence of scale-bearing Chrysophyceae in a eutrophic Danish lake. *Verh. Int. Ver. Limnol.* **22**: 2826–9.
Kristiansen, J. 1986. Silica-scale bearing chrysophytes as environmental indicators. *Br. Phycol. J.* **21**: 425–36.
Kristiansen, J. 1988. Seasonal occurrence of silica-scaled chrysophytes under eutrophic conditions. *Hydrobiologia* **161**: 171–84.
Kristiansen, J. & Takahashi, E. 1982. Chrysophyceae: introduction and bibliography. In: Rosowski, J. & Parker, B. (eds.) *Selected Papers in Phycology*, vol. II. Phycological Society of America, Lawrence, Kansas. pp. 698–704.
Likens, G.E. (ed.) 1985. *An Ecosystem Approach to Aquatic Ecology: Mirror Lake and its Environment*. Springer, Berlin. 516pp.
Munawar, M. & Munawar, I.F. 1986. The seasonality of phytoplankton in the North American Great Lakes: a comparative synthesis. *Hydrobiologia* **138**: 83–115.
Nalewajko, C. & Paul, B. 1985. Effects of manipulations of aluminum concentrations and pH on phosphate uptake and photosynthesis of planktonic communities in two Precambrian Shield lakes. *Can. J. Fish. Aquat. Sci.* **42**: 1946.
Nicholls, K.H. 1987. The distinction between *Mallomonas acaroides* v. *acaroides*

and *Mallomonas acaroides* v. *muskokana* var. nova (Chrysophyceae). *Can. J. Bot.* **65**: 1779–84.

Nicholls, K.H. 1988. Descriptions of three new species of *Mallomonas* (Chrysophyceae): *M. hexagonis* sp. nov., *M. liturata* sp. nov., and *M. galeiformis* sp. nov. *Br. Phycol. J.* **23**: 159–66.

Olsson, H. 1983. Origin and production of phosphatases in the acid Lake Gardsjon. *Hydrobiologia* **101**: 49–58.

Ostrofsky, M.L. & Duthie, H.C. 1975. Primary productivity and phytoplankton of lakes on the Eastern Canadian Shield. *Verh. Int. Ver. Limnol.* **19**: 732–8.

Philipose, M.T. 1953. Contribution of our knowledge of Indian Algae I. Chrysophyceae: on the occurrence of *Mallomonas* Perty, *Synura* Ehr. and *Dinobryon* Ehr. in India. *Proc. Indian Acad. Sci.* **37**: 232–48.

Pick, F.R. & Lean, D.R.S. 1984. Diurnal movements of metalimnetic phytoplankton. *J. Phycol.* **20**: 430–6.

Pick, F.R., Nalewajko, C. & Lean, D.R.S. 1984. The origin of a metalimnetic peak. *Limnol. Oceanogr.* **29**: 125–34.

Pick, F.R. & Cuhel, R.L. 1986. Light quality effects on carbon and sulfur uptake of a metalimnetic population of the colonial chrysophyte *Chrysosphaerella longispina*. In: Kristiansen, J. & Andersen, R.A. (eds.) *Chrysophytes: Aspects and Problems.* Cambridge University Press, Cambridge. pp. 197–206.

Prowse, G.A. 1962. Further Malayan freshwater flagellata. *Gardens Bull., Singapore* **19**: 105–40.

Reinertsen, H. 1982. The effect of nutrient addition on the phytoplankton community of an oligotrophic lake. *Holarctic Ecol.* **5**: 225–52.

Roijackers, R.M.M. 1981. Chrysophyceae from freshwater localities near Nijmegen, The Netherlands. *Hydrobiologia* **76**: 179–89.

Roijackers, R.M.M. 1986. Development and succession of scale-bearing Chrysophyceae in two shallow freshwater bodies near Nijmegen, The Netherlands. In: Kristiansen, J. & Andersen, R. (eds.) *Chrysophytes: Aspects and Problems.* Cambridge University Press, Cambridge. pp. 241–58.

Roijackers, R.M. & Kessels, H. 1986. Ecological characteristics of scale-bearing Chrysophyceae from the Netherlands. *Nord. J. Bot.* **6**: 373–83.

Rosén, G. 1981. Phytoplankton indicators and their relations to certain chemical and physical factors. *Limnologica* **13**: 263–90.

Saha, L.C. & Wujek, D.E. 1990. Scale-bearing chrysophytes from tropical Northeast India. *Nord. J. Bot.* **10**: 343–54.

Sandgren, C.D. 1986. Effects of environmental temperature on the vegetative growth and sexual life history of *Dinobryon cylindricum* Imhof. In: Kristiansen, J. & Andersen, R.A. (eds.) *Chrysophytes: Aspects and Problems.* Cambridge University Press, Cambridge. pp. 207–25.

Sandgren, C.D. 1988. The ecology of chrysophyte flagellates: their growth and perennation strategies as freshwater phytoplankton. In: Sandgren, C.D. (ed.) *Growth and Reproductive Strategies of Freshwater Phytoplankton.* Cambridge University Press, Cambridge. pp. 9–104.

Sandgren, C.D. & Flanagin, J. 1986. Heterothallic sexuality and density dependent encystment in the chrysophycean alga *Synura petersenii* Korsh. *J. Phycol.* **22**: 206–16.

Schindler, D.W. & Holmgren, S.K. 1971. Primary production and phytoplankton in the Experimental Lakes Area, northwestern Ontario, and other low-carbonate waters, and a liquid scintillation method for determining ^{14}C activity in photosynthesis. *J. Fish. Res. Bd. Can.* **28**: 189–201.

Distribution of chrysophytes along environmental gradients 267

Schindler, D.W., Kling, H., Schmidt, R.V., Prokopowich, J., Frost, V.E., Reid, R.A. & Capel, M. 1973. Eutrophication of Lake 227 by addition of phosphate and nitrate: the second, third and fourth years of enrichment, 1970, 1971 and 1972. *J. Fish. Res. Bd. Can.* **30**: 1415–40.

Schofield, C.L., Galloway, J.N. & Hendry, G.R. 1985. Surface water chemistry in the ILWAS basins. *Water Air Soil Pollut.* **26**: 403–23.

Sheath, R.G. 1986. Seasonality of phytoplankton in northern tundra ponds. *Hydrobiologia* **138**: 75–83.

Siver, P.A. 1988a. Distribution of scaled chrysophytes in 17 Adirondack (New York) lakes with special reference to pH. *Can. J. Bot.* **66**: 1391–403.

Siver, P.A. 1988b. Morphology and ecology of *Mallomonas galeiformis* (Chrysophyceae), a potentially useful paleolimnological indicator. *Trans. Am. Microsc. Soc.* **107**: 152–61.

Siver, P.A. 1989a. The distribution of scaled chrysophytes along a pH gradient. *Can. J. Bot.* **67**: 2120–30.

Siver, P.A. 1989b. The separation of *Mallomonas acaroides* v. *acaroides* and v. *muskokana* (Synurophyceae) along a pH gradient. *Beih. Nova Hedwigia* **95**: 111–17.

Siver, P.A. 1991. *The Biology of* Mallomonas: *Morphology, Taxonomy and Ecology.* Kluwer Academic, Dordrecht. 230pp.

Siver, P.A. 1992. A critical literature review on the usefulness of chrysophytes to detect lake chronic or episodic acidification and/or recovery in the context of a long term monitoring program. U.S. Environmental Protection Agency, Corvallis, OR.

Siver, P.A. 1993. Inferring the specific conductance of lake water with scaled chrysophytes. *Limnol. Oceanogr.* **38**: 1480–92.

Siver, P.A. & Chock, J.S. 1986. Phytoplankton dynamics in a chrysophycean lake. In: Kristiansen, J. & Andersen, R.A. (eds.) *Chrysophytes: Aspects and Problems.* Cambridge University Press, Cambridge. pp. 165–83.

Siver, P.A. & Hamer, J.S. 1989. Multivariate statistical analysis of the factors controlling the distribution of scaled chrysophytes. *Limnol. Oceanogr.* **34**: 368–81.

Siver, P.A. & Hamer, J.S. 1990. Use of extant populations of scaled chrysophytes for the inference of lakewater pH. *Can. J. Fish. Aquat. Sci.* **47**: 1339–47.

Siver, P.A. & Hamer, J.S. 1992. Seasonal periodicity of Chrysophyceae and Synurophyceae in a small New England lake: implications for paleolimnological research. *J. Phycol.* **28**: 186–98.

Siver, P.A. & Smol, J.P. 1993. The use of scaled chrysophytes in long term monitoring programs for the detection of changes in lakewater acidity. *Water Air Soil Pollut.* (in press).

Siver, P.A., Hamer, J.S. & Kling, H. 1990. The separation of *Mallomonas duerrschmidtiae* sp. nov. from *M. crassisquama* and *M. pseudocoronata*: implications for paleolimnological research. *J. Phycol.* **26**: 728–40.

Smol, J.P. 1986. Chrysophycean microfossils as indicators of lakewater pH. In: Smol, J.P., Battarbee, R.W., Davis, R.B. & Meriläinen, J. *Diatoms and Lake Acidity.* Junk, The Netherlands. pp. 275–87.

Smol, J.P. 1990. Paleolimnology: recent advances and future challenges. *Mem. Ist. Ital. Idrobiol.* **47**: 253–76.

Smol, J.P., Charles, D.F. & Whitehead, D.R. 1984. Mallomonadacean (Chrysophyceae) assemblages and their relationships with limnological characteristics in 38 Adirondack (New York) lakes. *Can. J. Bot.* **62**: 911–23.

Stokes, P. 1983. Responses of freshwater algae to metals. In: Round, F.E. &
    Chapman, D.J. (eds.) *Progress in Phycological Research*, vol. 2. Elsevier,
    Amsterdam, p. 87.
Takahashi, E. 1967. Studies on genera *Mallomonas, Synura* and other plankton
    in fresh-water with the electron microscope. VI. Morphological and
    ecological observations on genus *Synura* in ponds and lakes in Yamagata
    Prefecture. *Bull. Yamagata Univ. (Agric. Sci.)* **5**: 99–118.
Takahashi, E. 1978. *Electron Microscopical Studies of the Synuraceae
    (Chrysophyceae) in Japan: Taxonomy and Ecology.* Tokai University Press,
    Tokyo.
Takahashi, E. & Hayakawa, T. 1979. The Synuraceae (Chrysophyceae) in
    Bangladesh. *Phykos* **18**: 129–47.
ter Braak, C.J.F. 1986. Canonical correspondence analysis: a new eigenvector
    method for multivariate direct gradient analysis. *Ecology* **67**: 1167–79.
ter Braak, C.J.F. 1989. CANOCO: an extension of DECORANA to analyze
    species–environment relationships. *Hydrobiologia* **184**: 169–70.
Wee, J.L. & Gabel, M. 1989. Occurrences of silica-scaled chromophyte algae in
    predominantly alkaline lakes and ponds in Iowa. *Am. Midl. Nat.* **121**:
    32–40.
Wetzel, R.G. 1983. *Limnology.* W.B. Saunders, Philadelphia. 753 pp.
Wujek, D.E. 1984a. Chrysophyceae (Mallomonadaceae) from Florida. *Fla. Sci.*
    47: 161–70.
Wujek, D.E. 1984b. Scale-bearing Chrysophyceae (Mallomonadaceae) from
    north-central Costa Rica. *Brenesia* **22**: 309–13.
Wujek, D.E. 1986. Scale bearing Chrysophyceae from the Panama Canal. *Jpn.
    J. Phycol.* **34**: 83–6.
Wujek, D.E. & Asmund, B.C. 1979. *Mallomonas cyathellata* sp. nov. and
    Mallomonas cyathellata va. *kenyana* var. nov. (Chrysophyceae) studied by
    means of scanning and transmission electron microscopy. *Phycologia* **18**:
    115–19.
Wujek, D.E. & Gardiner, W.E. 1985. Chrysophyceae (Mallomonadaceae) from
    Florida. II. New species of *Paraphysomonas* and the prymnesiophyte
    Chrysochromulina. Fla. Sci. **48**: 59–64.
Wujek, D.E. & Bland, R.G. 1991. Chrysophyceae (Mallomonadaceae and
    Paraphysomonadaceae) from Florida. III. Additions to the flora. *Fla. Sci.*
    54: 41–8.
Yan, N.D. 1979. Phytoplankton community of an acidified, heavy
    metal-contaminated lake near Sudbury, Ontario: 1973–1977. *Water Air
    Soil Pollut.* **11**: 43–55.
Zeeb, B.A. & Smol, J.P. 1991. Paleolimnological investigation of the effects of
    road salt seepage on scaled chrysophytes in Fonda Lake, Michigan.
    J. Paleolimnol. **5**: 263–6.

12

The influence of zooplankton herbivory on the biogeography of chrysophyte algae

CRAIG D. SANDGREN & WILLIAM E. WALTON

Introduction

Chrysophyte algae: ecology and biogeographic distribution

The great majority of planktonic chrysophytes (algal class Chrysophyceae *sensu* Hibberd 1976; incl. Synurophyceae, *sensu* Andersen 1987) are rather delicate, golden-colored flagellates. Both unicellular and colonial chrysomonads are common in lake plankton and they exhibit three distinct types of cell coverings: 'naked' cells (cell membrane only), cells in expanded organic loricas, and cells covered with ornamented siliceous scales and/or bristles. This morphological diversity may affect their palatability for herbivores or may increase the effective diameter of chrysophyte cells as zooplankton 'food particles'. Chrysophytes range in natural particle size from a few micrometers to several hundred micrometers in diameter; larger colonies are mostly spherical (*Synura, Uroglena, Chrysosphaerella*), but some are dendroid (*Dinobryon*) or linear (*Chrysidiastrum*). Chrysophyte algae demonstrate seasonally restricted population cycles in lakes (Sandgren 1988); they produce siliceous resting cysts and probably recruit annually from sedimentary 'seed' populations of these cysts (Sandgren 1991).

Chrysophytes are among the most poorly studied freshwater phytoplankton with regard to their nutrition, physiology and ecology. Those genera of interest here are phototrophs, but many also have a facultative or obligate capacity for supplementary phagotrophic and osmotrophic feeding (Sanders 1991; reviewed in Sandgren 1988; also see Holen & Boraas, this volume). Chrysophyte algae are frequently biomass dominants, together with other algal flagellates, in the myriad of small, softwater, and largely oligotrophic lakes of the north-temperate regions of North America and Scandinavia (as summarized in Sandgren 1988).

However, chrysophytes are often absent or of minor importance in meso- or eutrophic lakes where most limnological studies have been conducted. In fact, a strong biogeographic trend described for chrysophyte algae suggests an inverse relationship between chrysophyte importance in lakes and various measures of lake productivity – the most widely available of which is mean seasonal phosphorus concentration. This trend was first emphasized by Nicholls (Nicholls 1976; Nicholls *et al.* 1986; also see Nicholls, this volume; Eloranta, this volume) and was supported by additional data ('Chrysophyte Database': Sandgren 1988: fig. 2-6a); similar trends also exist in independent databases (e.g. Eloranta 1986). Sandgren further demonstrated (Sandgren 1988: fig. 2-6b) that this inverse coupling of lake phosphorus availability and chrysophyte importance had a consistent dynamic attribute: the relative contribution of chrysophytes in the phytoplankton changed predictably through time as individual lakes became progressively more productive in response to cultural eutrophication or less productive in response to pollution abatement measures. Such dynamics suggest that this correlation reflects an underlying cause-and-effect relationship.

This empirical inverse correlation between chrysophyte importance and lake productivity is as strong as other published distributional relationships for phytoplankton which have been demonstrated to reflect a physiological basis (e.g. correlation between N:P supply ratios and cyanophyte abundance in lakes: Smith 1983). However, no physiological mechanism is obvious in this case. Existing experimental studies and distributional analyses suggest that the physiological tolerances of freshwater chrysophytes to temperature, pH, water color and major cation concentrations are, in general, broad and very similar to those of other groups of eukaryotic phytoplankton (reviewed in Sandgren 1988). Although individual chrysophyte species may have restrictive physiological tolerances (e.g. Siver & Hamer 1989; Smol 1990; Siver 1991; Cumming *et al.* 1992; also see Smol, this volume; Siver, this volume), the group as a whole does not. There is evidence that chrysomonad populations are frequently nutrient-limited in lakes, particularly by phosphorus, iron or a mixture of micrometals (reviewed in Sandgren 1988; also see Raven, this volume). But, evaluation of physiological characteristics such as potential growth rates, phosphorus uptake kinetics and storage capacity (Kennedy 1984; Sandgren 1986, 1988; Wiesinger 1992; Sandgren & Wiesinger, unpublished data) suggests that chrysophytes are fairly typical of freshwater phytoplankton in general. And, simulations of competition under a wide array of nutrient supply conditions using this extensive physiological database (Wiesinger 1992; Wiesinger & Sandgren, unpublished data) suggest that growth potential and competition under conditions of *constant, common mortality rates* cannot explain the observed inverse correlation among lakes between chrysophyte importance and lake productivity.

Sandgren (1988) hypothesized that differential loss rates to herbivore grazing, particularly strong, selective herbivory on chrysophytes by large species of the cladoceran *Daphnia*, might contribute significantly to this fairly dramatic biogeographic trend. Here we reconsider this evidence with: (1) expanded interrogation of biogeographic trends in the 'Chrysophyte Lakes Database' (see 'Methods'), (2) expanded analysis of chrysophyte responses in *in situ* herbivore grazing experiments, (3) laboratory grazing studies using morphologically diverse chrysophytes and herbivorous zooplankton, and (4) an experimental analysis of the efficacy of chrysophyte siliceous scale and bristle layers as herbivore avoidance adaptations. Our objective is to investigate further the mechanisms and consequences of predator–prey interactions between chrysophytes and diverse herbivorous zooplankton, with emphasis on the cladoceran genus *Daphnia*. We begin with brief reviews of herbivorous zooplankton feeding biology and of changes in zooplankton–phytoplankton coupling along trophic gradients in lakes.

Selective feeding by zooplankton on freshwater algae

Dietary preferences are evident among the major zooplankton groups (Gliwicz 1969, 1980; Porter 1973, 1977; Ferguson *et al.* 1982; Gilbert & Bogdan 1984), and selectivity is largely dependent upon the size and shape of the phytoplankton (Burns 1968; Ferguson *et al.* 1982; Gophen & Geller 1984; Reynolds 1984; Infante & Litt 1985). Selectivity is also affected by the nature of the cell covering or the presence of a gelatinous sheath (Porter 1973, 1977; Geller 1975; McNaught *et al.* 1980; Horn 1985; DeMott 1989), armature (Schnack 1979), motility/flagella (Gilbert & Bodgan 1981; DeMott 1982), taste/presence of toxins (Poulet & Marsot 1978; DeMott 1986, 1988; Venderploeg 1990), surface characteristics such as wettability and surface charge (Gerritsen & Porter 1982; Gerritsen *et al.* 1988) and nutritional quality (Cowles *et al.* 1988; Butler *et al.* 1989).

Rotifers

Rotifers exhibit low clearance rates (0.007–0.11 ml per individual per day: Pourriot 1977; Starkweather *et al.* 1979) and are generally restricted to ingesting food particles < 15 µm GALD (greatest axial linear dimension) (Gliwicz 1969; Reynolds 1984). Feeding habits are related to the morphologies of the ciliated corona and mouthparts (Pourriot 1977; Gilbert & Bogdan 1984).

Copepods

Calanoid copepods are behaviorally flexible omnivores (Vanderploeg 1990). Freshwater calanoids typically consume phytoplankton between 1 and 35 µm GALD (Gliwicz 1969; Richman *et al.* 1980; Reynolds 1984). More recent studies have found that calanoids also ingest microzooplankton, particularly soft-bodied rotifers (Williamson & Butler 1986; Williamson & Vanderploeg 1988). *Diaptomus* is the best studied freshwater genus. Because *Diaptomus* spp. exhibit three capture modes – passive, active and thrust – they can utilize a broad range of prey; selectivity (W') increases with increasing size of alga or rotifer prey (Vandenploeg & Paffenhöfer 1985; Vanderploeg *et al.* 1988; Williamson & Vanderploeg 1988; Vanderploeg 1990). Although *Diaptomus* spp. can potentially consume a wide variety of prey, community grazing studies run during 1990 and 1991 (Sandgren & Walton, unpublished data) have consistently demonstrated that Lake Michigan calanoids (*D. sicilis*, *D. ashlandi* and *D. minutus*) significantly affect phytoplankton ≤ 10 µm. The net growth rate of algae ≤ 10 µm (based on chlorophyll *a*) declined as a function of increasing calanoid biomass in these studies while the net growth rates of phytoplankton between 10–50 µm and > 50 µm did not change or increased, respectively, across a gradient of increasing calanoid biomass. Cyclopoids, a second group of freshwater copepods, are also omnivorous; but adult cyclopoids consume a broader range of animal prey (rotifers, copepods, cladocerans and fish larvae: Li & Li 1979; Hartig *et al.* 1982; Stemberger 1986) than does *Diaptomus*.

Clearance rates of zooplankton are temperature dependent and filtration rates of adult copepods (ml per individual per unit time), at typical summer water temperatures between approximately 15 and 25 °C, are generally lower than those of cladocerans of comparable body size (comparisons based on average body length (mm): literature values summarized by Wetzel 1983 and Reynolds 1984). Copepods are thought to be less 'efficient' grazers than are cladocerans because copepods feed at substantially lower weight-specific rates (Lampert & Muck 1985). In addition, copepods require relatively longer times to reach the age of first reproduction than do cladocerans and, consequently, dioecious copepods do not exhibit the rapid numerical response to increased resource abundance shown by parthenogenetic cladocerans (Allan 1976). Because copepods undergo marked morphological changes during development, prey selectivity probably changes during ontogeny. Ontogenetic shifts in selectivity for phytoplankton prey have not been studied in most copepods.

Cladocerans

Filter-feeding cladocerans can be placed in two general categories: single-mode feeders (e.g., *Daphnia*) and dual-mode feeders (e.g., *Bosmina*) (Vanderploeg 1990). In the filter-feeding, single mode species, particles are captured on relatively homogeneous filter combs (Vanderploeg 1990). Selectivity and weight-specific clearance rates are related to the intersetule distances on the filter combs (DeMott 1985). Large-sized cladocerans such as *Daphnia* exhibit high filtration rates (~ 100 ml per individual per day in *D. magna*: Porter *et al.* 1982) and, among the three major zooplankton groups, consume the broadest size range of phytoplankton. Some species can also efficiently utilize bacteria (DeMott 1985). The GALDs of most particles in *Daphnia* guts are $< 50\,\mu$m (Reynolds 1984), but *Daphnia* can also consume large-sized phytoplankton. For example, large *Daphnia magna* can consume phytoplankton up to $225\,\mu$m in length and intermediate-sized *D. hyalina* ingest algae up to $120\,\mu$m in length (Nadin-Hurley & Duncan 1976; Ferguson *et al.* 1982).

Clearly, algae differ in their nutritional value for *Daphnia* (Infante & Litt 1985). But, compared with calanoid copepods and dual-mode feeding cladocerans, single-mode feeding cladocerans have limited abilities to discriminate between large and small particles that differ in nutritional quality (Vanderploeg 1990). *Daphnia* species *can* discriminate to some extent, however. Knisely & Geller (1986) found that phytoplankton species in mixed, natural assemblages were differentially grazed by *Daphnia hyalina* and *Daphnia galeata* during 3 hour incubations. We might therefore expect to see elevated loss rates experienced by preferred prey in mixed prey assemblages.

Dual-mode feeders utilize both filter feeding and raptorial feeding. Selectivity increases with increasing algal size (Bleiwas & Stokes 1985); however, *Bosmina* does not readily filter small particles such as bacteria (DeMott 1985; Vanderploeg 1990). Filtration rates of small-sized dual-mode feeders (average body length: *Bosmina*, 0.4–0.6mm; *Chydorus*, 0.1–0.2 mm) are approximately 5–50 times lower than those of *Daphnia* spp. (average body length 0.7–1.9 mm) (average clearance rate in ml per individual per day: *Bosmina,* 0.44–3.0; *Chydorus,* 0.18–2.6; *Daphnia,* 1.0–63; Wetzel 1983, Reynolds 1984).

If chrysophyte algae are a preferred food item, then we would expect comparatively higher selectivities and filtration rates for chrysophytes by large *Daphnia* than by rotifers, copepods and small-sized cladocerans. Secondly, chrysophytes should not attain a refuge from *Daphnia* in cell size or growth form (i.e. coloniality), but larger unicells and colonies may be comparatively immune to herbivory by small-bodied cladocerans, calanoid copepods and rotifers.

Zooplankton community changes along lake trophic gradients

Zooplankton community structure changes along trophic gradients. Rotifers, copepods and small cladocerans tend to dominate the zooplankton communities of oligotrophic lakes and are replaced as community dominants in more eutrophic systems by large-bodied cladocerans such as *Daphnia* (Elster & Schwoerbel 1970; Gannon & Stemberger 1978; Einsle 1983). Copepods survive better than cladocerans at low food densities characteristic of oligotrophic lakes because copepods store lipids rather than using them immediately for reproduction (McNaught 1975; Lampert & Muck 1985). With regard to cladocerans, Romanovsky (1985) suggested that lake trophic status selects for different life history strategies among cladocerans, giving small forms such as *Bosmina* a competitive advantage in oligotrophic habitats and large forms such as *Daphnia* an advantage in eutrophic habitats. Competitive dominance of *Daphnia* over *Bosmina* has been demonstrated at the elevated levels of phytoplankton food availability characteristic of eutrophic lakes (DeMott & Kerfoot 1982; Kerfoot *et al.* 1985; Vanni 1986). And large *Daphnia* species also strongly suppress rotifer abundances in productive lakes through a combination of competitive superiority and directly inflicted mortality (Gilbert 1988).

Abundances and productivity of phytoplankton and zooplankton are correlated among lakes. Wetzel (1983) has summarized literature evidence describing a positive correlation between rates of primary and secondary production in lakes. Although much of the autotrophic production is not used by herbivores, Cyr & Pace (1993) have recently also documented a positive correlation between community-level rates of aquatic primary production and zooplankton herbivory. Empirical evidence also exists for a positive correlation between total zooplankton and total phytoplankton biomass among lakes (McCauley & Kalff 1981). It does not follow, however, that predation pressure on algae increases directly with increasing lake trophy because, in addition to changes in species assemblages, food web complexity and connectance are related to lake trophic state (Carney 1990). Food webs of very oligotrophic lakes are comparatively diverse with a large number of weak interactions within the food web. Such systems are often dominated by copepods which have lower per capita filtering rates and recycle less nutrients and different ratios of nutrients than do the large cladocerans (Carney 1990; Sterner 1990). Mesotrophic lakes have higher concentrations of edible and nutritious algae than do oligotrophic lakes and are dominated by cladocerans, such as *Daphnia*, with high per capita filtering rates and high nutrient regeneration abilities. Phytoplankton–zooplankton coupling is strong because of herbivore control of both algal birth and death rates (Elser & Goldman 1991). The food webs of eutrophic lakes are comparatively simple and less diverse than are the food webs in oligotrophic and mesotrophic lakes (Carney 1990). The strength of phytoplankton–zooplankton interactions is comparatively weaker in eutrophic systems (McQueen *et al.* 1986; Carney 1990; Richman *et al.* 1990), the phytoplankton populations are dominated by inedible, unpalatable and less nutritious species, and grazing pressure and nutrient regeneration rates are lower than in mesotrophic systems. Selective grazing on subdominant, palatable algae must occur to support the summer *Daphnia* populations.

On the basis of this evidence, chrysophyte algae should experience comparatively weak predation pressure and a refuge from predation by large *Daphnia* in oligotrophic lakes. In mesotrophic lakes, chrysophytes should experience generally strong predation pressure and strong interactions with *Daphnia*, particularly in summer; chrysophytes should be of minor overall importance and should be present only during periods of time (spring or autumn) when populations of large *Daphnia* are very small or entirely absent. In eutrophic lakes, predation pressure on phytoplankton generally may decline, but heavy selection for chrysophytes may exist so long as large *Daphnia* persist in the plankton. Because of the strong coupling between planktivorous fish and *Daphnia*, fish community structure may indirectly have a strong impact on both total phytoplankton abundance and chrysophyte relative importance in eutrophic lakes (Liebold 1989; Carpenter & Kitchell 1993; others in DeMelo *et al.* 1992; but also see McQueen *et al.* 1992; Turner & Mittelbach 1992). If chrysophyte–*Daphnia* interactions are the most significant factor excluding chrysophytes from productive lakes, then a clear biogeographic dichotomy should be apparent: in eutrophic lakes with large *Daphnia*, chrysophytes will be absent; in eutrophic lakes where large *Daphnia* are *consistently* eliminated by intense predation from planktivorous fish, chrysophytes will be present.

Methods

Description of the Chrysophyte Lakes Database

The literature-based 'Chrysophyte Lake Database' was constructed for examination of general chrysophyte biogeographic trends in a previous review paper on chrysophyte ecology (Sandgren 1988). In all, approximately 200 published papers and reports were used to extract data on 142 lakes and 172 lake-years (multiple year data available for some lakes). Only lakes with some quantitative record of chrysophyte presence were included; there is therefore an inherent undersampling of non-chrysophyte, often eutrophic, lakes. The effect of this is discussed in the Results section. Lakes incorporated into the database are distributed worldwide. Although availability of data is prejudiced by the published emphasis on north-temperate lakes, a particular effort was made to consider latitudinal relationships by including tropical lakes and arctic ponds. The database contains measurements of 20 physicochemical variables (morphometry, mixing characteristics, water color, light penetration, conservative ion chemistry, phosphorus and nitrogen concentrations) and 15 biological variables (productivity, chlorophyll levels, standing crop of subgroups of phytoplankton and zooplankton). The data set is incomplete in the sense that each variable lacks values for some of the included lakes. Temporal resolution is low; algal and zooplankton abundance estimates are annual or seasonal means, usually for a single year. Zooplankton data are available for only about half of the lakes. Taxonomic resolution is also low; only dominant species are known and abundance estimates are categorical (i.e. 'total chrysophytes', 'total cladocerans', 'total *Daphnia*', etc.).

Laboratory grazing studies

The chrysophytes and zooplankton were grown in one chemically defined medium ('DYIV', a modification of DYIII: Lehman 1976). Algal populations used in feeding studies were conditioned in clonal, phosphorus-limited 'semi-continuous' cultures (chemostat mimic: Kilham 1978) to insure uniformity and reproducibility of morphology and nutritional value. Cultures received daily dilutions of low-phosphorus medium (2 μM

Na₂HPO₄) at a dilution rate of 0.125 v/v, which at steady state results in an instantaneous birth rate of 0.134 per day. Standard growth conditions were 150 µEin m^{-2} s^{-1} cool-white fluorescent lighting on a 16L:8D cycle at 17 °C. Zooplankton were maintained at low light intensity under otherwise identical conditions. Animals were preconditioned to chryso-phyte prey for several weeks prior to grazing experiments by maintaining them on a mixture of fresh discard medium from the algal cultures. Animal cultures were periodically culled to prevent behavioral modifications associated with food limitation.

Three medium to large-size flagellate chrysophytes were used as prey (Fig. 12.1): *Mallomonas pseudocoronata* (scale-covered unicell, 17.8 ± 3.0 µm; clone CDS-2), *Mallomonas caudata* (scale- and bristle-covered unicell, 29.1 ± 6.5 µm; clone CDS-5) and *Synura petersenii* (spherical colony of scale-covered cells, 21.2 ± 7.8 µm; clone CDS-7). These species/

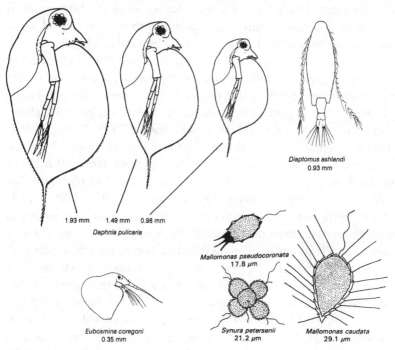

Figure 12.1. Illustrations of the three scale-covered chrysophytes and the five herbivore categories used in the laboratory feeding studies. Organisms are drawn to common scales (one scale for algal prey and one for herbivores) to facilitate size comparisons. Measurements are based on experimental means; algal sizes are for phosphorus-limited cells as determined by a particle counter (halo of siliceous bristles and scales not included).

clones were chosen because they are sufficiently large for differential selectivity and feeding rates by small versus large zooplankton to be expected; they would thus provide the most interesting test of predictions. Size estimates are mean values expressed as equivalent spherical diameters derived from particle counter measurements on equilibrium, phosphorus-limited cells just prior to grazing trials. The large variation in *Synura* size reflects an approximately equal mixture of unicells and 2-, 4- or 8-celled colonies; *M. caudata* unicells are consistently very variable in size in culture. Visual inspection (light microscope) confirmed that cells of all three cultures exhibited normal cell surface layers of scales and bristles. The scale layer contributes only 1–2 μm to the cell's diameter, but bristles, which emerge perpendicular to the cell surface, can add considerably to the cell's effective size as a food particle. It is apparent from comparison of the particle counter data and microscopical measurements that the particle counter did not detect the bristle layer surrounding *M. caudata* cells; the actual effective particle diameter including the bristle halo is 43 μm (visual measurements). *M. caudata*, with its covering of spiny bristles, is the largest planktonic unicellular chrysomonad.

Three common herbivorous zooplankton (Fig. 12.1) were used as predators: *Daphnia pulicaria* (large, single feeding mode cladoceran; three size classes of 0.98 ± 0.15 mm, 1.49 ± 0.15 mm, 1.93 ± 0.14 mm), *Eubosmina coregoni* (similar to *Bosmina*: small, dual feeding mode clado-ceran, 0.35 ± 0.055 mm) and *Diaptomus ashlandi* (small, omnivorous calanoid copepod, 0.93 ± 0.09 mm). Size classes were created by initial sieving and then selecting individual animals by pipet. Size measurements were made on preserved animals actually used in feeding trials ($n = 40$).

Short-term feeding experiments (Vanderploeg *et al.* 1984, 1988; Knisely & Geller 1986) were used to estimate predation rates for each pairing of zooplankton and chrysophyte species. Each experiment consisted of a single control (without zooplankton) and three experimental bottles (150 ml) containing 2500 cells ml^{-1} of the prey alga (only 1000 cells ml^{-1} for *M. caudata* because of low equilibrium abundance in the parent culture). These initial phytoplankton densities are below the incipient limiting concentration (ILC: Peters 1984); feeding rates should therefore be constant through time for short-term feeding studies. The number of animals enclosed in each experimental bottle and the duration of the feeding trial were varied among zooplankton trials in order to obtain measurable feeding rates (large *Daphnia*: 16 animals, 5 hours; medium-sized *Daphnia*: 24 animals, 4.3 hours; small *Daphnia*: 50 animals, 11 hours; *Eubosmina*: 75 animals, 4.5 hours; *Diaptomus* (adults and C-5 copepodites):

18 animals, 13.75 hours). Bottles were rotated on small Plexiglas plankton wheels under reduced (50 µEin m^{-2} s^{-1}), constant light during experiments. Changes in prey abundance were determined from preserved cell counts. Contamination of experimental trials with algae carried over from zooplankton stock cultures was minimal.

For analyses, it was assumed that: (1) algal growth rate in the control bottles and the rate of decline of phytoplankton concentrations in the experimental bottles was exponential, and (2) any enhancement of algal growth rate in the experimental vessels due to nutrient regeneration by zooplankton was negligible (Peters 1984). The equations of Frost (1972) were used to determine animal-specific filtration rates (= clearance rates) and ingestion rates.

Scales and bristles as herbivore avoidance adaptations

Silica-scale-covered chrysophytes can be converted to completely naked cells when maintained as actively growing populations under severely silica-limited conditions (Sandgren & Barlow 1989; Sandgren *et al.* unpublished data; Sutherland & Sandgren, unpublished data). Cell and colony morphology has been observed in these previous experiments to be otherwise normal for the genera and species studied. We used this silica starvation protocol here to generate populations of scale-free *Synura* and *Mallomonas* which were then used in feeding trials run in parallel to those employing phosphorus-limited, normally-scaled populations. Scale-free chrysophytes were produced in semi-continuous cultures identical to those described above with a low-silica (5 µM) supply medium. The scale-free status of these cells was confirmed with light microscopy. The measured equivalent spherical diameter of scale-free cells of the two *Mallomonas* species under severe silica-limitation was very similar to that of the phosphorus-limited cells (*M. caudata*, 34.4 ± 8.5 µm; *M. pseudocoronata*, 17.4 ± 3.0 µm). The effective particle diameter, i.e., what a feeding animal would encounter as a food particle, of these scale- and bristle-free *M. caudata* is, however, approximately 10 µm smaller than that measured from normally-scaled, phosphorus-limited cells. Mean size of *Synura* 'particles' was also considerably smaller (12.3 ± 4.5 µm), in this instance reflecting a change in colony size to a population composed almost entirely of naked unicells in culture. The implications of these size shifts will be considered in comparisons of grazing on scaled and unscaled cells.

In situ *herbivore gradient experiments*

The *in situ* herbivore gradient experiments measure net growth rate of individual prey species in short-term incubations (usually 3–5 days) conducted over an artificial gradient of herbivore abundances (Lehman & Sandgren 1985; Bergquist *et al.* 1985; Elser *et al.* 1987; Elser & Goldman 1991). Whole water is coarsely screened to separate algae from herbivores in a set of field enclosures (usually 4–20 l volume). Then herbivores are reintroduced in varying amounts to create a range of herbivore biomass among enclosures, usually ranging from ×0 to ×8 ambient, and enclosures are incubated in the lake. Algal growth rates over the course of the *in situ* incubation experiment are estimated from changes in cell abundance and plotted as a function of herbivore abundance. Regression analysis is used to estimate the density-dependent net herbivore effect (i.e. sum of grazing/mortality rates and enhanced birth rates from regenerated nutrients). The regression slope estimates the weight-specific clearance rate of the zooplankton (ml μg^{-1} dry wt d^{-1}) on the individual prey species. The magnitude and sign of the slope indicate the edibility of prey species: very edible species will exhibit large, negative net growth rates across the zooplankton gradient (very negative slopes), algae which are generally inedible and benefit from zooplankton-remineralized nutrients exhibit positive slopes, while insignificant ('0') slopes indicate no net effect of herbivores (i.e. losses balance gains over the range of herbivore abundance tested).

Results

We consider the following four exercises as tests of a hypothesis suggesting that herbivory, in particular severe losses to large *Daphnia*, strongly regulates chrysophyte biogeography.

Analysis of the 'Chrysophyte Lakes Database'

A positive correlation exists between total zooplankton and total phytoplankton biomass among the Chrysophyte Lakes Database (Sandgren 1988: fig. 2-17), as has been documented for other lake data sets (McCauley & Kalff 1981). However, the portion of phytoplankton biomass contributed by chrysophytes, i.e. total chrysophyte biomass, in these same lakes was not correlated to zooplankton biomass; it was variable but remained at low levels. Chrysophytes thus become progressively less important in the phytoplankton as zooplankton abundance increases among lakes. This could result from selective regulation of chrysophytes by zooplankton. Analysis of seasonal periodicity patterns of chrysophytes from the literature (Sandgren 1988: table 2-1) suggested that chrysophyte appearance in lakes becomes progressively more brief and is restricted to spring or autumn pulses as lakes become increasingly productive and zooplankton-rich. Such a pattern could result from predation pressure from herbivores such as *Daphnia* which have high food requirements and strongly temperature-dependent development. Repeated decimation of vegetative populations in productive lakes over successive years can eventually result in recruitment failure and local extirpation of species in individual lakes.

We considered the possibility that the strong inverse relationship between phosphorus loading in lakes and chrysophyte importance (discussed in 'Introduction') may reflect an underlying correlation with zooplankton composition, specifically the presence of large *Daphnia* species. The original figure from Sandgren (1988: Fig. 2-6a) is redrawn here (Fig. 12.2*a*); the same relationship was then derived for the subset of those lakes for which zooplankton composition data were available (Fig. 12.2*b*). Clearly, most chrysophyte-dominated lakes lack large-bodied *Daphnia*, while eutrophic lakes with very minor chrysophyte presence have large *Daphnia*. However, a great deal of the variability in this plot can obviously not be attributed to this simple dichotomy. Unfortunately, the data set contains too many missing values for other variables to run even descriptive multivariate statistical analyses.

The relationship between chrysophyte biomass and mean phosphorus concentration in the Chrysophyte Lakes Database more clearly demonstrates two alternative trends: chrysophyte biomass increases with phosphorus availability in some lakes while it remains very low in others (Fig. 12.3*a*). The first trend pertains to lakes with small-bodied zooplankton assemblages while the second trend pertains primarily to lakes with large-bodied *Daphnia* (Fig. 12.3*b*). Again, some individual lakes exhibit variability in chrysophyte biomass which is not attributable to this simple dichotomy in zooplankton composition, and we are unable to perform statistical analyses because of the lack of comparable data for all lakes. One obvious confounding factor is seasonal periodicity of species. Two lakes which appear to run counter to the general trends (circled asterisks) are *Daphnia*-dominated in summer when chrysophytes are absent, but develop significant chrysophyte blooms in autumn after the *Daphnia* populations wane. Changes in species' periodicity patterns along

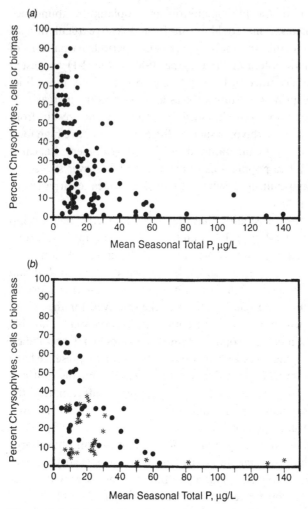

Figure 12.2. Relationships between mean chrysophyte importance in phyto-plankton and mean seasonal total phosphorus, an indicator of lake productivity. Data are from the 'Chrysophyte Lakes Database' (Sandgren 1988). (*a*) Scatterplot of all data available. (*b*) The subset of these same lakes for which information on zooplankton composition was also available. Asterisk, large *Daphnia* lakes; dot, lakes lacking *Daphnia*. Note that chrysophyte-dominated lakes lack *Daphnia* while most phosphorus-rich lakes have them.

the trophic gradient (i.e. summer chrysophytes in oligotrophic lakes lacking *Daphnia* shifting to spring/autumn chrysophytes in mesotrophic lakes with summer *Daphnia*) also largely explains data distribution in the lower left-hand portion of the scatterplot (Fig. 12.3*b*). We believe that the

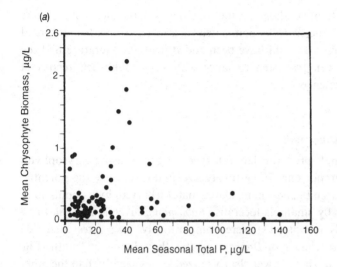

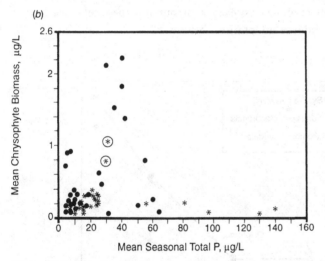

Figure 12.3. Relationship between mean seasonal phosphorus concentration and mean chrysophyte biomass in lakes from the Chrysophyte Lakes Database. (*a*) All data available. Note the two trends among lakes with increasing phosphorus: increasing chrysophytes in some lakes and a flat or decreasing response in other lakes. (*b*) The subset of these same lakes for which information on zooplankton composition was also available. Asterisk, large *Daphnia* lakes; dot, lakes lacking *Daphnia*. Note that lakes with high chrysophyte biomass lack *Daphnia* while most phosphorus-rich lakes have them. The two circled asterisks represent lakes in which large chrysophyte populations develop in autumn/winter after summer *Daphnia* populations decline.

presentation actually understates the consistency of the dichotomy. *Many* additional eutrophic, *Daphnia*-dominated lakes (asterisks) with *no* recorded chrysophyte biomass could have been added from the literature had the database not been restricted to lakes with some published record of chrysophyte presence.

Laboratory grazing studies

'Single prey–single predator' laboratory feeding trials using chrysophytes document the importance of herbivore size, and emphasize the potential importance of large *Daphnia* in regulating natural chrysophyte populations. Ingestion rates by small cladocerans (*Eubosmina*, juvenile *Daphnia*) and small copepods (*Diaptomus*) were comparatively low; rates increased several-fold with classes of *Daphnia* greater than 1 mm in length (Fig. 12.4). *Mallomonas caudata* was clearly grazed less efficiently than the other two chrysophyte prey species by all herbivores (ingestion rates consistently *c.* 50–70% lower). This is probably a function of greater prey cell size rather than an effect of the bristle layer surrounding these cells (see section on herbivore avoidance adaptations below).

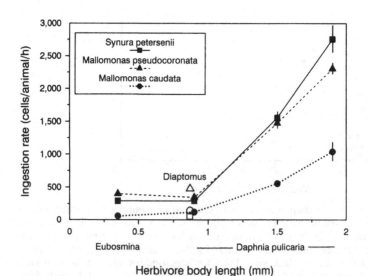

Figure 12.4. Results from one predator–one prey laboratory grazing studies: particle ingestion rates are plotted as a function of herbivore size for each algal prey species. Rates for cladocerans (*Eubosmina, Daphnia*) are indicated by filled symbols; those for *Diaptomus* by open symbols.

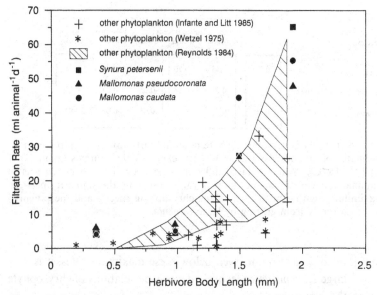

Figure 12.5. Results from one predator–one prey laboratory grazing studies: filtering rates are plotted as a function of herbivore size for each algal prey species. Comparisons are indicated with published studies using similar animals feeding on other freshwater algae.

When converted to filtration and compared with literature values for other phytoplankton species, the chrysophyte values were as high or higher than any others reported for freshwater algae (Fig. 12.5). Chrysophytes would appear to be extremely attractive food items for all of these herbivorous zooplankton. The only other published laboratory grazing rate data for a chrysophyte we found (*Dinobryon* sp.: Sanders & Porter 1990) support this conclusion.

Ingestion rates were also converted to estimated mortality rates for the algal population due to herbivory (G), assuming a very modest herbivore population density of 1 animal ml^{-1} (Fig. 12.6). Such loss rates were compared with estimates of the maximum potential birth rate of the prey algae (μ_{max}) expressed in the same units (from table 2-3 in Sandgren 1988). There was a profound difference between the potential impact of herbivory on chrysophytes by large *Daphnia* versus small-bodied zooplankters. Specific losses associated with *Daphnia* >1 mm in size were much higher for all three chrysophyte species; values for 1.9 mm *Daphnia* approached or exceeded the maximum potential growth rates of these species. Given that chrysophyte population growth is probably often

Species	μ_{MAX}(days^{-1})	G (days^{-1})				
		Daphnia pulicaria			Eubosmina	Diaptomus
		1.9mm	1.5mm	0.9mm	0.4mm	0.9mm
Synura petersenii	0.76	-0.45	-0.20	-0.04	-0.03	-0.01
Mallomonas pseudocoronata	0.50	-0.31	-0.18	-0.05	-0.04	-0.05
Mallomonas caudata	0.30	-0.38	-0.32	-0.04	-0.01	-0.02

Figure 12.6. Comparison of chrysophyte potential birth rates (μ_{max}) to conservative estimates of grazing loss rate (G) to the experimental animals (assuming 1 animal l^{-1}; 15 °C). Note the large differences between the two boxed sets of large-animal versus small-animal grazing rates; also note the similarity between the maximum potential birth rates of the chrysophyte species and the potential impact of mortality from grazing by large *Daphnia*.

nutrient-limited, and therefore well below these maximum rates, it is very likely that large *Daphnia* can effectively depress populations of chrysophyte species in this size range.

Scales and bristles as herbivore avoidance adaptations

Populations of normally-scaled and scale-free cells were fed to each of the five herbivore categories in parallel trials (Fig. 12.7). All three size classes of the generalist filter feeder *Daphnia* exhibited higher filtering rates on the normal, scale-covered cells of both *Mallomonas* species. The tendency is the same for the small *Synura* colonies tested, although the differences are not statistically significant in this case. *Synura* results are also confounded by the distinct size shift to smaller mean food particle sizes in the scale-free population (12.3 μm unicells vs. 21.2 μm small colonies); both sizes are, however, still near the center of the range for particle capture by *Daphnia*. *Eubosmina*, a small, dual-mode feeding cladoceran, exhibited a preference for scale-free *M. caudata* but no preference between scaled and unscaled cells of the other species. *M. caudata* cells with normal scale and bristle layers are apparently too large for this animal to handle (Fig. 12.4), but the scale-free cells are not. The copepod *Diaptomus* exhibited significant differential feeding preferences for *Synura* but not for *M. pseudocoronata* (*M. caudata* was not tested). Naked *Synura* cells were strongly preferred by *Diaptomus*, but these results are again confounded by the differences in particle size of the comparative prey populations.

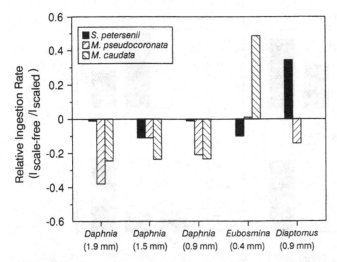

Figure 12.7. Comparison of ingestion rates on normally-scaled and scale-free chrysophytes. Results are presented as means of triplicate trials performed simultaneously. A positive ratio indicates preference for scale-free cells while a negative ratio indicates a preference for normal, scale-covered cells. The two prey morphotypes of each species were prepared by maintaining growing cell populations alternatively under phosphorus or silica limitation.

In situ *herbivore gradient experiments*

Sandgren (1988: table 2-4) summarized the results from 48 individual species' responses to this type of experiment as reported in two previous publications (Lehman & Sandgren 1985; Bergquist *et al.* 1985). We have added data from two more recent papers (Elser *et al.* 1987; Elser & Goldman 1991) and have summarized results in frequency histograms from a total of 90 chrysophyte species-specific responses to herbivore assemblages dominated by either small or large cladocerans (Fig. 12.8). Conclusions remain similar to those previously suggested; they only partially support our hypothesis suggesting consistently strong, negative effects of *Daphnia* on all chrysophytes. Small chrysomonads were grazed by both small and large cladocerans (top panels); rates of grazing are higher for large *Daphnia* (rates not shown). As predicted, larger chryso-monads and colonial forms demonstrated an apparent refuge from predation from small cladocerans (no ' − ' slopes) and occasionally benefited from the presence of such grazers (' + ' slopes). Against predictions, positive interactions were also recorded for larger chrysophytes (*Uroglena, Dinobryon, Synura*) in some experiments with large *Daphnia*.

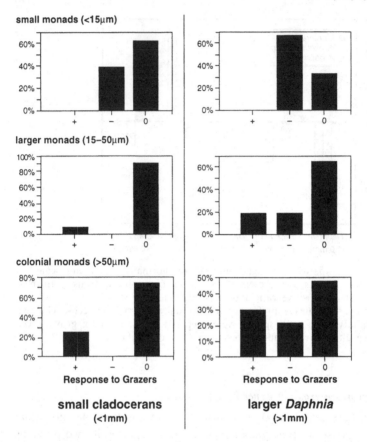

small monads (<15μm)

larger monads (15–50μm)

colonial monads (>50μm)

Response to Grazers

Response to Grazers

small cladocerans
(<1mm)

larger *Daphnia*
(>1mm)

Figure 12.8. Summary histograms of grazing slope trajectories in Lehmann–Sandgren type herbivore gradient experiments collected from the literature. Histogram categories: +, positive slope between herbivore abundance and algal species growth rate (i.e. net benefit to alga from presence of herbivores); −, negative slope (alga shows net losses); 0, no significant slope (net effect of interaction is zero).

This interpretation is dependent on the threshold chosen to designate 'large' animals: each of these same taxa was effectively grazed by 'large' *Daphnia* in other experiments included in this literature summary. We can say that no chrysophyte examined to date in this manner has a consistent refuge in large size from *Daphnia* greater than 2 mm in length. The many '0' responses in Fig. 12.8 for interactions with large *Daphnia* also seemingly fail to support our predictions by suggesting a balance between birth and loss rates during the incubation. However, such results are not definitive because they could also result from: (1) herbivore abundance gradients inadequate to resolve a functional prey response, (2) low growth rates of larger chrysophytes (i.e., no response over incubation time), or (3) poor enumeration accuracy at low algal abundances.

Discussion

Importance of research

One of the major challenges for the discipline of ecology is to explain the patterns of animal and plant distribution and abundances seen in nature (Hunter & Price 1992). Biogeographic studies have generally tended to emphasize abiotic factors (climate, topography, soil type, substrate pH, etc.) which interact with species' tolerance limits (e.g. Carlton 1990; Hengveld 1990; Duigan & Kovach 1991; Hill 1991) or to emphasize resource availability (MacArthur 1972; Diamond 1975). Most recent ecological studies, on the other hand, have most strongly emphasized biotic factors (competition, predation) in ordering biological communities. Most such ecological studies (thousands of papers!) have obvious biogeographic implications, but fail to evaluate results over biogeographically meaningful scales. Examples of biotic factors such as predator–prey interactions controlling species' biogeographic distribution patterns are therefore relatively rare. Those that do exist are often well known and dramatic, predominantly involving exotic (introduced) predators and naive prey (human-caused extinctions of food animals: Vermeij 1986; Nile perch and Lake Victoria cichlids: Witte *et al.* 1992; Sea lamprey and marine fish effects on Great Lakes native fishes: Christie 1974; introduced *Mysis* effects on daphnids in western US lakes: Richards *et al.* 1975; cactus moth control of invasive cactus in Australia: Monro 1967; introduced sea otter control of sea urchins in kelp forest management: Duggins 1980, Lubina & Levin 1988; efficacy of insect biological control agents: Waage & Greathead 1985, Julien 1982; decimation of native flora and fauna of islands by introduced herbivores and predators: Stone 1985, Vermeij 1986, Loope & Mueller-Dombois 1989). But how general are these dynamics? How important is predation/herbivory in controlling the biogeographic distributional patterns of prey species which have had the opportunity to evolve anti-predation mechanisms over evolutionary time scales? Existing examples fitting this criterion (e.g. Brooks & Dodson 1965; Vermeij 1978; Gilbert 1988) are too few to answer this basic ecological question.

   Our investigation of chrysophyte–*Daphnia* interactions thus provides a significant new test of a general hypothesis suggesting that strong predator–prey interactions can resolve as biogeographical distribution patterns of prey species. Although we recognize that many biotic and abiotic factors contribute to the distribution of these algae in nature, some factors are far more important than others. The problem is to assess the *relative contributions* of different factors whose interactions with species overlap in space and time (Bocard *et al.* 1992; Hunter & Price 1992). Sometimes a single factor can be identified that exerts tremendous influence. Examples of such situations are important because they identify strong interactions that dictate how natural communities are ordered. Identification of strong interactions also provides a 'short cut' to understanding the organism's ecology and can facilitate prediction of community responses to perturbations. This study represents the first focused attempt to test predation/herbivory as a pivotal factor in phytoplankton biogeography; it is also unique in suggesting that distributions of an *entire group* of related organisms (i.e. planktonic chrysophytes) is dependent to a large degree on a single keystone predator.

Strength of evidence and future considerations

Laboratory studies

The one prey–one predator grazing studies presented here support the predation-control hypothesis by providing evidence for the required mechanism: strong, selective predation pressure which increases with herbivore size (Figs. 12.4, 12.5). The high filtering rates on all three chrysophyte species tested in comparison with other freshwater algae suggest that mortality due to herbivory in general must be a very significant factor in chrysophyte population dynamics. The particularly strong potential impact of large *Daphnia* is confirmed by comparison of chrysophyte birth rates and grazing-based mortality rates (Fig. 12.6). Clearly, more data on herbivory are necessary to support this mechanism as a principal factor in regulating biogeography of chrysophytes in general. Most profitable would be additional one prey–one predator studies with a greater diversity of chrysophyte prey. For the predation-control hypothesis to be valid, all chrysophytes must suffer high grazer-mediated mortality compared with their potential birth rates. Morphologically very large chrysophytes, such as large colonial clones of *Synura*, *Dinobryon* and *Uroglena*, must be tested. Also important would be laboratory studies using single predators and mixed prey assemblages that include both chrysophyte and non-chrysophyte species. Herbivores, particularly large *Daphnia*, should exhibit higher selectivity for chrysophytes than other algae in mixed assemblages.

Results from the test of chrysophyte scales and bristles as anti-herbivore adaptations are surprising, particularly for *Daphnia*. Results for the calanoid copepod, a notoriously selective feeder, are equivocal. However, the small-bodied cladoceran *Eubosmina* exhibited the expected response: feeding on cells within its documented size range for particle capture but not on large, bristle-covered *Mallomonas caudata* cells. Surprisingly, *Daphnia* of all sizes consistently preferred the scale-covered particles over naked cells of the same prey species. We conclude that spines and bristles, may be an effective anti-herbivore mechanism if they make particles too large to be handled, but the added surface 'texture' would seem actually to increase mortality from *Daphnia*, which has a very broad particle capture size range. These data, therefore, support the proposed strong effect of *Daphnia* on chrysophytes and the proposed refuge in large particle size from small-bodied herbivores. Additional studies should certainly be done with larger colonial chrysophytes, particularly *Chrysosphaerella* and other clones or species of *Synura*.

In situ *grazing studies*

This type of experiment seems to be a critical link in the extrapolation of laboratory feeding studies to population interactions in whole lakes over longer periods of time. The existing data support the predictions of persistent heavy predation on small chrysophytes and of a differential response of morphologically large chrysophytes to assemblages of small versus large herbivores (Fig. 12.8). However, if the hypothesis of consistent negative effects of large *Daphnia* on *all* chrysophytes is to be supported, an explanation for the net '0' responses of some chrysophyte species in the presence of large *Daphnia* needs to be found. Clearly, a better analytical approach would be to replace the 'categorical' presentation used here with actual herbivore and prey size data so that relationships can be evaluated on the basis of size-dependent grazing functions (e.g. Figs. 12.3, 12.4). Additional experimental data of this type from lakes of different trophic states would help to expand the database. Also, an expansion in both spatial and temporal scale from these short-term experiments is critical to resolving the complex interactions that determine actual species abundances of chrysophytes in nature. Month- or season-long manipulations of herbivory and nutrient availability in very large artificial enclosures or in natural ponds and lakes are required.

One of the best documented sets of whole-lake manipulations that contain inter-annual data on chrysophyte responses are the Cascade Lakes experiments of Carpenter & Kitchell (1993). Fish manipulations in Tuesday Lake that resulted in a striking shift in zooplankton from dominance by small herbivores to large *Daphnia* (probably a novel event in this lake's history!) resulted in a clear negative impact on chrysophytes, including morphologically large colonial forms (Carpenter *et al.* 1993). This is consistent with the chrysophyte predation-control hypothesis. Attempts to remove large *Daphnia* in Peter Lake by stocking minnows was a less successful biomanipulation; *Daphnia* seasonal dynamics and abundance were altered but large forms were not eliminated from the lake. In this case, there was no significant negative correlation between large colonial chrysophytes and mean cladoceran length, just as might be expected on the basis of our hypothesis (Carpenter *et al.* 1993).

Lake database analyses

The Chrysophyte Lakes Database provides tantalizing evidence for a strong chrysophyte–*Daphnia* interaction with important biogeographic consequences. Particularly suggestive is the dichotomy in trends for chrysophyte biomass between lakes with and without large *Daphnia* (Fig. 12.3). However, the strength of these relationships, i.e. the relative significance of the herbivore factor alone in dictating chrysophyte distributions, cannot be statistically evaluated because of inadequacies of the database. Moreover, because this is a literature-based data set, these inadequacies cannot be alleviated; the missing data are not available. Future tests of our hypothesis must ultimately consider chrysophyte–*Daphnia* interactions over broad spatial scales if the predation-control mechanism we propose is to have biogeographic consequences. An alternative approach to testing our hypothesis, which retains the necessary broad geographic scale, is to use regional databases collected with a standardized sampling protocol and with quantitative assessment of phytoplankton and zooplankton. It would also obviously be attractive to utilize multi-lake data sets which incorporate periodicity information for the plankton. Additional emphasis on eutrophic lakes with and without large *Daphnia* would be of particular importance. Such comparisons could be made by experimental biomanipulation of the fish community (piscivore versus planktivore dominance), but the manipulation must be of sufficient (multi-year) duration to affect both the vegetative cell populations and the benthic propagule 'bank' (see below). An obvious alternative to water column databases would be to employ regional sets of surficial lake sediment samples, as has been done recently to generate regional 'calibration sets' for chrysophyte and diatom microfossils used in assessing the impact of acid precipitation and other limnological variables (Charles 1985; Charles & Smol 1988; Dixit *et al.* 1989, 1990; Rybak *et al.* 1991; reviewed in Smol, this volume).

The predation-control mechanism we invoke requires effective and persistent reduction of chrysophyte algal populations to such an extent that their perennation strategy fails: that is, few new cyst propagules are formed and the established cyst propagule banks in the sediment become depleted over years through unsuccessful excystment (recruitment) attempts or through burial. Reliance by chrysophytes on cyst propagules for perennation (i.e. general lack of vegetative overwintering), the density-dependent dynamics of chrysophyte cyst production (Sandgren 1986; Sandgren & Flanagin 1986; Sandgren 1991) and the relatively short survivorship of mature chrysophyte cysts (Sandgren 1988) all suggest that local extirpation of chrysophytes in individual lakes is a likely outcome of severe predation pressure on vegetative populations.

Paleolimnological investigations of sediment cores should provide pertinent data because the propagule banks of all species and vegetative cell coverings of scale-covered species are well preserved as siliceous microfossils. However, considerations of predator–prey or herbivore–prey cycles in long-term (decades, centuries) records afforded by lake sediments are very few (Kerfoot 1974; Kitchell & Kitchell 1980; Kitchell & Carpenter 1987; Leavitt *et al.* 1989; Carpenter & Leavitt 1991). Leavitt *et al.* (1989) used fossilized pigment profiles in the varved sediments of Peter and Paul Lakes to investigate the effect of phytoplankton community composition of oscillations between large *Daphnia* dominance and small herbivore dominance over a period of 40 years; oscillations were due to episodic, unsuccessful fish stocking efforts. Leavitt and coworkers (1989) detected a direct correlation between *Daphnia* and chrysophyte abundances: the opposite of what our hypothesis would predict. These authors hypothesized that *Daphnia* grazing in the epilimnion decreased phytoplankton abundance and increased light penetration, providing the opportunity for the development of large metalimnetic chrysophyte populations. From the perspective of the predator regulation hypothesis we advance here, it is crucial to re-examine such sedimentary records using a species-level metric such as chrysophyte cysts or scales that would allow us to distinguish between the dynamics of epilimnetic and metalimnetic chrysophytes. We would predict that *epilimnetic* chrysophyte abundance will be inversely correlated with episodes of *Daphnia* dominance. There is certainly a need for additional paleolimnological data sets that couple long-term dynamics of chrysophytes and zooplankton (see Smol, this volume).

Shifting emphasis between abiotic and biotic factors in the regulation of species growth and distributions has been an important aspect of the historical development of plankton ecology. These diverse views form part of the background that has fueled the recent 'bottom up' versus 'top down' debate regarding regulations of biomass and productivity in aquatic ecosystems (McQueen *et al.* 1989; DeMelo *et al.* 1992; Carpenter & Kitchell 1992; Power 1992). Several authors have pointed out that such perspectives are neither contradictory nor mutually exclusive (Sterner 1990; Carpenter & Kitchell 1992; Hunter & Price 1992; Power 1992). We concur, and regarding the *distributions* of planktonic organisms, we propose that the differences can largely be resolved as a disparity between *potential* species distributions, dictated by interaction of physiological tolerances with abiotic factors, and *realized* distributions, which represent further restrictions resulting from biotic interactions in many habitats (i.e. Vandermeer 1972; Connell 1975). These biotic interactions (competition, predation) require more serious consideration in phytoplankton biogeographic analyses.

Summary

The merits of a hypothesis suggesting that the resolved biogeographic distribution patterns of chrysophyte algae are primarily regulated by strong, selective herbivory is evaluated. This study represents the first focussed attempt to test predation/herbivory as a pivotal factor in phytoplankton biogeography; it is also unique in suggesting that distributions of an entire group of related organisms (i.e. planktonic chrysophytes) are dependent primarily on a single keystone predator. Existing data from a biogeographic database and *in situ* herbivore grazing studies are summarized, and new results from laboratory grazing studies are presented. The high filtering rates demonstrated in laboratory studies for herbivorous zooplankton on chrysophyte prey suggest that grazing-dependent mortality must be a very significant factor in natural chrysophyte population dynamics. Large individuals of *Daphnia* are particularly potent grazers on these algae, suggesting that the zooplankton composition of lakes can have a selective impact on the success of chrysophyte species. The siliceous scale and bristle layers of some chrysophytes were demonstrated to affect prey selectivity by a small cladoceran (*Eubosmina coregoni*) but to have no effect on prey selectivity of a calanoid copepod (*Diaptomus ashlandi*) or of three size classes of a large *Daphnia* species (*D. pulicaria*). Published results from *in situ* herbivore grazing studies demonstrate a refuge from grazing for morphologically large chrysophyte species in association with small-bodied herbivores, but no consistent refuge with size in the presence of large (>2 mm) *Daphnia*. Examination of a large database composed of lakes containing chrysophyte algae suggests: (1) that chrysophyte biomass patterns are consistent with expectations of strong regulation by herbivores, and (2) that chrysophyte biomass trends associated with increasing lake productivity are different for lakes with and without large *Daphnia*. The predation-regulation hypothesis remains a viable explanation for chrysophyte biogeographical distribution patterns. Suggestions are made for additional studies critical to further testing of this hypothesis.

Acknowledgment

This work was funded by the University of Wisconsin Sea Grant Institute under grants from the National Sea Grant College Program, National Oceanic and Atmospheric Administration, US Department of Commerce, and from the State of Wisconsin. Federal Grant NA90AA-D-SG469, project R/RL-39.

References

Allan, J.D. 1976. Life history patterns in zooplankton. *Am. Nat.* **110**: 165–80.
Andersen, R.A. 1987. Synurophyceae classic nov., a new class of algae. *Am. J. Bot.* **74**: 337–53.
Bergquist, A.M., Carpenter, S.R. & Latino, J.C. 1985. Shifts in phytoplankton size structure and community composition during grazing by contrasting zooplankton assemblages. *Limnol. Oceanogr.* **30**: 1037–45.
Bleiwas, A.H. & Stokes, P.M. 1985. Collection of large and small food particles by *Bosmina. Limnol. Oceanogr.* **30**: 1090–2.
Bocard, D., Legendre, P. & Drapeau, P. 1992. Partialling out the spatial component of ecological variation. *Ecology* **73**: 1045–55.
Brooks, J.L. & Dodson, S.I. 1965. Predation, body size, and the composition of plankton. *Science* **150**: 28–35.
Burns, C.W. 1968. Direct observation of mechanisms regulating feeding behavior of *Daphnia* in lake water. *Int. Rev. Ges. Hydrobiol.* **53**: 83–100.
Butler, N.M., Suttle, C.A. & Neill, W.E. 1989. Discrimination by freshwater zooplankton between single algal cells differing in nutritional status. *Oecologia* **78**: 368–77.

Carlton, T.J. 1990. Variation in terricolous bryophyte and macroliche vegetation along primary gradients in Canadian boreal forests. *J. Veg. Sci.* **1**: 585–94.

Carney, H.J. 1990. A general hypothesis for the strength of food web interactions in relation to trophic state. *Verh. Int. Ver. Limnol.* **24**: 487–492.

Carpenter, S.R. & Kitchell, J.F. 1992. Trophic cascade and biomanipulation: interface of research and management. A reply to the comment of DeMelo *et al. Limnol. Oceanogr.* **37**: 208–13.

Carpenter, S.R. & Kitchell, J.F. 1993. *The Trophic Cascade in Lakes.* Cambridge University Press, Cambridge.

Carpenter, S.C. & Leavitt, P.R. 1991. Temporal variation in a paleolimnological record arising from a trophic cascade. *Ecology* **72**: 277–85.

Carpenter, S.R., Morrice, J.A., Elser, J.J., St. Amand, A. & McKay, N.A. 1993. Phytoplankton community dynamics. In: Carpenter, S.R. & Kitchell, J.F. (eds.) *The Trophic Cascade in Lakes.* Cambridge University Press, Cambridge. pp. 189–209.

Charles, D.F. 1985. Relationships between surface sediment diatom assemblages and lake water characteristics in Adirondack lakes. *Ecology* **66**: 994–1011.

Charles, D.F. & Smol, J.P. 1988. New methods for using diatoms and chrysophytes to infer past pH of low-alkalinity lakes. *Limnol. Oceanogr.* **33**: 1451–62.

Christie, W.J. 1974. Changes in fish species composition of the Great Lakes. *J. Fish. Res. Bd. Can.* **31**: 827–54.

Connell, J.H. 1975. Some mechanisms producing structure in natural communities: a model and evidence from field experiments. In: Cody, M.L. & Diamond, J.M. (eds.) *Ecology and Evolution of Communities.* Belknap Press, Cambridge, Mass. pp. 460–90.

Cowles, T.J., Olsen, R.J. & Chisholm, S.W. 1988. Food selection by copepods: discrimination on the basis of food quality. *Mar. Biol.* **100**: 41–9.

Cumming, B.F., Smol, J.P. & Birks, H.J.B. 1992. Scaled chrysophytes (Chrysophyceae and Synurophyceae) from Adirondack (N.Y., USA) drainage lakes and their relationship to measured environmental variables, with special reference to lakewater pH and labile monomeric aluminum. *J. Phycol.* **28**: 162–78.

Cyr, H. & Pace, M.L. 1993. Magnitude and patterns of herbivory in aquatic and terrestrial ecosystems. *Nature* **361**: 148–50.

DeMelo, R., France, R. & McQueen, D.J. 1992. Biomanipulation: hit or myth? *Limnol. Oceanogr.* **37**: 192–207.

DeMott, W.R. 1982. Feeding selectivities and relative ingestion rates of *Daphnia* and *Bosmina. Limnol. Oceanogr.* **27**: 518–27.

DeMott, W.R. 1985. Relations between filter mesh-size, feeding mode, and capture efficiency for caldocerans feeding on ultrafine particles. *Ergeb. Limnol.* **21**: 125–34.

DeMott, W.R. 1986. The role of taste in food selection by freshwater zooplankton. *Oecologia* **69**: 334–40.

DeMott, W.R. 1988. Discrimination between algae and artificial particles by freshwater and marine copepods. *Limnol. Oceanogr.* **33**: 397–408.

DeMott, W.R. 1989. Optimal foraging behavior as a predictor of chemically mediated food selection by suspension-feeding copepods. *Limnol. Oceanogr.* **34**: 140–54.

DeMott, W.R. & Kerfoot, W.C. 1982. Competition among cladocerans: nature of the interaction between *Bosmina* and *Daphnia. Ecology* **63**: 1949–66.

Diamond, J.M. 1975. Assembly of species communities. In: Cody, M.L. & Diamond, J.M. (eds.) *Ecology and Evolution of Communities.* Belknap Press, Cambridge, Mass. pp. 342–443.

Dixit, S.S., Dixit, S.A. & Smol, J.P. 1989. Relationship between chrysophyte assemblages and environmental variables in seventy-two Sudbury Lakes as examined by Canonical Correspondence Analysis (CCA). *Can J. Fish. Aquat. Sci.* **46**: 1667–76.

Dixit, S.S., Smol, J.P., Anderson, D.S. & Davis, R.B. 1990. Utility of scaled chrysophytes in predicting lakewater pH in northern New England lakes. *J. Paleolimnol.* **3**: 269–86.

Duggins, D.O. 1980. Kelp beds and sea otters: an experimental approach. *Ecology* **61**: 447–53.

Duigan, C.A. & Kovach, W.L. 1991. A study of the distribution and ecology of littoral freshwater chydorid (Crustaceae, Cladocera) communities in Ireland using multivariate analyses. *J. Biogeogr.* **18**: 267–80.

Einsle, U. 1983. Long-term changes in planktonic associations of crustaceans in Lake Constance and adjacent waters and their effects on competitive situations. *Hydrobiologia* **106**: 127–43.

Eloranta, P. 1986. Phytoplankton structure in different lake types in central Finland. *Holarctic Ecol.* **9**: 214–24.

Elser, J.J. & Goldman, C.R. 1991. Zooplankton effects on phytoplankton in lakes of contrasting trophic status. *Limnol. Oceanogr.* **36**: 64–90.

Elser, J.J., Goff, N.C., MacKay, N.A., St. Amand, A.L., Elser, M.M. & Carpenter, S.R. 1987. Species-specific algal responses to zooplankton: experimental and field observations in three nutrient-limited lakes. *J. Plankton Res.* **9**: 699–717.

Elster, H.-J. & Schwoerbel, I. 1970. Beiträge zur Biologie und Populationsdynamik der Daphnien in Bodensee. *Arch. Hydrobiol. Beih.* **38**: 18–72.

Ferguson, A.J.D., Thompson, J.M. & Reynolds, C.S. 1982. Structure and dynamics of zooplankton communities maintained in closed systems, with special reference to the algal food supply. *J. Plankton Res.* **4**: 523–43.

Frost, B.W. 1972. Effects of the size and concentration of food particles on the feeding behavior of the marine planktonic copepod *Calanus pacificus. Limnol. Oceanogr.* **17**: 805–15.

Gannon, J.E. & Stemberger, R.S. 1978. Zooplankton (especially crustaceans and rotifers) as indicators of water quality. *Trans. Am. Microsc. Soc.* **97**: 16–35.

Geller, W. 1975. Food ingestion of *Daphnia pulex* as a function of food concentration, temperature, animal's body length and hunger. *Arch. Hydrobiol. (Suppl.)* **48**: 47–107.

Gerritsen, J. & Porter, K.G. 1982. The role of surface chemistry in filter feeding by zooplankton. *Science* **216**: 1225–7.

Gerritsen, J., Porter, K.G. & Strickler, J.R. 1988. Not by sieving alone: observations of suspension-feeding *Daphnia. Bull. Mar. Sci.* **43**: 366–76.

Gilbert, J.J. 1988. Suppression of rotifer populations by *Daphnia*: a review of the evidence, the mechanisms, and the effects on zooplankton community structure. *Limnol. Oceangr.* **33**: 1286–303.

Gilbert, J.J. & Bogdan, K.G. 1981. Selectivity of *Polyarthra* and *Keratella* for flagellate and aflagellate cells. *Verh. Int. Ver. Limnol.* **21**: 1515–21.

Gilbert, J.J. & Bogdan, K.G. 1984. Rotifer grazing: *in situ* studies on selectivity and rates. In: Meyers, D.J. & Strickler, J.R. (eds.) *Trophic Interactions with Aquatic Ecosystems.* AAAS Special Symposium 85.

298   *Chrysophyte algae*

Gliwicz, Z.M. 1969. The food sources of lake zooplankton. *Ekol. Pol., Ser. B* **15**: 205–23.

Gliwicz, Z.M. 1980. Filtering rates, food size selection and feeding rates in cladocerans: another aspect of interspecific competition in filter-feeding zooplankton: In: Kerfoot, W.C. (ed.) *Evolution and Ecology of Zooplankton Communities.* University Press of New England, Hanover, N.H. pp. 282–91.

Gophen, M. & Geller, W. 1984. Filter mesh size and food particle uptake by *Daphnia. Oecologia* **64**: 408–12.

Hartig, J.H., Jude, D.J. & Evans, M.S. 1982. Cyclopoid predation on Lake Michigan fish larvae. *Can. J. Fish. Aquat. Sci.* **39**: 1563–8.

Hengeveld, R. 1990. *Dynamic Biogeography.* Cambridge University Press, Cambridge.

Hibberd, D.J. 1976. The ultrastructure and taxonomy of the Chrysophyceae and Prymnesiophyceae (Haptophyceae): a survey with some new observations on the ultrastructure of the Chrysophyceae. *Bot. J. Linn. Soc.* **72**: 55–80.

Hill, M.O. 1991. Patterns of species distributions in Britain elucidated by canonical correspondence analysis. *J. Biogeogr.* **18**: 247–55.

Horn, W. 1985. Investigations into the food selectivity of the planktonic crustaceans *Daphnia hyalina, Eudiaptomus gracilis* and *Cyclops vicinus. Int. Rev. Ges. Hydrobiol.* **70**: 603–12.

Hunter, M.D. & Price, P.W. 1992. Playing chutes and ladders: heterogeneity and the relative roles of bottom-up and top-down forces in natural communities. *Ecology* **73**: 724–32.

Infante, A. & Litt, A.H. 1985. Differences between two species of *Daphnia* in the use of 10 species of algae in Lake Washington. *Limnol. Oceangr.* **30**: 1053–9.

Julien, M.H. 1982. *Biological Control of Weeds: A World Catalogue of Agents and their Target Weeds.* Commonwealth Agricultural Bureau.

Kennedy, K.C. 1984. The influence of cell size and the colonial habit on phytoplankton growth and nutrient uptake kinetics: a critical evaluation using a phylogenetically related series of volvocene green algae. MS thesis, University of Texas-Arlington. 109pp.

Kerfoot, W.C. 1974. Net accumulation rates and the history of cladoceran communities. *Ecology* **55**: 51–61.

Kerfoot, W.C., DeMott, W.R. & DeAngelis, D.L. 1985. Interactions among cladocerans: food limitation and exploitive competition. *Arch. Hydrobiol. Beih.* **21**: 431–52.

Kilham, S.S. 1978. Nutrient kinetics of freshwater phytoplankton algae using batch and semicontinuous methods. *Mitt. Int. Ver. Limnol.* **21**: 147–57.

Kitchell, J.F. & Carpenter, S.R. 1987. Piscivores, planktivores, fossils, and phorbins. In: Kerfoot, W.C. & Sih, A. (eds.) *Predation: Direct and Indirect Impacts on Aquatic Communities.* University Press of New England, Hanover, N.H. pp. 132–46.

Kitchell, J.A. & Kitchell, J.F. 1980. Size-selective predation, light transmission, and oxygen stratification: evidence from the recent sediments of manipulated lakes. *Limnol. Oceangr.* **25**: 389–402.

Knisely, K. & Geller, W. 1986. Selective feeding of four zooplankton species on natural lake phytoplankton. *Oecologia* **69**: 86–94.

Lampert, W. & Muck, P. 1985. Multiple aspects of food limitation in zooplankton communities: the *Daphnia–Eudiaptomus* example. *Arch. Hydrobiol. Beih.* **21**: 311–22.

Leavitt, P.R., Carpenter, S.C. & Kitchell, J.F. 1989. Whole lake experiments: the annual record of fossil pigments and zooplankton. *Limnol. Oceanogr.* **34**: 700–17.

Lehman, J.T. 1976. Ecological and nutritional studies on *Dinobryon*: seasonal periodicity and the phosphorus toxicity problem. *Limnol. Oceanogr.* **21**: 646–58.

Lehman, J.T. & Sandgren, C.D. 1985. Species-specific rates of growth and grazing loss among freshwater algae. *Limnol. Oceanogr.* **30**: 34–46.

Li, J.L. & Li, H.W. 1979. Species-specific factors affecting predator–prey interactions of the copepod *Acanthocyclops vernalis* with its natural prey. *Limnol. Oceangr.* **24**: 613–26.

Liebold, M.A. 1989. Consumer–resource interactions in environmental gradients. *Am. Nat.* **134**: 922–49.

Loope, L.L. & Mueller-Dombois, D. 1989. Characteristics of invaded islands, with special reference to Hawaii. In: Drake, J.A. *et al.* (eds.) *Biological Invasions: A Global Perspective.* Wiley, New York. pp. 257–80.

Lubina, J.A. & Levin, S.A. 1988. The spread of a reinvading species: range expansion in the California sea otter. *Am. Nat.* **131**: 526–43.

MacArthur, R.H. 1972. *Geographical Ecology.* Harper and Row, New York.

McCauley, E. & Kalff, J. 1981. Empirical relationships between phytoplankton and zooplankton biomass in lakes. *Can. J. Fish. Aquat. Sci.* **38**: 458–63.

McNaught, D.C. 1975. A hypothesis to explain the succession from calanoids to cladocerans during eutrophication. *Verh. Int. Ver. Limnol.* **19**: 724–31.

McNaught, D.C., Griesmer, D. & Kennedy, M. 1980. Resource characteristics modifying selective grazing in copepods. In: Kerfoot, W.C. (ed.) *Evolution and Ecology of Zooplankton Communities.* University Press of New England, Hanover, N.H. pp. 292–8.

McQueen, D.J., Post, J.R. & Mills, E.L. 1986. Trophic relationships in freshwater pelagic food webs. *Can. J. Fish. Aquat. Sci.* **43**: 1571–81.

McQueen, D.J., Johannes, M.R.S., Post, J.R., Stewart, T.J. & Lean, D.R.S. 1989. Bottom-up and top-down impacts on freshwater community structure. *Ecol. Monogr.* **59**: 289–309.

McQueen, D.J., Mills, E.L., Forney, J.L., Johannes, M.R.S. & Post, J.R. 1992. Trophic level relationships in pelagic food webs: comparisons derived from long-term data sets for Oneida Lake, New York (USA) and Lake St George, Ontario (Canada). *Can. J. Fish. Aquat. Sci.* **49**: 1588–96.

Monro, J. 1967. The exploitation and conservation of resources by populations of insects. *J. Animal Ecol.* **36**: 531–47.

Nadin-Hurley, C.M. & Duncan, A. 1976. A comparison of daphnid gut particles with sestonic particles in two Thames Valley reservoirs throughout 1970 and 1971. *Freshwater Biol.* **6**: 109–23.

Neill, W.E. 1984. Regulation of rotifer densities by crustacean zooplankton in an oligotrophic montane lake in British Columbia. *Oecologia* **48**: 164–77.

Nicholls, K.H. 1976. The phytoplankton of the Kawartha Lakes (1972 and 1976). In: *Kawartha Lakes Management Study.* Ontario Ministry of the Environment.

Nicholls, K.H., Heintsch, L., Carney, E., Beaver, J. & Middleton, D. 1986. Some effects of phosphorus loading reductions on the phytoplankton of the Bay of Quinte, Lake Ontario. *Can. Spec. Publ. Fish. Aquat. Sci.* **86**.

Peters, R.H. 1984. Methods for the study of feeding, filtering and assimilation of zooplankton. In: Downing, J.A. & Rigler, F.H. (eds.) *A Manual for the Assessment of Secondary Production in Fresh Waters.* Blackwell Scientific, Oxford. pp. 336–412.

Porter, K.G. 1973. Selective grazing and differential digestion of algae by zooplankton. *Nature* **244**: 179–80.

Porter, K.G. 1977. The plant–animal interface in freshwater systems. *Am. Sci.* **65**: 159–70.

Porter, K.G., Gerritsen, J. & Orcutt, J.D., Jr 1982. The effect of food concentration on swimming patterns, feeding behavior, ingestion, assimilation and respiration by *Daphnia. Limnol. Oceanogr.* **27**: 935–49.

Poulet, S.A. & Marsot, P. 1978. Chemosensory grazing by marine calanoid copepods (Arthropoda: Crustacea). *Science* **200**: 1403–5.

Pourriot, R. 1977. Food and feeding habits of Rotifera. *Ergeb. Limnol.* **8**: 243–60.

Power, M.E. 1992. Top-down and bottom-up forces in food webs: do plants have primacy? *Ecology* **73**: 733–46.

Reynolds, C.S. 1984. *The Ecology of Freshwater Phytoplankton.* Cambridge University Press, Cambridge.

Richards, P.J., Goldman, C.R., Frantz, T.C. & Wickwire, R. 1975. Where have all the *Daphnia* gone? The decline of a major cladoceran in Lake Tahoe, California. *Verh. Int. Ver. Limnol.* **19**: 835–42.

Richman, S., Bohon, S.A. & Robbins, S.E. 1980. Grazing interactions among freshwater calanoid copepods. In: Kerfoot, W.C. (ed.) *Evolution and Ecology of Zooplankton Communities.* University Press of New England, Hanover, N.H. pp. 219–33.

Richman, S., Branstrator, D.K. & Huber-Villegas, M. 1990. Impact of zooplankton grazing on phytoplankton along a trophic gradient. In: Tilzer, M.M. & Serruya, C. (eds.) *Large Lakes: Ecological Structure and Function.* Springer, Berlin. pp. 592–614.

Romanovsky, Y.E. 1985. Food limitation and life history strategies in cladoceran crustaceans. *Ergeb. Limnol.* **21**: 363–72.

Rybak, M., Rybak, I. & Nicholls, K.H. 1991. Sedimentary chrysophycean cyst assemblages as paleoindicators in acid sensitive lakes. *J. Paleolimnol.* **5**: 19–72.

Sanders, R.W. 1991. Trophic strategies among heterotrophic flagellates. In: Patterson, D.J. & Larsen, J. (eds.) *The Biology of Heterotrophic Flagellates.* Systematics Association Special Volume 45. Oxford University Press, Oxford. pp. 21–38.

Sanders, R.W. & Porter, K.G. 1990. Bactivorous flagellates as food resources for the freshwater crustacean zooplankter *Daphnia ambigua. Limnol. Oceanogr.* **35**: 188–91.

Sandgren, C.D. 1986. Effects of environmental temperature on the vegetative growth and sexual life history of *Dinobryon cylindricum* Imhof. In: Kristiansen, J. & Andersen, R. (eds.) *Chrysophytes: Aspects and Problems.* Cambridge University Press, Cambridge. pp. 207–25.

Sandgren, C.D. 1988. The ecology of chrysophyte flagellates: their growth and perennation strategies as freshwater phytoplankton. In: Sandgren, C.D. (ed.) *Growth and Reproductive Strategies of Freshwater Phytoplankton.* Cambridge University Press, Cambridge. pp. 9–104.

Sandgren, C.D. 1991. Chrysophyte reproduction and resting cysts: a paleolimnologist's primer. *J. Paleolimnol.* **5**: 1–9.

Sandgren, C.D. & Barlow, S.B. 1989. Siliceous scale production in chrysophyte algae. II. SEM observations regarding the effects of metabolic inhibitors on scale regeneration in laboratory population of scale-free *Synura petersenii* cells. *Beih. Nova Hedwigia* **95**: 27–44.

Sandgren, C.D. & Flanagin, J. 1986. Heterothallic sexuality and density-dependent encystment in the chrysophycean flagellate *Synura petersenii* Korshikov. *J. Phycol.* **22**: 206–16.

Schnack, S.B. 1979. Feeding of *Calanus helgolandicus* on phytoplankton mixtures. *Mar. Ecol. Prog. Ser.* **1**: 41–7.

Siver, P.A. 1991. *The Biology of* Mallomonas: *Morphology, Taxonomy and Ecology.* Kluwer Academic Press, Dordrecht.

Siver, P.A. & Hamer, J.S. 1989. Multivariate statistical analysis of the factors controlling the distribution of scaled chrysophytes. *Limnol. Oceanogr.* **34**: 368–81.

Smith, V.H. 1983. Low nitrogen to phosphorus ratios favor dominance by bluegreen algae in lake phytoplankton. *Science* **221**: 669–71.

Smol, J.P. 1990. Diatoms and chrysophytes: a useful combination in paleolimnological studies. Report of a workshop and a working bibliography. In: Simola, H. (ed.) *Proceedings of the Tenth International Diatom Symposium.* Koeltz, Koenigstein. pp. 582–92.

Starkweather, P.L., Gilbert, J.J. & Frost, T.M. 1979. Bacteria feeding by *Brachionus calyciflorus*: clearance and ingestion rates, behavior and population dynamics. *Oecologia* **44**: 26–30.

Stemberger, R.S. 1986. The effects of food deprivation, prey density and volume on clearance rates and ingestion of *Diacyclops thomasi*. *J. Plankton Res.* **8**: 243–51.

Sterner, R.W. 1990. The ratio of nitrogen to phosphorus resupplied by herbivores: zooplankton and the algal competitive arena. *Am. Nat.* **136**: 209–29.

Stone, C.P. 1985. Alien animals in Hawaii's native ecosystem: towards controlling the adverse effects of introduced vertebrates. In: Stone, C.P. & Scott, J.M. (eds.) *Hawaii's Terrestrial Ecosystems: Preservation and Management.* University of Hawaii, Honolulu, pp. 252–97.

Turner, A.M. & Mittelbach, G.C. 1992. Effects of grazer community composition and fish on algal dynamics. *Can. J. Fish. Aquat. Sci.* **49**: 1908–15.

Vandermeer, J.H. 1972. Niche theory. *Annu. Rev. Ecol. Syst.* **3**: 107–32.

Vanderploeg, H.A. 1990. Feeding mechanisms and particle selection in suspension-feeding zooplankton. In: Wotton, R.S. (ed.) *The Biology of Particles in Aquatic Systems.* CRC Press, Boca Raton, Fla. pp. 183–212.

Vanderploeg, H.A. & Paffenhöfer, G.-A. 1985. Modes of algal capture by the freshwater copepod *Diaptomus sicilis* and their relation to food-size selection. *Limnol. Oceanogr.* **30**: 871–85.

Vanderploeg, H.A., Scavia, D.A. & Liebig, J.R. 1984. Feeding rate of *Diaptomus sicilis* and its relation to selectivity and effective food concentration in algal mixtures and in Lake Michigan. *J. Plankton Res.* **6**: 919–41.

Vanderploeg, H.A., Paffenhöfer, G.-A. & Liebig, J.R. 1988. *Diaptomus* vs. net phytoplankton: effects of algal size and morphology on selectivity of a behaviorally flexible, omnivorous copepod. *Bull. Mar. Sci.* **43**: 377–94.

Vanni, M.J. 1986. Competition in zooplankton communities: suppression of small species by *Daphnia pulex*. *Limnol. Oceanogr.* **31**: 1039–56.

Vermeij, G.J. 1978. *Biogeography and Adaptation.* Harvard Press, Cambridge, Mass.

Vermeij, G.J. 1986. The biology of human-caused extinctions. In: Norton, B.G. (ed.) *The Preservation of Species*. Princeton University Press, Princeton, NJ. pp. 28–49.

Waage, J.K. & Greathead, D.J. 1985. Biological control: challenges and opportunity. *Phil. Trans. R. Soc. Lond., Ser. B.* **318**: 111–28.

Wetzel, R.G. 1983. *Limnology*. W.B. Saunders, Philadelphia.

Wiesinger, J. 1992. Competition for phosphorus among eight phylogenetically diverse phytoplankton species under conditions of non-continuous supply. MS thesis, University of Wisconsin-Milwaukee, 105pp.

Williamson, C.E. & Butler, N.M. 1986. Predation on rotifers by the suspension-feeding calanoid copepod *Diaptomus pallidus*. *Limnol. Oceanogr.* **31**: 393–402.

Williamson, C.E. & Vanderploeg, H.A. 1988. Predatory suspension feeding in *Diaptomus*: prey defenses and the avoidance of cannibalism. *Mar. Ecol. Prog. Ser.* **43**: 568–77.

Witte, F., Goldschmidt, T., Wanink, J., van Oijen, M., Goudswaard, K., Witte-Maas, E. & Bouton, N. 1992. The destruction of an endemic species flock: quantitative data on the decline of haplochromine cichlids of Lake Victoria. *Env. Biol. Fish.* **34**: 1–28.

13
Application of chrysophytes to problems in paleoecology

JOHN P. SMOL

Introduction

Chrysophytes have long been recognized to be powerful indicators of environmental conditions (e.g. Siver, this volume); however, only recently have they been used extensively in paleoecological studies. In many ways, the increased use of chrysophyte microfossils has closely tracked the heightened interest and application of paleoecological approaches to the study of environmental change (see Davis 1989; Smol 1990a, b, 1992; Battarbee 1991; Smol & Glew 1992; and Charles et al. 1994 for recent reviews and commentaries). Historical perspectives, such as those that can be gleaned from lake sediments using paleoecological approaches, have now been melded into a wide array of studies that cover both theoretical and applied aspects of limnology (Smol 1990a). With these proxy data, long-term environmental conditions and variability can be assessed, hypotheses can be generated and tested, and models can be verified.

This chapter highlights some of the recent advances in the use of chrysophytes as paleolimnological markers, and stresses examples that I believe are most relevant to phycologists. Because of the increased volume of literature, it is no longer possible to provide a thorough synthesis in an article of this size. The reader is referred to reviews by Adam & Mahood (1981), Cronberg (1986a, b), Kristiansen (1986) and Smol (1987, 1988a), who summarize much of the historical literature on chrysophyte microfossils and provide a point of departure for this chapter. Smol (1990c) compiled a bibliography of most work on chrysophyte-based paleolimnology published in the 1980s.

Chrysophyte microfossil indicators

The previously cited papers summarize the historical development of chrysophytes in paleoecological studies, as well as the major features, advantages and disadvantages of the chrysophyte indicators commonly used in paleoecological studies. Below, I briefly summarize some of the salient features.

The siliceous scales, and to a lesser extent the bristles, of the Synurophyceae and some Chrysophyceae (i.e. the 'scaled' chrysophytes, such as taxa in the genera *Mallomonas* and *Synura*), have now been used in many paleolimnological reconstructions. Chrysophyte scales have the chief advantage that individual scales (i.e. the fossils that a paleolimnologist encounters in lake sediments, often in great quantities) are generally taxon-specific, and so past populations of scaled chrysophytes can be reconstructed from a stratigraphic analysis of scales preserved in dated sediment cores. Because different taxa possess varying numbers of scales per cell, simply enumerating disarticulated scales will provide only an indirect measure of past chrysophyte populations. To remedy this problem, Siver (1991) published correction tables that allow paleolimnologists to convert scale numbers back into cell numbers, and Cumming & Smol (1993) discussed how these corrections and other transformations affected some interpretations.

Many scaled chrysophyte taxa have well-defined and quantifiable (see below) ecological optima and tolerances, and so they are powerful biomonitors of environmental change. Nonetheless, scaled chrysophytes represent only a subset of past chrysophyte populations, as most taxa do not possess siliceous scales (e.g. *Dinobryon*, *Uroglena*, *Ochromonas*).

Fortunately, all chrysophytes produce a second group of microfossils. These are the endogenously-formed siliceous stomatocysts (also called statospores or resting cysts), which are widely believed to be taxon-specific, although this has recently elicited some controversy (Simola 1991; Sandgren & Smol 1991). Sandgren (1991*a*) provides a useful primer on the use of chrysophyte cysts in paleolimnological research; clearly much has changed since Nygaard's (1956) pioneering and inspiring studies on Lake Gribsø.

Stomatocysts are usually spheroids, although a range of shapes from ovate to pyramidal has been observed. All cysts are distinguished by the presence of a single pore, through which the germinating cell emerges. The pore may be surrounded by an elevated margin referred to as a collar. Although the mature or immature cysts produced by some taxa are unsculpted (thus making taxonomic designations difficult), many are highly ornamented. Ornamentation may consist of siliceous thickenings in the form of scabrae, verrucae, spines or reticula, or sculpturing in the form of psilae, depressions or fossae. This variety of ornamentation, as well as collar morphology, allows for differentiation of individual cyst morphs using scanning electron microscopy and, in most cases, light microscopy. Even unornamented cysts can often be distinguished on the basis of wall thickness and pore morphology (Zeeb & Smol 1993a). A major advantage of using cysts is that they are generally thickly silicified and therefore well preserved in many types of lake environments.

One drawback to the full exploitation of chrysophyte cysts is that relatively few cyst morphotypes (less than 10%) have been linked to the taxa that produce them. Identification of cysts has been hampered by problems associated with inducing cyst development in culture, and the paucity of lake studies that focus on encysting chrysophyte populations (Sandgren 1988). Hence, paleolimnological analyses have relied upon the assignment of temporary numbers or names to the cyst morphs (reviewed in Cronberg 1986a). Recent advances in this area include the development of a standardized terminology and guidelines for cyst numbering, under the auspices of the International Statospore Working Group (ISWG; see Cronberg & Sandgren 1986). According to the ISWG guidelines, each researcher or research group maintains a numerical list of all the cyst morphotypes that they observe. A complete description is based on scanning electron microscopy (SEM) and includes a characteristic SEM specimen, a written description and any associated ecological information. Light micrographs and/or line drawings may also be included. An active area of research is the critical description of stomatocyst floras from numerous habitats according to ISWG standards (e.g. Sandgren 1991b), and the use of regression/calibration techniques (see below) to characterize the range of environmental variables under which each cyst type is found (e.g. Rybak et al. 1991). Much of this work has been synthesized in the *Atlas of Chrysophyte Cysts*, which my laboratory has compiled.

There is no doubt that research attempting to link cyst morphotypes to the taxa that produce them should be encouraged, and each year more linkages are published. However, from a paleoecological perspective, the fact that cyst morphs can, at this time, often be given only temporary names is not the insurmountable problem that some might suspect. Given that most of our ecological information is gleaned from the analyses of surface-sediment calibration sets (see below), it makes little difference (from a statistical inference point of view) that cyst morphs are identified by, for example, numbers instead of species names, provided that the criteria used to differentiate cysts are sound and consistent. Clearly, paleoecologists see this as a temporary measure, and it is to be hoped that all cyst numbers will be replaced eventually by species names;

nonetheless, much can still be learned using this temporary nomenclature.

Although used less often than the morphological fossils described above, past chrysophyte populations can also be estimated from biogeochemical indicators. For example, Leavitt *et al.* (1989) used the pigments chlorophyll *c*, fucoxanthin and β-carotene in sediment cores to indicate past deepwater chrysophyte blooms. Although fossil pigments are a very useful paleolimnological tool, to date they cannot be used to differentiate past chrysophyte populations below the class level. However, it may eventually be possible to differentiate taxa further using other minor carotenoids (P. Leavitt, personal communication). Specific long-chain unsaturated ketones preserved in lake sediments may also track past chrysophyte populations (Cranwell 1985). Finally, estimates of biogenic silica concentrations in lake sediments (e.g. Conley 1988), although primarily designed to infer past diatom concentrations, will also reflect chrysophyte scale, bristle and cyst concentrations. These biogeochemical approaches have considerable potential, but for the purposes of this review I will restrict my discussion to the siliceous, morphological fossils found in lake sediments.

Calibration (training) sets

The development of surface-sediment calibration sets (also called reference and/or training sets) and the application of powerful multivariate statistical techniques have done much to strengthen and quantify paleolimnological inferences based on biological indicators. Calibration sets are constructed by examining the statistical relationship between taxa abundances in the recent sediments of a large number of lakes, and the present-day physical, chemical and biological characteristics of those lakes. Birks *et al.* (1990*a*), Battarbee (1991), Dixit *et al.* (1992*d*), Smol & Glew (1992) and Charles & Smol (1994) provide reviews on some of the most commonly used approaches, and Charles (1990) provides a checklist of the information that should be included in calibration sets.

Briefly, surface-sediment calibration sets consist of a series of lakes (e.g. 50 or so in number, but generally the more the better) from a particular region where ecological inferences are desired. Lakes are selected to span a large gradient of the environmental variables of interest. For example, if a study required reconstruction of past lakewater nutrient levels, such as for a eutrophication study, the lake set should span a large spectrum of lake trophy. Contemporaneous limnological data are required for each of the study lakes, preferably over several years. This data matrix is regressed on the distributional data of the biological indicators (in this case chrysophyte scales and/or cysts, usually expressed as relative percentages: see below) from the calibration lakes. This regression infers the species parameters from the distributional data of the biological indicators.

In a paleo-data set, the biological distributional data come from microfossils preserved in the surface sediments (e.g. top 1 cm or so, usually representing the last 2 or 3 years of sediment accumulation), which are usually collected using a simple surface coring device (e.g. Wright 1990; Glew 1991). Herein lies one of the major advantages of this approach: surface sediments represent an annually integrated sample of chrysophyte populations which have collected from the entire water column, 24 hours a day, every day of the year. In short, a surface-sediment assemblage represents the 'average chrysophyte assemblage' for that lake for its recent history. Since chrysophytes often form short-lived populations, at discrete levels in a water column, standard phycological sampling protocols may often miss important chrysophyte populations (Sandgren 1988). Moreover, if a large number of lakes are to be sampled (e.g. 50), it would be logistically impractical to study each lake, say every week, for all depth intervals, for several years, to obtain an 'average' or integrated chrysophyte assemblage.

Once the biological and environmental data have been gathered from all the calibration lakes, one can then determine which variables exert a statistically significant influence on the distribution of chrysophyte taxa. Canonical Correspondence Analysis (CCA: see ter Braak 1986; Jongman *et al.* 1987) is one ecologically realistic and statistically robust approach for these types of analyses (see examples described below), and one that my research group has used most extensively. Once the significant environmental variables are identified (e.g. pH, salinity, phosphorus), then species parameters can be estimated using a variety of techniques including weighted averaging (Line & Birks 1990) and Gaussian logit regression (Jongman *et al.* 1987). Comparison of these reveals that the simpler weighted averaging techniques appear to be a good approximation of the more complex logit regression methods (e.g. Cumming *et al.* 1992*a*).

Refinements to the design of calibration sets continue to be an area of active concern; for example, Siver & Hamer (1992) have recently suggested the inclusion of seasonal data to improve paleolimnological inferences based on chrysophytes. Further advances in inference models and user-friendly computer programs are also in development.

Implicit in the calibration set approach is the belief that a surface-sediment sample taken from near the center of a lake is a faithful surrogate for living chrysophyte populations. These assumptions have been confirmed by, for example, Siver & Hamer (1990), who found a good relationship between scaled chrysophyte assemblages recorded in phytoplankton collections and those preserved in the surface sediments of the same lakes, and by Charles *et al.* (1991), who showed that relatively little between-sample variability occurred in deepwater, surface-sediment samples.

Calibration sets have provided phycologists with a wealth of autecological data, particularly on scaled chrysophytes, which it would not have been feasible to gather in any other way. For example, Smol *et al.* (1984*a*), Charles & Smol (1988) and Cumming *et al.* (1992*a*) used surface-sediment calibration sets to document the ecological optima and tolerances of scaled chrysophytes in Adirondack Park (New York), with respect to factors related to lake acidification (e.g. lakewater pH and monomeric aluminum concentrations). Dixit *et al.* (1988*b*, 1989*b*), Dixit & Dixit (1989), Christie *et al.* (1988), Hartmann-Zahn (1991) and Cumming *et al.* (1991) similarly used surface sediments to gather ecological data on scaled chrysophytes for lakes in Sudbury (Ontario), Quebec, northern New England, Finland, Germany and Norway, respectively. These studies all documented the overriding importance of lakewater pH on chrysophyte distributions, but also noted the secondary importance of pH-related variables such as metal or dissolved organic carbon (DOC) concentrations. Plots of observed lakewater pH versus the pH inferred from chrysophyte scales preserved in the surface sediments of four calibration sets attest to the strength of these relationships (Fig. 13.1).

To date, the only calibration sets available for stomatocysts are by Rybak *et al.* (1991), who showed that lakewater alkalinity, total dissolved solids and trophic status were important variables determining the distribution of cysts in 50 central Ontario lakes; and by Carney *et al.* (1992) who studied cyst distributions in the surface sediments of 33 California lakes from the Sierra Nevada mountains. Duff *et al.* (1992) compared chrysophyte cyst distributions preserved in the surface sediments of 36 shallow ponds in the Canadian High Arctic, but did not develop transfer functions. Many other calibration sets are now being developed for other lake regions (for both chrysophyte cysts and scales), and undoubtedly the wealth of data these studies will produce will be of interest to many phycologists, in addition to paleoecologists.

Application of chrysophytes to problems in paleoecology

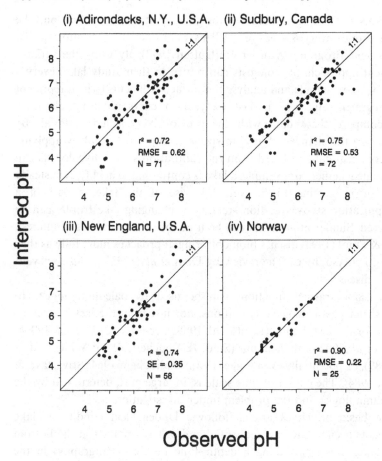

Figure 13.1. Plots of observed (i.e. measured) lakewater pH versus the inferred pH using four chrysophyte predictive models. Details on each of these calibration sets can be found in Cumming *et al.* (1992*a*), Dixit *et al.* (1989*b*), Dixit *et al.* (1990*b*) and Cumming *et al.* (1991), respectively. RMSE, root mean squared error; SE, standard error. The reader is referred to the original papers for details.

Applications

Once the environmental optima and tolerances of chrysophyte taxa are estimated, and statistically and ecologically realistic transfer functions are constructed, the phycologist and the paleolimnologist possess powerful tools to monitor environmental change. For example, if a phycologist is involved in a long-term monitoring program, where a large number of lakes must be sampled, logistic concerns will, in all likelihood, preclude detailed phycological investigations. In fact, in some large-scale monitoring programs now being initiated in the United States, lakes will only be sampled once every 4 years (e.g. EPA's Environmental Monitoring and Assessment Program: Whittier & Paulsen 1992). By using the surface-sediment approach, phycologists can return to their study lakes every 4 years or so, re-sample and analyse the surface sediments (which represent an integrated sample of the last few years of chrysophyte populations), and compare these data with the previously collected material. By comparing the changes (if any) in species composition, and by applying the ecological data gathered from the calibration steps, phycologists can determine whether, for example, a lake is continuing to acidify, is in steady state, or is starting to recover. Likewise, if the lake is part of a eutrophication study, questions related to changing lake trophy can be answered. Similar approaches can be used for other variables of interest. Siver & Smol (1993) discuss biomonitoring approaches more fully as they relate to chrysophytes. The review by Dixit *et al.* (1992*b*) is also relevant to this discussion.

The same transfer functions can be used by paleolimnologists to reconstruct past ecological conditions, not just from the last 4 years or so but for the last decade (Dixit *et al.* 1989*c*), century (Birks *et al.* 1990*b*; Dixit *et al.* 1992*a*), millennium (Smol & Boucherle 1985; Whitehead *et al.* 1989), or potentially even longer (VanLandingham 1964; Srivastava & Binda 1984). The choice of time scale is, in large part, determined by the paleolimnologist and the problem under investigation.

The basic principles are as follows. Under ideal conditions, lake sediments accumulate in a sequential and orderly manner at the bottom of a lake, thus establishing a depth–time profile (stratigraphy) in the sediments. A sedimentary stratum can be dated using a variety of techniques, such as ^{210}Pb or ^{14}C accelerator mass spectrometry (AMS) geochronology. Preserved in these stratigraphic sequences is a wealth of physical, chemical and biological (including chrysophyte scales and cysts) information (reviewed in Smol & Glew 1992) that can be used by paleolimnologists to reconstruct past lake environments, and thereafter these data can be used to test hypotheses and validate models. Implicit in all such studies is the belief that the sediments are not greatly disturbed by bioturbation or other factors, and that the information entombed in sediments is a faithful record of past limnological conditions. Using a variety of approaches, paleolimnologists have satisfied many critics that the above assumptions are reasonable and/or that problem lakes can generally be identified. The continued study of sample variability (e.g. Charles *et al.* 1991) and the establishment of defendable quality assurance and quality control protocols (Kingston *et al.* 1992*a*), coupled with strong statistical methodology (Birks *et al.* 1990*a*), has done much to alleviate the concerns of many sceptics.

Trends in lake eutrophication

Work done in the late 1970s and early 1980s tended to focus on using chrysophytes as paleolimnological markers of lake trophic status. For example, Battarbee *et al.* (1980), Munch (1980, 1985) and Smol (1979, 1980) completed three independent studies, all showing that fossil chryso-phyte scales can be used to track eutrophication trends in the recent histories of Finnish, United States and Canadian lakes, respectively. Haworth (1983, 1984) included scaled chrysophytes in her paleolimno-logical studies of Blelham Tarn (England) and Smol *et al.* (1983) and Smol & Boucherle (1985) tracked fluctuations in lake trophic status over the last century and since deglaciation, respectively, in Little Round Lake (Ontario). Smol (1979, 1982) presents additional post-glacial stratig-raphies of chrysophyte scales for other Ontario lakes.

Chrysophyte cysts were also mentioned in some of the above studies but (because of taxonomic difficulties) relatively few workers attempted to differentiate the cysts into temporary names (such as those proposed by Nygaard 1956). Because of the reluctance of many workers (including the author!) to attempt to differentiate chrysophyte cysts, Smol (1983, 1985) proposed that a simple ratio comparing the chrysophyte cysts and diatom frustules preserved in lake sediments would provide an approxi-mate measure of the relative abundances of these two important algal groups. For example, in many temperate lakes chrysophytes are often less abundant following nutrient additions (Sandgren 1988; Sandgren & Walton, this volume), and the ratio appears to follow these patterns (Smol 1985). Paleolimnological studies on the Great Lakes have shown that chrysophytes were more abundant prior to cultural eutrophication (e.g. Stoermer *et al.* 1991). The near extirpation of chrysophytes from Hamilton Harbor coincided with the severe pollution this Lake Ontario harbor experienced (Yang *et al.* 1993). Risberg (1990) similarly interpreted the decline of cysts in the recent sediments of the Baltic Sea as a consequence of eutrophication, and overall shifts in cyst abundance were one part of the paleolimnological evidence used to assess the limnological effects of past ungulate grazing in Yellowstone National Park (Engstrom *et al.* 1991). Comparing the abundance of cysts and diatoms may also be useful in interpreting long-term successional patterns in other types of systems, such as high montane (Marciniak 1988), sub-alpine (Hickman & Schweger 1991) or arctic (Smol 1983; Lemmen *et al.* 1988; Blake *et al.* 1992) lakes. Clumping all chrysophytes into one category, however, greatly reduces the information content of the sample from which environmental inferences can be made.

Several studies have begun to differentiate cysts into morphological categories, first using Nygaard's (1956) temporary nomenclature and then using ISWG guidelines. For example, Carney (1982), Carney & Sandgren (1983) and Sandgren & Carney (1983) incorporated cysts into their paleolimnological study of lake eutrophication patterns of Frains Lake (Michigan). Rybak (1986, 1987) and Rull (1986, 1991) used chrysophyte cysts to interpret paleoenvironmental conditions in two Polish lakes and a Spanish pond, respectively. Zeeb *et al.* (1994) recorded striking shifts in cyst and scale assemblages in response to the artificial fertilization in 1969 of Lake 227 (Experimental Lakes Area, Ontario). Rybak *et al.* (1987) and Zeeb *et al.* (1990) studied long-term eutrophication patterns in meromictic lakes in southern Ontario.

A major hindrance to the full exploitation of quantitative inferences of lake trophic status from chrysophytes is that, at the time of writing, a surface-sediment calibration set has not been completed (although some are in progress) relating chrysophyte species assemblages to trophic variables. As a result, trophic inferences are still qualitative, based primarily on ecological comments that are often only sparsely available in the phycological literature. However, once such calibration sets are complete, as they are for pH gradients, then these inferences will un-doubtedly become more quantitative, and therefore more useful to limnologists and lake managers.

Trends in lake acidification

During the 1980s, much paleolimnological work in parts of North America and Europe shifted to a new environmental problem: acid rain. Because chrysophyte distributions closely track lakewater pH and related variables, chrysophyte microfossils played an important role in the scientific and often political debates on lake acidification.

Charles *et al.* (1989) reviewed the ways in which diatoms and chryso-phytes were used in paleolimnological studies of lake acidification. Limnologists could fairly easily determine what percentage of lakes were presently acidic; however, because of the lack of historical limnological data, they could not determine if and when and how much lakes had acidified. These were crucial scientific, political and policy questions. Paleolimnology played a pivotal role in these important debates.

As noted previously, surface-sediment calibration sets have frequently shown the importance of lakewater pH in determining the distribution of scaled chrysophytes, and soon chrysophytes were incorporated into several large-scale paleo-acidification projects, such as PIRLA-I (Charles & Whitehead 1986), SWAP (Battarbee *et al.* 1990) and PIRLA-II (Charles & Smol 1990). The first attempt at using scaled chrysophytes to infer recent acidification trends was from the presently acid and fishless Deep Lake (Adirondack Park, New York), where Smol *et al.* (1984*b*) interpreted the striking increase in *Mallomonas hindonii* and *M. hamata* in the recent sediments as indicative of acidification. The Adirondack Park became the site of many other such investigations (e.g. Smol 1986; Christie & Smol 1986; Douglas & Smol 1988; Smol & Dixit 1990; Cumming 1991), as did northern New England (Gibson *et al.* 1987; Smol & Dixit 1990; Davis *et al.* 1990), southern New England (Marsicano & Siver 1993), Quebec (Dixit *et al.* 1988*a*), Germany (Hartmann & Steinberg 1986; Steinberg *et al.* 1988; Hartmann-Zahn 1991), Sweden and Scotland (Cronberg 1990), Norway (Birks *et al.* 1990*b*; Cronberg 1990), Finland (Tolonen *et al.* 1986), Czech Republic (Vesely *et al.* 1993), Italy (Marchetto 1992), and the acid and metal stressed lakes of Sudbury, Ontario (Dixit *et al.* 1989*a, b, c*; 1992*a, b, c, e*). In addition to the above investigations which used scaled chrysophytes, Duff & Smol (1991) also identified (using ISWG guidelines) and enumerated a total of 66 cyst morphotypes from the sediments of a recently acidified Adirondack lake, and showed that cyst assemblages changed with acidification.

With the continued development and refinement of surface-sediment calibration sets, paleolimnological inferences became increasingly more accurate. Moreover, with new developments in sediment coring and close-interval sectioning methods (e.g. Glew 1988), the temporal resolution attainable from sedimentary profiles increased greatly.

One example which demonstrates the power of paleolimnological analyses is a core taken from a hill-top lake in Norway. As reviewed in Birks *et al.* (1990*b*), two competing hypotheses had been proposed to explain the present acidity levels of some Norwegian lakes. One was that the lakes had recently acidified as a result of atmospheric deposition of strong acids, whereas the 'land-use hypothesis' postulated that land use and associated soil changes were a major cause of lake acidification. Choosing lakes that are perched on hill-tops (i.e. lakes that have very small catchments, and so would be little affected by land use) ensured

HOLETJØRN

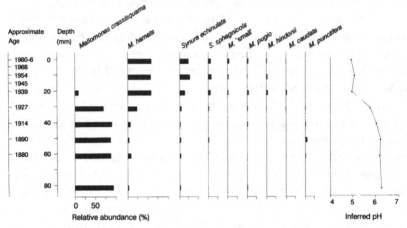

Figure 13.2. Stratigraphic changes in chrysophyte scales from the recent sediments of Holetjørn, Norway. The approximate ages of the sediment intervals (shown on the left) were calculated using ^{210}Pb chronology. Chrysophyte taxa are expressed as a percentage of total scales counted. The inferred pH profile (shown on the right) was determined using the transfer functions developed by Cumming *et al.* (1991), also shown in Fig. 13.1(iv). This core is described in detail in Birks *et al.* (1990*b*).

that the water chemistry would be primarily influenced by the chemical composition of precipitation and by the underlying bedrock geology. Birks *et al.* (1990*b*) used several paleolimnological indicators to show that, although the lakes were naturally acidic, they showed additional acidification during this century, consistent with the acid-deposition hypothesis.

A summary diagram of the scaled chrysophytes from the sediments of one of these hill-top lakes (Holetjørn) illustrates some of the species changes that occurred in its recent history (Fig. 13.2: see Birks *et al.* (1990*b*) for a complete discussion). In the nineteenth century the scaled chrysophyte flora was dominated by *Mallomonas crassisquama* (NB: this category also contains *M. duerrschmidtiae*, which at the time of the study was not described and differentiated from the morphologically similar *M. crassisquama* scales). The relative abundance of this taxon declined markedly and almost disappeared in the first part of the twentieth century and was replaced by more acidophilic species such as *M. hamata*, *Synura echinulata* and other minor taxa. These changes coincided with periods of increased acidic deposition, and not land-use activities.

Because of the acid-rain debate, an additional hypothesis that required evaluation was that present lakewater pHs are the result of long-term (thousands of years) natural acidification processes. To answer this question, sediment cores were collected from presently acidic lakes. However, in contrast to the previous studies, the cores not only spanned the lakes' recent histories but included their full post-glacial development (i.e. dating back about 12 000 years). Fossil chrysophyte analyses of these sediments (e.g. Christie & Smol 1986; Whitehead *et al.* 1989) helped ascertain that, although the lakes may have slowly acidified over long time frames, their rapid recent acidification has been unprecedented in the lakes' histories, as regards both the extent and rate of acidification.

By slicing sediment cores into thin sections (e.g. using an instrument such as a Glew (1988) extruder to section at 0.25 cm intervals, which in some lakes represents only a few years per section) many important limnological questions have been answered. For example, in the Sudbury region of Ontario, which supports the largest nickel smelting operation and point source of sulphur emissions in the world, chrysophyte indicators have been used to trace the acidification and recovery (following recent dramatic declines in local deposition) of a large number of lakes. Paleolimnological studies of some Sudbury lakes also provided the opportunity to test the reliability of fossil inferences, as long-term (e.g. over the last 20 years) lakewater pH measurements were available for a few of the study lakes, and so paleolimnological inferences could be compared with actual pH measurements collected by the Ontario Ministry of the Environment. The similarity between the fossil-inferred pHs from dated sediment cores and from archived lakewater pH readings was often striking (Dixit *et al.* 1989*c*, 1992*e*), and attested to the veracity of the paleolimnological method for tracking acidification and recovery patterns. Interestingly, lakes situated as far as 70 km southwest of the smelters also showed marked acidification and recovery trajectories (Dixit *et al.* 1992*a*). Chrysophyte scales have also been used to study the effects of remedial actions, such as lake liming and other manipulations (Dixit *et al.* 1990*a*).

In order to determine whether lakewater pH and monomeric aluminum concentrations have been changing in Adirondack Park lakes over the last 20 years, concomitant with declines in sulphur deposition, Cumming (1991) and Cumming *et al.* (1994) studied stratigraphic changes in chrysophyte scales from thinly sectioned (0.25 cm intervals) sediment cores from 18 Adirondack lakes. They found that all three possible scenarios were occurring in the Park: since 1970, some lakes were continuing to acidify, some were recovering, and some were in a steady state.

Scaled chrysophytes may indicate other environmental variables, in addition to lakewater pH. For example, Hartmann-Zahn (1991) interpreted the decline in *Mallomonas paludosa* scales, and other taxonomic shifts, in the recent sediments of a presently acid German lake to a decline in lakewater DOC as a result of recent anthropogenic acidification. Kingston *et al.* (1992*b*) included inferences regarding pH and monomeric aluminum in their paleolimnological study of fisheries loss in Adirondack Park lakes. Although in general agreement, chrysophyte species changes often pre-date those recorded in fossil diatom stratigraphies, and the magnitude of inferred pH and monomeric aluminum changes tend to be higher with the chrysophytes. Uutala *et al.* (1994) noted similar trends in Southern Ontario lakes. It is possible that the often vernal-blooming, euplanktonic chrysophytes may respond more rapidly to environmental changes, and may be especially sensitive indicators of acid-shock conditions, characteristic of spring snowmelt periods.

The above techniques are time-consuming, but they are the only way in which paleoecologists can answer questions concerning the timing and the rate of environmental change. However, for some lake management questions it is more important to determine what the pre-disturbance conditions were for a large number of lakes, and the magnitude of the environmental changes. For example, in northeastern North America three major policy questions were: (1) what percentage of Adirondack lakes have acidified since AD 1850? (2) How much did they acidify? and (3) Were any lakes naturally acidic? As an effective and efficient way to answer such questions in a large number of lakes, paleolimnologists use what has come to be the called the 'top and bottom' approach, which compares environmental inferences from biological indicators in the surface (i.e. 'tops' of sediment cores, representing present-day conditions) with inferences from sediments representing pre-impact lake conditions (i.e. the 'bottoms' of cores; in the Adirondacks these were sediments that accumulated before anthropogenically caused acidification was possible, or pre-1850).

Figure 13.3 summarizes the chrysophyte data from one such 'top and bottom' study completed for 34 Adirondack lakes, the full details of which are presented in Cumming *et al.* (1992*b*). These study lakes were selected as probabilistic samples from a subset of Adirondack study lakes, and so represent a 'target population' of relatively undisturbed, low alkalinity lakes. Chrysophyte scales were identified and enumerated from the surface sediments (i.e. the 'tops') and from sediments usually at greater than 25 cm depth in the cores (i.e. the 'bottoms'). Environmental inferences for

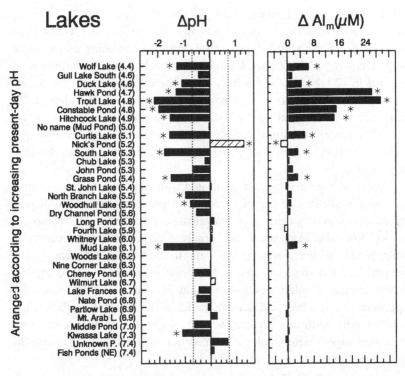

Figure 13.3. Chrysophyte-based estimates of historical change (pre-industrial to the present) in pH (ΔpH) and total monomeric aluminum (ΔAl_m) from the 34 Adirondack study lakes. The estimates are presented as the differences in inferred water chemistry between the top (0–1 cm) and bottom (usually >25 cm) sediment core intervals. The lakes are arranged according to increasing present-day measured pH. Lakes indicated by hatched bars were limed within 5 years before coring. The dotted lines represent $\pm RMSE_{boot}$ of prediction for pH; asterisks denote changes that are $> RMSE_{boot}$ of prediction for pH. The reader is referred to Cumming *et al.* (1992*b*) for full details. (From Cumming *et al.* (1992*b*) with permission.)

lakewater pH and monomeric aluminum were then obtained using the calibration equations developed by Cumming *et al.* (1992*b*), using a calibration set of 71 Adirondack Park lakes.

As indicated in Fig. 13.3, all but one (see below) of the presently acidic lakes have acidified and have increased their monomeric aluminum concentrations since the nineteenth century. The latter is an important variable because elevated monomeric aluminum concentrations can result in fishery declines and recruitment failures. The one lake that seems to be an exception to the above (i.e. Nick's Pond, shown by a hatched bar in Fig. 13.3) was actually limed a few years before coring – further attesting to the veracity of the approach. Cumming *et al.* (1992*b*) present these data in a variety of ways, including spatial presentations using maps of the park. Dixit *et al.* (1992*c*) used similar approaches to infer limnological changes in 72 lakes in the Sudbury area. These data are very important to lake managers because they provide surrogates for the unavailable historical data required for the estimation of realistic background (i.e. pre-disturbance) conditions, which in turn are necessary for planning remedial action as well as providing estimates of damage to lake populations.

Because of the uncertainties associated with empirical and/or dynamic computer modelling of complex ecosystem processes, paleolimnological data are increasingly being used to evaluate the performance of computer models. Instead of using the models simply to predict future environmental changes (for which there are no data to assess the accuracy of the predictions), the model output is 'hindcast', and then verified against paleolimnological inferences. For example, Sullivan *et al.* (1992) used this approach to assess lake acidification models in the Adirondack Park lakes. Undoubtedly, with new challenges such as global climate and environmental change, paleolimnology will continue to be used to evaluate the myriad of models being developed.

Trends in climatic change

Paleoclimatology has always been an important research area, but with heightened interest in possible future climate modifications, reconstructing past climate variables has taken on a new sense of urgency. Reliable instrument readings documenting past trends in climatic variables, such as temperature, are generally only available for, at the most, the last two centuries or so. Longer-term time frames are required to distinguish actual trends from natural variability, and to evaluate climate models. Paleolimnologists can help provide these data (Smol *et al.* 1991).

Chrysophyte microfossils have not yet been used extensively in paleoclimatic reconstructions, but they have considerable potential. Although direct temperature reconstructions for chrysophytes may in fact be possible (Siver & Hamer 1992), it is likely that most climatic inferences will be based on indirect data (i.e. tracking environmental changes that are only indirectly related to climate). For example, in many closed basin lakes where evaporation exceeds precipitation, lakewater salinity is governed to the greatest extent by the balance between precipitation and evaporation. Many algal taxa, such as diatoms, can be used to reconstruct past lakewater salinity levels, from which past inferences on some climatic variables can be made (e.g. Fritz *et al.* 1991). Although chrysophytes are not often important components of saline lake systems, some preliminary data suggest that identifying and enumerating cyst morphotypes may track past salinity changes (Pienitz *et al.* 1992). The species composition of scaled chrysophytes can also be used to track specific conductivity (Siver 1993). In general, it seems that chrysophyte abundance decreases as salinity increases, which in itself can be informative in paleoenviron-mental studies (Cumming *et al.* 1993).

Chrysophyte species composition may also be responding to a suite of other environmental variables that are related to climate. For example, Zeeb & Smol (1993*a*, *b*) studied chrysophyte cysts from the laminated sediments of Elk Lake (Minnesota), as part of a multidisciplinary paleoclimate study of this lake. In general, shifts in chrysophyte cyst assemblages coincided with presumed changes in past climate, as inferred from a suite of other proxy data available for this important reference site. The most striking taxon shifts occurred about 8500 years ago, which coincided with the shift to prairie vegetation dominance in the lake's drainage. Interestingly, the ratio of chrysophyte cysts to diatom frustules was exceptionally high during the Little Ice Age. Duff & Smol (1988) speculated that changes in cyst assemblages over the last 9000 years in a high arctic lake were related to climatic changes. Further studies are clearly warranted.

Several workers have noted that some of the morphological variability noted in chrysophytes may be temperature dependent. For example, Siver & Skogstad (1986) showed that bristle morphology in *Mallomonas crassisquama* was directly related to temperature, and Sandgren (1983) found that variations in the length and definition of both spines and the cyst collar of *Dinobryon cylindricum* were temperature dependent. These structures are preserved in lake sediments, and it is potentially possible to refine paleoclimatic inferences by quantifying this morphological variability in fossils.

Other applications

Most of the examples used in this chapter have focused on recent environmental problems, such as lake acidification. Nonetheless, chryso-phyte microfossils can be used in many other challenging applications. For example, Rhodes (1991) included chrysophyte scales in his paleolimnological assessment of past fires and windstorms near Mud Pond, Maine. Most chrysophyte taxa responded sharply to these disturbances, with increases in *Mallomonas caudata*, *Synura spinosa* and *S. petersenii*. Zeeb & Smol (1991) recorded past chrysophyte species changes in a Michigan lake that has undergone marked salinification due to salt intrusion from an adjacent salt-storage facility. Moreover, Siver (1993) used the species composition of the chrysophyte scales from a southern New England calibration set to show that, in addition to pH, many taxa were significantly correlated with concentrations of dissolved salt. He used these data to construct a transfer function which allowed him to infer conductivity levels in a small Connecticut lake over the last 170 years. The inferred increase in conductivity coincided with known changes in the watershed (e.g. the operation of a silex mine).

Because many chrysophyte taxa are known to exhibit marked seasonal changes, Battarbee (1981) and Peglar *et al.* (1984), amongst others, used chrysophyte microfossils in individual bands of laminated sediments to establish the annual nature of these deposits. J. Webb (unpublished) used cysts to delineate the lacustrine/marine boundary in the historical development of a low arctic lake. Smith & White (1985) observed different cyst assemblages in lakes affected by the Mount St Helens volcanic eruptions. Chrysophytes often contribute to taste and odor problems in lake systems (e.g. Nicholls, this volume); paleoecological studies of these lakes can help determine the causes of these blooms (Smol 1992). Paleolimnology can also be used for more theoretical pursuits, such as questions related to biogeography (Smol 1988*b*; Sandgren & Walton, this volume).

In addition to lake sediments, the above approaches can also be applied to other types of deposits. For example, Brown *et al.* (1994) identified and enumerated chrysophyte cysts to help interpret the ontogeny of a high arctic peat deposit, and Harwood (1986) used cysts as part of his paleoenvironmental study of the Greenland Camp Century ice core. Although this chapter has focused on freshwater systems, there is considerable potential for using chrysophyte microfossils in marine systems as well (e.g. Mitchell & Silver 1986). Many other types of applications are possible.

Summary

This review focuses on recent developments in the use of chrysophyte microfossils. Over the last decade, chrysophyte microfossils have played important roles in the study of many environmental problems. New challenges will undoubtedly lead to new applications. I find it especially exciting to look back and see how much progress has been made since the first reviews were published on fossil chrysophytes, just a few years ago. Clearly, there have been some striking advances, many of which were fostered by the increased recognition and acceptance that paleolimnology has enjoyed. These advances have been driven, in part, by new developments and designs in coring and sampling equipment, sediment dating techniques, new protocols and powerful statistical treatments. Moreover, the continued demonstration of reproducibility and the rigorous application of quality assurance and quality control protocols have done much to bolster confidence in the accuracy and precision of paleolimnological inferences. Phycologists interested in chrysophytes can take advantage of many of these developments.

Acknowledgments

Research in my laboratory on chrysophycean microfossils was made possible largely by grants from the Natural Sciences and Engineering Research Council of Canada and the US Environmental Protection Agency. Many helpful comments on this manuscript were provided by PEARL scientists as well as H.J.B. Birks, C. Sandgren and P. Siver.

References

Adam, D.P. & Mahood, A.D. 1981. Chrysophycean cysts as potential environmental indicators. *Geol. Soc. Am. Bull.* **92**: 839–44.
Battarbee, R.W. 1981. Diatom and Chrysophyceae microstratigraphy of the annually laminated sediments of a small meromictic lake. *Striae* **14**: 105–9.
Battarbee, R.W. 1991. Recent paleolimnology and diatom-based environmental reconstruction. In: Shane, L.C.K. & Cusing, E.J. (eds.) *Quaternary Landscapes.* University of Minnesota Press, Minneapolis. pp. 129–74.
Battarbee, R.W., Cronberg, G. &. Lowry, S. 1980. Observations on the occurrence of scales and bristles of *Mallomonas* spp. (Chrysophyceae) in the micro-laminated sediments of a small lake in Finnish North Karelia. *Hydrobiologia* **71**: 225–32.
Battarbee, R.W., Mason, J., Renberg, I. & Talling, J.F. 1990. *Palaeolimnology and Lake Acidification.* The Royal Society, London, 219pp.
Birks, H.J.B., Line, J.M., Juggins, S., Stevenson, A.C. & ter Braak, C.J.F. 1990*a*. Diatoms and pH reconstruction. *Phil. Trans. Soc. Lond. B.* **327**: 263–78.
Birks, H.J.B., Berg, F., Boyle, J.F. & Cumming, B.F. 1990*b*. A palaeoecological test of the land-use hypothesis for recent lake acidification in south-West Norway using hill-top lakes. *J. Paleolimnol.* **4**: 69–85.

Blake, W., Boucherle, M.M., Fredskild, B., Janssens, J.A. & Smol, J.P. 1992.
The geomorphological setting, glacial history and Holocene development
of Kap Inglefield Sø, Inglefield Land, north-west Greenland. *Medd.
Groenl., Geosci.* **27**, 42pp.

Brown, K., Douglas, M.S.V. & Smol, J.P. 1994. Siliceous microfossils in a
Holocene, High Arctic peat deposit (Nordvestø, Northwest Greenland).
Can. J. Bot. **72**: 208–16.

Carney, H.J. 1982. Algal dynamics and trophic interactions in the recent history
of Frains Lake, Michigan. *Ecology* **63**: 1814–26.

Carney, H.J. & C.D. Sandgren. 1983. Chrysophycean cysts: indicators of
eutrophication in the recent sediments of Frains Lake, Michigan, USA.
Hydrobiologia **101**: 195–202.

Carney, H.J., Whiting, M.C., Duff, K.E. & Whitehead, D.R. 1992.
Chrysophycean cysts in Sierra Nevada (California) lake sediments:
paleoecological potential. *J. Paleolimnol.* **7**: 73–94.

Charles, D.F. 1990. A checklist for describing and documenting diatom and
chrysophyte calibration data sets and equations for inferring water
chemistry. *J. Paleolimnol.* **3**: 175–8.

Charles, D.F. & Smol, J.P. 1988. New methods for using diatoms and
chrysophytes in reconstructions of lakewater pH in low alkalinity lakes.
Limnol. Oceanogr. **33**: 1451–62.

Charles, D.F. & Smol, J.P. 1990. The PIRLA II project: regional assessment of
lake acidification trends. *Verh. Int. Ver. Limnol.* **24**: 474–80.

Charles, D.F. & Smol, J.P. 1994. Long-term chemical changes in lakes:
quantitative inferences using biotic remains in the sediment record. In:
Baker, L. (ed.) *Environmental Chemistry of Lakes and Reservoirs.* Advances
in Chemistry Series 237. American Chemical Society Books, Washington,
DC pp. 3–31.

Charles, D.F. & Whitehead, D.R. 1986. The PIRLA project: paleoecological
investigations of recent lake acidification. *Hydrobiologia* **143**: 13–20.

Charles, D.F. Battarbee, R.W., Renberg, I., van Dam, H. & Smol, J.P. 1989.
Paleoecological analysis of lake acidification trends in North America
and Europe using diatoms and chrysophytes. In: Norton, S.A., Lindberg,
S.E. & Page, A.L. (eds.) *Acid Precipitation*, vol. 2. Springer, Berlin.
pp. 207–76.

Charles, D.F., Dixit, S.S., Cumming, B.F. & Smol, J.P. 1991. Variability in
diatom and chrysophyte assemblages and inferred pH: paleolimnological
studies of Big Moose L., N.Y. *J. Paleolimnol.* **5**: 267–84.

Charles, D.F., Smol, J.P. & Engstrom, D.R. 1994. Paleolimnological approaches
to biomonitoring. In: Loeb, S. & Spacie, A. (eds.) *Biological Monitoring of
Freshwater Ecosystems.* Lewis Publishers, Ann Arbor, Michigan. pp. 233–93.

Christie, C.E. & Smol, J.P. 1986. Recent and long-term acidification of Upper
Wallface Pond (N.Y.) as indicated by mallomonadacean microfossils.
Hydrologia **143**: 355–60.

Christie, C.E., Smol, J.P., Meriläinen, J. & Huttunen, P. 1988. Chrysophyte
scales recorded in lake sediments from eastern Finland. *Hydrobiologia* **161**:
237–43.

Conley, D.J. 1988. Biogenic silica as an estimate of siliceous microfossil
abundance in Great Lakes sediments. *Biogeochemistry* **6**: 161–79.

Cranwell, P.A. 1985. Long-chain unsaturated ketones in recent lacustrine
sediments. *Geochim. Cosmochim. Acta* **49**: 1545–51.

Cronberg, G. 1986a. Chrysophycean cysts and scales in lake sediments: a review. In: Kristiansen, J. & Andersen, R.A. (eds.) *Chrysophytes: Aspects and Problems*. Cambridge University Press, Cambridge. pp. 281–315.

Cronberg, G. 1986*b*. Blue-green algae, green algae and chrysophyceae in sediments. In: Berglund, B.E. (ed.) *Handbook of Holocene Palaeoecology and Palaeohydrology*. Wiley, Chichester, pp. 507–26.

Cronberg, G. 1990. Recent acidification and changes in the subfossil chrysophyte flora of lakes in Sweden, Norway and Scotland. *Phil. Trans. R. Soc. Lond. B*. **327**: 289–93.

Cronberg, G. & Sandgren, C.D. 1986. A proposal for the development of standardized nomenclature and terminology for chrysophycean statospores. In: Kristiansen, J. & Andersen, R.A. (Eds). *Chrysophytes: Aspects and Problems*. Cambridge University Press, Cambridge. pp. 317–28.

Cumming, B.F. 1991. Paleoecological assessment of changes in Adirondack (New York, USA) lake-water chemistry: historical (pre-1850) and recent (post-1970) trends. PhD thesis. Queen's University. 183pp.

Cumming, B.F. & Smol, J.P. 1993. Scaled chrysophytes and pH inference models: the effects of converting scale counts to cell counts and other species transformations. *J. Paleolimnol.* **9**: 17–153.

Cumming, B.F., Smol, J.P. & Birks, H.J.B. 1991. The relationship between sedimentary chrysophyte scales (Chrysophyceae and Synurophyceae) and limnological characteristics in 25 Norwegian lakes. *Nord. J. Bot.* **11**: 231–42.

Cumming, B.F., Smol, J.P. & Birks, H.J.B. 1992*a*. Scaled chrysophytes (Chrysophyceae and Synurophyceae) from Adirondack (N.Y., USA) drainage lakes and their relationship to measured environmental variables, with special reference to lakewater pH and labile monomeric aluminum. *J. Phycol.* **28**: 162–78.

Cumming, B.F., Smol, J.P., Kingston, J.C., Charles, A.F., Birks, H.J.B., Camburn, K.E., Dixit, S.S., Uutala, A.J. & Selle, A.R. 1992*b*. How much acidification has occurred in Adirondack region (New York, USA) lakes since pre-industrial times? *Can. J. Fish. Aquat. Sci.* **49**: 128–41.

Cumming, B.F., Davey, K., Smol, J.P. & Birks, H.J.B. 1994. When did Adirondack Mountain lakes begin to acidify and are they still acidifying? *Can. J. Fish. Aquat. Sci.*. (in press).

Cumming, B.F., Wilson, S.E. & Smol, J.P. 1993. Paleolimnological potential of chrysophyte cysts and scales, and sponge spicules as indicators of lakewater salinity. *Int. J. Salt Lake Res.* **2**(1): 87–92.

Davis, R.B. 1989. The scope of Quaternary paleolimnology. *J. Paleolimnol.* **2**: 263–83.

Davis, R.B., Anderson, D.S., Whiting, M.C., Smol, J.P. & Dixit, S.S. 1990. Alkalinity and pH of 3 lakes in northern New England, USA over the past 3000 years. *Phil. Trans. R. Soc. Lond. B* **327**: 413–21.

Dixit, A.S. & Dixit, S.S. 1989. Surface sediment chrysophytes from 35 Quebec lakes and their usefulness in reconstructing lakewater pH. *Can. J. Bot.* **67**: 2071–6.

Dixit, S.S., Dixit, A.S. & Evans, R.D. 1988*a*. Chrysophyte scales in lake sediments provide evidence of recent acidification in two Quebec (Canada) lakes. *Water Air Soil Pollut.* **38**: 97–104.

Dixit, S.S., Dixit, A.S. & Evans, R.D. 1988*b*. Scaled chrysophytes (chrysophyceae) as indicators of pH in Sudbury, Ontario, lakes. *Can. J. Fish. Aquat. Sci.* **45**: 1411–21.

Dixit, S.S., Dixit, A.S. & Evans, R.D. 1989*a*. Paleolimnological evidence for trace metal sensitivity in scaled chrysophytes. *Environ. Sci. Technol.* **23**: 110–15.

Dixit, S.S., Dixit, A.S. & Smol, J.P. 1989*b*. Relationship between chrysophyte assemblages and environmental variables in 72 Sudbury lakes as examined by Canonical Correspondence Analysis (CCA). *Can. J. Fish. Aquat. Sci.* **46**: 1667–76.

Dixit, S.S., Dixit, A.S. & Smol, J.P. 1989*c*. Lake acidification recovery can be monitored using chrysophycean microfossils. *Can. J. Fish. Aquat. Sci.* **46**: 1309–12.

Dixit, S.S., Dixit, A.S. & Smol, J.P. 1990*a*. Paleolimnological investigation of three manipulated lakes from Sudbury, Canada. *Hydrobiologia* **214**: 245–52.

Dixit, S.S., Smol, J.P., Anderson, D.S. & Davis, R.B. 1990*b*. Utility of scaled chrysophytes in predicting lakewater pH in northern New England lakes. *J. Paleolimnol.* **3**: 269–86.

Dixit, A.S., Dixit, S.S. & Smol, J.P. 1992*a*. Long-term trends in lake water pH and metal concentrations inferred from diatoms and chrysophytes in three lakes near Sudbury, Ontario. *Can. J. Fish. Aquat. Sci.* **49** (Suppl. 1): 17–24.

Dixit, S.S., Smol, J.P., Kingston, J.C. & Charles, D.F. 1992*b*. Diatoms: powerful indicators of environmental change. *Environ. Sci. Technol.* **26**: 22–33.

Dixit, S.S., Dixit, A.S. & Smol, J.P. 1992*c*. Assessment of pre-industrial changes in lakewater chemistry in Sudbury area lakes. *Can. J. Fish. Aquat. Sci.* **49** (Suppl. 1): 8–16.

Dixit, S.S., Cumming, B.F., Smol, J.P. & Kingston, J.C. 1992*d*. Monitoring environmental changes in lakes using algal microfossils. In: McKenzie, D., Hyatt, D.E. & MacDonald, V.S. (eds.) *Ecological Indicators*, vol. 2, Elsevier Applied Science, Amsterdam. pp. 1135–55.

Dixit, A.S., Dixit, S.S. & Smol, J.P. 1992*e*. Algal microfossils provide high temporal resolution of environmental trends. *Water Air Soil Pollut.* **62**: 75–87.

Douglas, M.S.V. & Smol, J.P. 1988. Siliceous protozoan and chrysophycean microfossils from the recent sediments of *Sphagnum* dominated Lake Colden, N.Y. *Verh. Int. Ver. Limnol.* **23**: 855–59.

Duff, K.E. & Smol, J.P. 1988. Chrysophycean stomatocysts from the postglacial sediments of a high arctic lake. *Can. J. Bot.* **66**: 1112–28.

Duff, K.E. & Smol, J.P. 1991. Morphological descriptions and stratigraphic distributions of the chrysophycean stomatocysts from a recently acidified lake (Adirondack Park, N.Y.). *J. Paleolimnol.* **5**: 73–113.

Duff, K., Douglas, M.S.V.D. & Smol, J.P. 1992. Chrysophyte cysts in 36 Canadian high arctic ponds. *Nord. J. Bot.* **12**: 471–99.

Engstrom, D.R., Whitlock, C., Fritz, S.C. & Wright, H.E. Jr. 1991. Recent environmental changes inferred from the sediments of small lakes in Yellowstone's northern range. *J. Paleolimnol* **5**: 139–74.

Fritz, S., Juggins, S., Battarbee, R.W. & Engstrom, D.R. 1991. Reconstruction of past changes in salinity and climate using a diatom-based transfer function. *Nature* **352**: 706–8.

Gibson, K.N., Smol, J.P. & Ford, J. 1987, Chrysophycean microfossils and the recent history of a naturally acidic lake (Cone Pond, N.H.). *Can. J. Fish. Aquat. Sci.* **44**: 1584–8.

Glew, J.R. 1988. A portable extruding device for close interval sectioning of unconsolidated core samples. *J. Paleolimnol.* **1**: 235–9.

Glew, J.R. 1991. Miniature gravity corer for recovering short sediment cores. *J. Paleolimnol.* **5**: 285–7.

Hartmann, H. & C. Steinberg, 1986. Mallomonadacean (Chrysophyceae) scales: Early biotic paleoindicators of lake acidification. *Hydrobiologia* **143**: 87–91.

Hartmann-Zahn, H. 1991. Scaled Chrysophytes as indicators of lake acidification in West Germany. *Verh. Int. Ver. Limnol.* **24**: 800–5.

Harwood, D.M. 1986. Do diatoms beneath the Greenland ice sheet indicate interglacials warmer than present? *Arctic* **39**: 304–8.

Haworth, E.Y. 1983. Diatom and chrysophyte relict assemblages in the sediments of Blelham Tarn in the English Lake District. *Hydrobiologica* **103**: 131–4.

Haworth, E.Y. 1984. Stratigraphic changes in algal remains (diatoms and chrysophytes) in the recent sediments of Blelham Tarn, English Lake District. In: Haworth, E.Y. & Lund, J.W.G. (eds.) *Lake Sediments and Environmental History*. University of Minnesota Press. pp. 165–90.

Hickman, M. & Schweger, C.E. 1991. A palaeoenvironmental study of Fairfax Lake, a small lake situated in Rocky Mountain Foothills of west-central Alberta. *J. Paleolimnol.* **6**: 1–15.

Jongman, R.H.G., ter Braak, C.J.F. & van Tongeren, O.F.R. 1987. *Data Analysis in Community and Landscape Ecology*. Pudoc, Wageningen. 299pp.

Kingston, J.C., Cumming, B.F., Uutala, A.J., Smol, J.P., Camburn, K.E., Charles, D.F., Dixit, S.S. & Kreis, R.G. Jr 1992a. Biological quality control and quality assurance: a case study in paleolimnological biomonitoring. In: McKenzie, D., Hyatt, D.E. & MacDonald, V.S. (eds) *Ecological Indicators*, vol. 2. Elsevier Applied Science, Amsterdam. pp. 1542–4.

Kingston, J.C., Birks, H.J.B., Uutala, A.J., Cumming, B.F. & Smol, J.P. 1992b. Assessing trends in fishery resources and lake water aluminum for paleolimnological analyses of siliceous algae. *Can. J. Fish. Aquat Sci.* **49**: 116–27.

Kristiansen, J. 1986. Silica-scale bearing chrysophytes as environmental indicators. *Br. Phycol. J.* **21**: 425–36.

Leavitt, P.R., Carpenter, S.R. & Kitchell, J.F. 1989. Whole-lake experiments: the annual record of fossil pigments and zooplankton. *Limnol. Oceanogr.* **34**: 700–17.

Lemmen, D.S., Gilbert, R., Smol, J.P. & Hall, R.I. 1988. Holocene sedimentation in glacial Tasikutaaq lake, Baffin Island. *Can. J. Earth Sci.* **25**: 810–23.

Line, J.M. & Birks, H.J.B. 1990. WACALIB version 2.1: a computer program to reconstruct environmental variables from fossil assemblages by weighted averaging. *J. Paleolimnol.* **3**: 170–3.

Marchetto, A. 1992. Valutazione paleolimnologica dei fenomeni di acidificazione dei laghi alpini. PhD thesis, University of Parma, Italy. 223 pp.

Marciniak, B. 1988. Late glacial *Fragiliaria* flora from lake sediments of the Tatra Mts, and the Alps. In: Round, F.E. (ed.) *Proceedings of the Ninth International Diatom Symposium*. Koeltz Scientific Books, Koenigstein. pp. 233–43.

Marsicano, L.J. & Siver, P.A. 1993. A paleolimnological assessment of lake acidification in five Connecticut lakes. *J. Paleolimnol.* **9**: 209–22.

Mitchell, J.G. & Silver, M.W. 1986. Archaeomonad (Chrysophyta) cysts: ecological and paleoecological significance. *BioSystems* **19**: 289–98.

Munch, C.S. 1980. Fossil diatoms and scales of Chrysophyceae in the recent history of Hall Lake, Washington. *Freshwater Biol.* **10**: 61–6.

Munch, C.S. 1985, Chrysophycean scales as paleoindicators in the sediments of Hall Lake, Washington, USA. *Nord. J. Bot.* **5**: 505–10.

Nygaard, G. 1956. Ancient and recent flora of diatoms and Chrysophyceae in Lake Gribsø. *Folia Limnol. Scand.* **8**: 32–262.

Peglar, S.M., Fritz, S.C., Alapieti, T., Saarnisto, M. & Birks, H.J.B. 1984. Composition and formation of laminated sediments in Diss Mere, Norfolk, England. *Boreas* **13**: 13–28.

Pienitz, R., Walker, I.R., Zeeb, B.A., Smol, J.P. & Leavitt, P.R. 1992. Biomonitoring past salinity changes in an athalassic sub-Arctic lake. *Int. J. Salt Lake Res.* **1**(2): 91–123.

Rhodes, T. 1991. Effects of late Holocene forest disturbance on an acidic Maine lake. PhD thesis, University of Maine. 241pp.

Risberg, J. 1990. Siliceous microfossil stratigraphy in a superficial sediment core from the northern part of the Baltic proper. *Ambio* **19**: 167–72.

Rull, V. 1986. Diatomeas y crisofíceas en los sedimentos acuáticos de una depresión cárstica del Pirineo catalán. *Oecol. Aquat.* **8**: 11–24.

Rull, V. 1991. Palaeoecological significance of chrysophycean stomatocysts: a statistical approach. *Hydrolobiologia* **220**: 161–5.

Rybak, M. 1986. The chrysophycean paleocyst flora of the bottom sediments of Kortowskie Lake (Poland) and its ecological significance. *Hydrobiologia* **140**: 67–84.

Rybak, M. 1987. Fossil chrysophycean cyst flora of Racze Lake, Wolin Island (Poland) in relation to paleoenvironmental conditions. *Hydrobiologia* **150**: 257–72.

Rybak, M., Rybak, I. & Dickman, M. 1987. Fossil chrysophycean cyst flora in a small meromictic lake in southern Ontario, and its paleoecological interpretation. *Can. J. Bot.* **65**: 2425–40.

Rybak, M., Rybak, I. & Nicholls, K. 1991. Sedimentary chrysophycean cyst assemblages as paleoindicators in acid sensitive lakes. *J. Paleolimnol.* **5**: 19–72.

Sandgren, C.D. 1983. Morphological variability in populations of Chrysophycean cysts. I. Genetic (interclonal) and encystment temperature effects on morphology. *J. Phycol.* **19**: 64–70.

Sandgren, C.D. 1988. The ecology of chrysophyte flagellates: their growth and perennation strategies as freshwater phytoplankton. In: Sandgren, C.D. (ed.) *Growth and Reproductive Strategies of Freshwater Phytoplankton.* Cambridge University Press, Cambridge, pp. 9–104.

Sandgren, C.D. 1991a. Chrysophyte reproduction and resting cysts: a paleolimnologist's primer. *J. Paleolimnol.* **5**: 1–9.

Sandgren, C.D. (ed.). 1991b. Application of chrysophyte stomatocysts in paleolimnology. *J. Paleolimnol.* **5**: 1–113.

Sandgren, C.D. & Carney, H.J. 1983. A flora of fossil Chrysophycean cysts from the recent sediments of Frains Lake, Michigan, USA. *Nova Hedwigia* **38**: 129–63.

Sandgren, C.D. & Smol, J.P. 1991. Reply to Dr Simola's comment on the taxonomy of 'snowflakes' (chrysophyte cysts). *J. Paleolimnol.* **6**: 261–3.

Simola, H. 1991. Chrysophyte stomatocyst taxonomy: classification of snowflakes? *J. Paleolimnol.* **6**: 257–60.

Siver, P.A. 1991. Implications for improving paleolimnological inference models utilizing scale-bearing siliceous algae: transforming scale counts to cell counts. *J. Paleolimnol.* **5**: 219–25.

Siver, P.A. 1993. Inferring specific conductivity of lake water using scaled chrysophytes. *Limnol. Oceanogr.* **30**: 1480–92.

Siver, P.A. & Hamer, J.S. 1990. Use of extant populations of scaled chrysophytes for the inference of lakewater pH. *Can. J. Fish. Aquat. Sci.* **47**: 1339–47.

Siver, P.A. & Hamer, J.S. 1992. Seasonal periodicity of Chrysophyceae and Synurophyceae in a small New England lake: implications for paleolimnological research. *J. Phycol.* **28**: 186–98.

Siver, P.A. & Skogstad, A. 1988. Morphological variation and ecology of *Mallomonas crassisquama. Nord. J. Bot.* **8**: 99–107.

Siver, P.A. & Smol, J.P. 1993. The use of scaled chrysophytes in long term monitoring programs for the detection of changes in lakewater acidity. *Water Air Soil Pollut.* **73**: 357–76.

Smith, M.A. & White, M.J. 1985. Observations on lakes near Mount St Helens: phytoplankton. *Arch. Hydrobiol.* **104**: 345–62.

Smol, J.P. 1979. Paleolimnology of selected Precambrian Shield lakes. MSc thesis, Brock University. 225pp.

Smol, J.P. 1980. Fossil synuracean (Chrysophyceae) scales in lake sediments: a new group of paleoindicators. *Can. J. Bot.* **58**: 458–65.

Smol, J.P. 1982. Postglacial changes in fossil algal assemblages from three Canadian lakes. PhD thesis, Queen's University. 151pp.

Smol, J.P. 1983. Paleophycology of a high arctic lake near Cape Herschel, Ellesmere Island. *Can. J. Bot.* **61**: 2195–204.

Smol, J.P. 1985. The ratio of diatom frustules to chrysophycean statospores: a useful paleolimnological index. *Hydrobiologia* **123**: 199–208.

Smol, J.P. 1986. Chrysophycean microfossils as indicators of lakewater pH. In: Smol, J.P. *et al.* (eds.) *Diatoms and Lake Acidity.* Junk, Dordrecht. pp. 275–87.

Smol, J.P. 1987. Methods in Quaternary ecology: freshwater algae. *GeoScience Canada* **14**: 208–17.

Smol, J.P. 1988*a*. Chrysophycean microfossils in paleolimnological studies. *Palaeogeog. Palaeoclim. Palaeoecol.* **62**: 287–97.

Smol, J.P. 1988*b*. The North American 'endemic' *Mallomonas pseudocoronata* (Mallomonadacea, Chrysophyta) in an Austrian lake. Phycologia **27**: 427–9.

Smol, J.P. 1990*a*. Paleolimnology: recent advances and future challenges. *Mem. Ist. Ital. Idrobiol.* **47**: 253–76.

Smol, J.P. 1990*b*. Are we building enough bridges between paleolimnology and aquatic ecology? *Hydrobiologia* **214**: 201–6.

Smol, J.P. 1990*c*. Diatoms and chrysophytes: a useful combination in palaeolimnological studies: report of a workshop and a working bibliography. In: Simola, H. (ed.) *Proceedings of the Tenth International Diatom Symposium.* Koeltz Scientific Books. Koenigstein. pp. 585–92.

Smol, J.P. 1992. Paleolimnology: an important tool for effective ecosystem management. *J. Aquat. Ecosystem Health* **1**: 49–58.

Smol, J.P. & Boucherle, M.M. 1985. Postglacial changes in algal and cladoceran assemblages in Little Round Lake, Ontario. *Arch. Hydrobiol.* **103**: 25–49.

Smol, J.P. & Dixit, S.S. 1990. Patterns of pH change inferred from chrysophycean microfossils in Adirondack and New England lakes. *J. Paleolimnol.* **4**: 31–41.

Smol, J.P. & Glew, J.R. 1992. Paleolimnology. In: Nierenberg, W.A. (ed.) *Encyclopedia of Earth System Science*, vol. 3. Academic Press, New York. pp. 551–64.

Smol, J.P., Brown, S.R. & McNeely, R.N. 1983. Cultural disturbances and trophic history of a small meromictic lake from Central Canada. *Hydrobiologia* 103: 125–30.

Smol, J.P., Charles, D.F. & Whitehead, D.R. 1984*a*. Mallomonadacean (Chrysophyceae) assemblages and their relationships with limnological characteristics in 38 Adirondack (N.Y.) lakes. *Can. J. Bot.* 62: 911–23.

Smol, J.P., Charles, D.F. & Whitehead, D.R. 1984*b*. Mallomonadacean microfossils provide evidence of recent lake acidification. *Nature* 307: 628–30.

Smol, J.P., Walker, I.R. & Leavitt, P.R. 1991. Paleolimnology and hindcasting climatic trends. *Verh. Int. Ver. Limnol.* 24: 1240–6.

Srivastava, S.K. & Binda, P.L. 1984. Siliceous and silicified microfossils from the Maastrichtian Battle Formation of Southern Alberta, Canada. *Paleobiolog. Cont.* 14: 1–24.

Steinberg, C. & Hartmann, H. 1986. A biological paleoindicator for early lake acidification: Mallomonadacean (Chrysophyceae) scale abundance in sediments. *Naturwissenschaften* 73: 37–9.

Steinberg, C., Hartmann, H., Arzet, K. & Krause-Dellin, D. 1988. Paleoindication of acidification in Kleiner Abersee (Federal Republic of Germany, Bavarian Forest) by chydorids, chrysophytes, and diatoms. *J. Paleolimnol.* 1: 149–57.

Stoermer, E.F., Kociolek, J.P., Schelske, C.L. & Andersen, N.A. 1991. Siliceous microfossil succession in the recent history of Green Bay, Lake Michigan. *J. Paleolimnol.* 6: 123–40.

Sullivan, T.J., Turner, R.S., Charles, D.F., Cumming, B.F., Smol, J.P., Schofield, C.L., Driscoll, C.T., Birks, H.J.B., Uutala, A.J., Kingston, J.C., Dixit, S.S., Bernert, J.A., Ryan, P.F. & Marmorek, D.F. 1992. Use of historical assessment for evaluation of process-based model projections of future environmental change: lake acidification in the Adirondack Mountains, New York, USA *Environ. Pollut.* 77: 253–62.

ter Braak, C.J.F. 1986. Canonical correspondence analysis: a new eigenvector technique for multivariate direct gradient analysis. *Ecology* 67: 1167–79.

Tolonen, K., Liukkonen, M., Harjula, R. & Pätilä, A. 1986. Acidification of small lakes in Finland documented by sedimentary diatom and chrysophycean remains. In: Smol, J.P. *et al.* (eds.) *Diatoms and Lake Acidity*. Dordrecht. pp. 169–99.

Uutala, A.J., Yan, N., Dixit, A.S., Dixit, S.S. & Smol, J.P. 1994. Paleolimnological assessment of declines in fish communities in three acidic Canadian shield lakes. *Fish. Res.* 19: 157–77.

VanLandingham, S.L. 1964. Chrysophyta cysts from the Yakima basalt (Miocene) in south-central Washington. *J. Paleontol.* 38: 729–39.

Vesely, J., Almquist-Jacobson, H., Miller, L., Norton, S., Appleby, P., Dixit, A. & Smol, J.P. 1993. The history and impact of air pollution at Lake Certova, southwestern Czech Republic. *J. Paleolimnol.* 8: 211–31.

Whitehead, D.R., Charles, D.F., Jackson, S.T., Smol, J.P. & Engstrom, D.R. 1989. The developmental history of Adirondack (N.Y.) lakes. *J. Paleolimnol.* 2: 185–206.

Whittier, T.R. & Paulsen, S.G. 1992. The surface waters component of the Environmental Monitoring and Assessment Program (EMAP): an overview. *J. Aquat Ecosystem Health* 1: 119–26.

Wright, H.E. Jr 1990. An improved Hongve sampler for surface sediments. *J. Paleolimnol.* **4**: 91–2.

Yang, J.-R., Duthie, H.C. & Delorme, L.D. 1993. Reconstruction of the recent environmental history of Hamilton Harbour (Lake Ontario, Canada) from analysis of siliceous microfossils. *J. Great Lakes Res.* **19**: 55–71.

Zeeb, B.A. & Smol, J.P. 1991. Paleolimnological investigation of the effects of road salt seepage on scaled chrysophytes in Fonda Lake, Michigan. *J. Paleolimnol.* **5**: 263–6.

Zeeb, B.A. & Smol, J.P. 1993*a*. Chrysophycean stomatocyst flora from Elk Lake, Minnesota. *Can. J. Bot.* **71**: 737–56.

Zeeb, B.A. & Smol, J.P. 1993*b*. Postglacial chrysophycean cyst record from Elk Lake, Minnesota. In: Bradbury, J.P. & Dean, W. (eds.) *Elk Lake, Minnesota: Evidence for Rapid Climate Change in the North-Central United States.* Geological Society of America Special Paper 276. pp. 239–49.

Zeeb, B.A., Duff, K.E. & Smol, J.P. 1990. Morphological descriptions and stratigraphic profiles of chrysophycean stomatocysts from the recent sediments of Little Round Lake, Ontario. *Nova Hedwigia* **51**: 361–80.

Zeeb, B.A., Christie, C.E., Smol, J.P., Findlay, D., Kling, H. & Birks, H.J.B. 1994. Responses of diatom and chrysophyte assemblages in Lake 227 to experimental eutrophication. *Can. J. Fish. Aquat Sci.* (in press).

Part IV
Contributed original papers

14

Mallomonas variabilis, sp. nov. (Synurophyceae) with stomatocysts found in Lake Konnevesi, Finland

GERTRUD CRONBERG

Introduction

At the seventh workshop of the International Association for Phytoplankton Taxonomy and Ecology (IAP) held on 5–15 May 1989 at Konnevesi Research Station, University of Jyväskylä, Finland, phytoplankton was collected in nearby Lake Konnevesi. At that time of year the lake was extremely rich in scaled chrysophytes (Table 14.1) and dominated by different species of *Mallomonas* and *Synura*. In the samples investigated, a *Mallomonas* species similar to *M. alpina* Pasch. & Ruttn. and *M. tonsurata*, Teil, was observed. Between 6 and 11 May this *Mallomonas* species developed stomatocysts (statospores, cysts). Preserved plankton samples were later examined with electron microscopy (EM). The *alpina–tonsurata*-like *Mallomonas* was found to be a new species and has been given the name *Mallomonas variabilis* sp. nov.

M. variabilis had, in fact, been recorded earlier from several Finnish lakes, even from Lake Konnevesi. However, no complete cells, only isolated scales were found; thus not enough information was available to make a new description of the species (Hällfors & Hällfors 1988). This species has also been recorded under the name *M. tonsurata* fa. in large lakes in northwestern Ontario, central Northwest Territories, and in the Alaskan Toolik Lake area (H. Kling, unpublished data).

Methods

Locality

Lake Konnevesi is situated in the northeastern part of central Finland (62°40'N, 26°30'E). It is an oligotrophic lake, with low nutrient concentrations (see Eloranta, this volume). However, during the past 15 years

333

Table 14.1. *The algal species associated with* Mallomonas variabilis,
Lake Konnevesi, Finland, 6–10 May 1989

CYANOPHYTA
Planktothrix mougeotii (Bory et Gom.) Anagn. & Kom.
Woronichinia compacta (Lemm.) Kom. & Hind.
W. naegeliana (Ung.) Elenk.

CHLOROPHYCEAE
Botryococcus braunii Kütz.
Chlorogonium maximum Skuja
Monoraphidium sp.
Kolliella longiseta (Visch.) Hind.
Pediastrum biradiatum Meyen
P. duplex Meyen
Pseudosphaerocystis lacustris (Lemm.) Nov.

CHROMOPHYTA

Diatomophyceae
Asterionella formosa Hass.
Cyclotella comensis Grun.
Diatoma elongatum Agardh
Fragilaria sp.
Melosira sp.
Rhizosolenia longiseta Zach.
Surirella elegans Ehrenb.
Synedra acus Kütz.
S. ulna (Nitzsch) Ehrenb.
Tabellaria fenestrata (Lyng.) Kütz.
T. fenestrata var. *asterionelloides* Grun.

Chrysophyceae and Synurophyceae
Dinobryon bavaricum Imh.
D. bavaricum var. *medium* (Lemm.) Krieg.
D. borgei Lemm.
D. cylindricum Imh.
D. sociale var. *stipiatum* (Stein) Lemm.
D. suecicum Lemm.
Chrysastrella furcata (Dolg.) Defl.
C. paradoxa Chod.
Chrysococcus cf. *diaphanus* Skuja
Chrysolykos planctonicus Mack
Chrysosphaerella brevispina Korsch.
Chrysostephanosphaera globulifera Scherff.
Mallomonas akrokomos Ruttn.
M. alpina Pasch. & Ruttn.
M. caudata Ivan.
M. crassisquama (Asm.) Fott

Table 14.1 (*continued*)

M. elongata Rev.
M. eoa Tak.
M. hamata Asm.
M. heterospina Lund
M. punctifera Korsch.
M. tonsurata Teil.
M. torquata Asm. & Cronb.
M. transsylvanica Péterfi & Mom.
M. variabilis Cronb.
Paraphysomonas imperforata Lucas
Spiniferomonas bourrellyi Tak.
S. trioralis Tak.
Synura leptorrhabda (Asm.) Nich.
S. petersenii Korsch.
S. spinosa Korsch.
S. uvella Ehrenb.

Cryptophyceae
Cryptomonas sp.
Rhodomonas sp.

Dinophyceae
Ceratium hirundinella (O.F.M.) Schrank
Gymnodinium helveticum Pen.
Peridinium cf. *umbonatum* Stein
Peridiniopsis berolinense (Lemm.) Bourr.
Woloszynskia sp.

Xanthophyceae
Tetradiella cf. *jovetii* (Bourr.) Bourr.

the eastern part of the lake has become eutrophied due to fish farming (Eloranta & Palomäki 1986). The lake is surrounded mainly by woods consisting of pine and birch. Some data on its hydrology and water chemistry are given in Table 14.2.

Sample preparation and examination

Phytoplankton samples were collected with 10 µm mesh plankton nets. The sampling took place between 6 and 11 May 1989, three times per day (at about 08:00, 13:00 and 20:00 hours). Live samples were studied with light microscopy (LM). Some of the material was preserved in Lugol's solution and some in an approximately 2% formalin solution.

336 *Chrysophyte algae*

Table 14.2. *Hydrographic and water chemistry data from Lake Konnevesi, Finland (Eloranta & Palomäki 1986) and lakes in North America from which* M. variabilis *has been collected (H. Kling, unpublished data)*

	Konnevesi	North America
Surface area (km^2)	187	0.07–66 000
Mean depth (m)	10	<5 to >100
Maximum depth (m)	54	7–219
Theoretical retention time	442 days	Month to several years
Temperature (°C)	5–10	<13
Conductivity at 25 °C (mS m^{-1})	4.4	5.7–25
pH	7.4–7.5	7.3–8.5
Color (mg Pt l^{-1})	30	–
Total N (μg l^{-1})	380–460	160–240
Total P (μg l^{-1})	5–8	<2–4
Chlorophyll a (mg m^{-3})	1.9–2.2	0.26–3.6

For scanning electron microscopy (SEM) unpreserved samples were filtered through Nuclepore filters (pore size 0.45 μm) and air dried. The filters were glued onto specimen stubs and coated with gold/palladium. Samples rich in *Mallomonas* cells with stomatocysts were treated with 30% hydrogen peroxide for about 24 h in order to oxidize the organic matter and to release the stomatocysts from the cells. The samples were then rinsed repeatedly with distilled water. A drop of the solution was applied to Formvar-coated copper grids or specimen stubs for EM studies.

The samples prepared for scanning electron microscopy (SEM) were studied with a Jeol JSM-25SII scanning electron microscope, operated at 12.5–15 kV. Those prepared for transmission electron microscopy (TEM) were studied with a Zeiss EM10C transmission electron microscope, operated at 60 kV, at the Department of Crop Genetics and Breeding, Swedish University of Agricultural Sciences in Svalöf.

The description of *Mallomonas variabilis* follows the terminology of Asmund & Kristiansen (1986), and the description of the stomatocyst follows the guidelines of the International Statospore Working Group (ISWG: Cronberg & Sandgren 1986). All measurements have been made on preserved material. The cells and stomatocysts of *Mallomonas variabilis* were measured with a Zeiss Standard light microscope; the scales and bristles were measured using SEM and TEM micrographs.

Description of *Mallomonas variabilis* nov. sp.

Latin diagnosis

Cellula ellipsoides, (13–)18–20 × (7–)11–14 µm magna, setis dimorphis praecipue anticis et spinis brevibus posticis armata.

Squamae anticae et saepe nonnullae mediae cupulatae, subrhombicae vel ovales, saepe asymmetricae, 7 × 4 µm magnae; laminae basales earum poris irregulariter distributis, in angulis cristarum V-formium minoribus, in limbis posterioribus majoribus perforatae, in scutis cristas breves sparsas praebentes, nunc satis multas, nunc paucas vel nullas (unde nomen specificum). Squamae mediae plurimae non cupulatae, 5 × 3 µm magnae. Squamae posticae ovales vel subcirculares, asymmetricae, 4 × 3 µm magnae, saepe spinigerae, spina e costa submarginali sinistra surgente, ad 0.3 µm longa. 6–8 setae apicales 10–13 µm longae, validae, 6–8 cuique dentibus serratae. 10–12 setae posteriores 16–20 µm longae, graciles, dentibus cuique 8–12.

Stomatocysta ellipsoides vel obovoides, 14–16 × 9–13 µm magna, laevis, ostio 1 µm diam. Prope polum latiorem sito, in cellula pariente semper postico.

Die Maji anni 1989 in lacu fennico Konnevesi (62°40′ lat. bor., 26°30′ long. orient.) ab auctore lecta, Fig. 11B typifica monstrata.

Cell and scale morphology

The cells are small and oval in shape. They possess two types of bristles (Figs. 14.1–14.2), which are mostly confined to the apical part of the cell. There are also posterior scales with short spines. The scales are rhomboid to oval and have a base plate perforated with irregularly distributed pores (Figs. 14.7–14.10). Four types of scales can be recognized: domed apical scales, domed body scales, domeless body scales and domeless rear scales with spines.

The dome-bearing scales are concentrated at the apical part of the cell. A few can also be found irregularly distributed further down on the cell (Figs. 14.1, 14.2 and 14.11). The shape of the domed scales is slightly rhomboid to oval and often asymmetric (Fig. 14.9). The proximal part of the scale is surrounded by a thin, partly upturned rim. The dome is upraised and has thin ridges. The V-rib is well developed, sometimes hooded, and connected to submarginal ribs, which fuse with the dome. In the angle of the V-rib there is a field with small pores. Pores larger than those on the shield are clearly visible on the posterior flanges. The

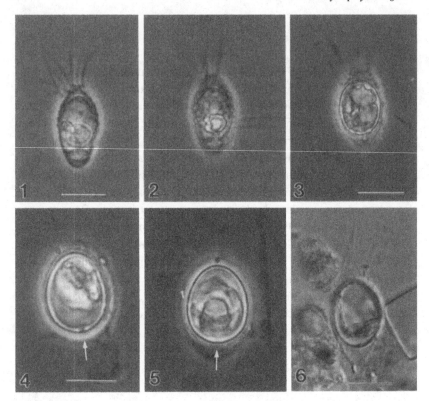

Figures 14.1–14.6. *Mallomonas variabilis* sp. nov. Figures 14.1, 14.2. Normal cell with the characteristic bristle conformation (the same cell with different focus). Figures 14.3–14.6. Cells with stomatocysts developed inside. The pore is directed towards the proximal end of the cell. Figure 14.4 shows a stomatocyst obovate in shape with asymmetrically positioned pore (arrow). Figure 14.5 shows the same stomatocyst but with different focus. Scale bars represent 10 μm.

body scales are oval and in most cases lack a dome (Figs. 14.7–14.10). The V-rib is also connected to the submarginal ribs, which fuse at the distal end of the scale. Basal scales are small, oval to almost circular, asymmetric, and the left submarginal rib often supports a short spine, length ≤ 0.3 μm (Figs. 14.7, 14.9).

The most characteristic feature of the scales is the presence of short, irregularly distributed ridges on the shield. However, some scales have only a few ridges and others none at all (Figs. 14.7–14.14).

Bristles are primarily concentrated at the apical part of the cell, with a few isolated bristles associated with scales in the middle of the cell (Figs. 14.1, 14.2, 14.7, 14.11, 14.13 and 14.14). There are two types of bristles.

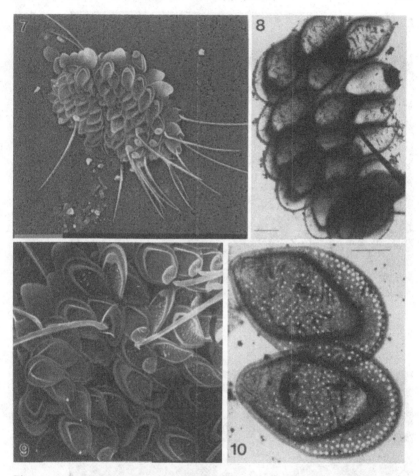

Figures 14.7–14.10. *Mallomonas variabilis* sp. nov. Figure 14.7. A cell with the different types of scales and bristles. Figure 14.8. Body scales with and without the characteristic ridges on the shield. Figure 14.9. Apical, domed scales, body scales and basal scales with short spines. Figure 14.10. Body scales. Scale bars represent 10 μm (Figs 14.7, 14.9) or 1 μm (Figs. 14.8, 14.10).

The short, stout, apical bristles are serrated, with about 6–8 evenly distributed teeth (Figs. 14.13, 14.14). The more posteriorly positioned bristles are long, slender and serrated, with about 8–12 teeth. Both types of bristles are slightly bent with the teeth on the convex side. There are 6–8 apical bristles and about 10–12 body bristles on each cell. Most body bristles are of the slender, long type, but some of the shorter stouter apical bristles can also be found on the body (Fig. 14.13).

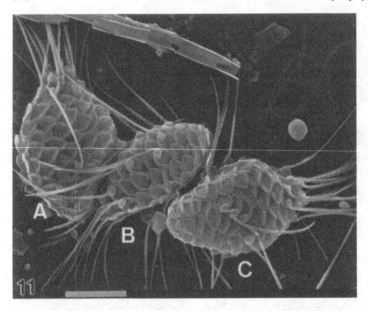

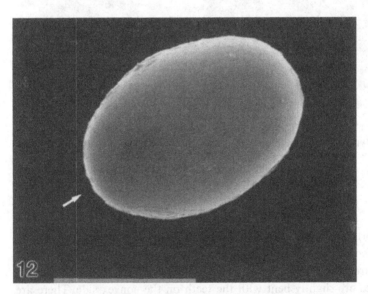

Figures 14.11, 14.12. *Mallomonas variabilis* sp. nov. Figure 14.11. Three cells. Cell B is the type specimen. Cells A and C have stomatocysts developed inside. Scale bar represents 10 μm. Figure 14.12. The stomatocyst. The pore region is indicated. Scale bar represents 10 μm.

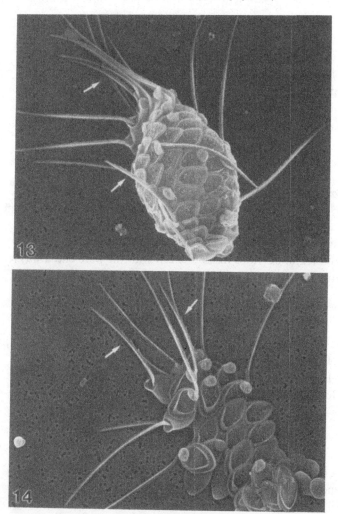

Figures 14.13, 14.14. *Mallomonas variabilis* sp. nov. Figure 14.13. Cell with both types of bristles. Figure 14.14. Scales and bristles. The short, stout bristles are indicated.

Dimensions   Cells (13)–18–20 × (7)–11–14 µm, domed scales 7 × 4 µm; domeless body scales 5 × 3 µm; basal scales 4 × 3 µm, apical bristles 10–13 µm; body bristles 16–20 µm; cysts 14–16 × 9–13 µm.

Type   Figure 14.11*B*.

Type locality   Lake Konnevesi, Finland (62°40′N, 26°30′E).

Date   6 May 1989.

Epithet   The epithet refers to the varying ultrastructure of the scales. The distribution of the ridges on the shield varied considerably within the scales on a single cell as well as among cells and populations. North American populations had a less dense distribution of ridges than seen in the Finnish material (H. Kling unpublished data).

Stomatocyst

The stomatocyst (statospore, cyst) is oval to obovate, with a smooth cyst wall (Figs. 14.3–14.6, 14.12). It has a minute pore (diameter 1 μm) without any structural differentiation in the pore area. The pore is positioned more or less asymmetrically at the broader end of the cyst body. The stomatocyst develops inside the cell with the pore end always directed towards the posterior end of the cell (Figs. 14.4–14.6).

Description of the stomatocyst of Mallomonas variabilis *according to ISWG*

Stomatocyst of *Mallomonas variabilis* Cronberg
Morphotype: Fig. 14.12; negative number Cronberg 910802-S42/9
Morphotype locality: Lake Konnevesi, Finland
Date: 11 May 1989

The stomatocyst is oval to obovate (14–16 μm × 9–13 μm) with a smooth cyst wall. It has a minute pore (diameter 1 μm) without any structural differentiation on the pore or in the pore area. The pore is positioned at the broader end of the cyst body.

Discussion

Using LM, it is impossible to distinguish *Mallomonas variabilis* from *M. alpina* or *M. tonsurata*, as the shape of the cells and the arrangement of the bristles of the three species look similar. However, with EM the differences are obvious. *M. tonsurata* has bristles concentrated at the apical part of the cell. The anterior bristles are short, stout, strongly curved and serrated. The body bristles are long, slender, without serration and with a subapical tooth on each. The bristles of *M. alpina* are distributed over a larger part of the cell than in *M. tonsurata*. The apical bristles are short, stout and serrated, while the body bristles are long, slender and also serrated. *M. variabilis* has the bristles confined to the apical part of the cell, with only a few bristles associated with the central portion of the cell. Bristle morphology is similar to that of *M. alpina*.

The shield of scales of *M. variabilis* has a secondary silica layer with irregularly distributed ridges. In *M. alpina*, no secondary layer is developed on the shield of the scales. However, *M. tonsurata* has a thick secondary layer with pits plus a pore in the bottom of the pits. In the angle of the V-rib there is a 'window' where the secondary layer is missing and the base plate with pores can be seen. Such pits are not present on the scales of *M. variabilis*, only the conspicuous ridges.

Both *M. alpina* and *M. tonsurata* have stomatocysts with well-developed spines distributed on the cyst body (Asmund & Kristiansen 1986; G. Cronberg unpublished data). The cyst wall of *M. variabilis* is smooth.

During the IAP workshop, sampling was undertaken in about 15 lakes surrounding the Konnevesi Research Station, but *Mallomonas variabilis* was recorded only from Lake Konnevesi. Although many investigations of chrysophytes in Scandinavian waterbodies have been carried out, *M. variabilis* has hitherto been recorded only from Finnish lakes. It thus seems to be a species with a restricted distribution within Scandinavia. However, in North America it has been recorded from large shield lakes and small arctic lakes (H. Kling unpublished data).

During the period 5–15 May 1989, the phytoplankton communities in the lakes investigated were nearly all dominated by chrysophytes, especially species belonging to the genera *Dinobryon*, *Mallomonas* and *Synura*. The plankton of Lake Konnevesi was very diverse and many scaled chrysophytes were present (Table 14.2). During the investigation period at Konnevesi the temperature of the water rose from about 5 °C to 10 °C and many chrysophytes started to develop stomatocysts. Species with recorded cyst formation were: *Mallomonas variabilis*, *M. hamata*, *M. heterospina*, *M. torquata*, *Dinobryon bavaricum* var. *medium* and *Chrysolykos plantonicus*. Also the chrysophytes *Chrysastrella furcta* and *C. paradoxa* were frequent. Thus, *M. variabilis* probably belongs to the vernal chrysophyte flora. When the lake water became warmer, the chrysophyte population developed cysts and then settled out of the water.

Summary

A new species of *Mallomonas*, *M. variabilis* (Synurophyceae), was found in Lake Konnevesi, Finland. It is similar to *Mallomonas alpina* Pasch. & Ruttn. and *M. tonsurata* Teil., but is distinguished from these species by different bristles and scale morphology. It is placed in the Section *Mallomonas*, Series Alpinae, fam. Mallomonadaceae. During the investigation period, 6–11 May 1989, *M. variabilis* developed stomatocysts, which are described here according to the ISWG guidelines.

Acknowledgments

I would like to thank Prof. Pertti Eloranta for arranging the seventh IAP workshop at Konnevesi Research Station and for putting all the facilities of the station at our disposal. Dr Tyge Christensen has kindly written the Latin diagnosis and Mrs Karin Ryde has corrected the English language. I am greatly indebted to H. Kling, who reviewed this paper and added valuable data on *Mallomonas variabilis* in North America. Technical assistance was kindly given by the laboratory assistants Kerstin Brismar and Gunnel Karlsson at the EM laboratory, Department of Crop Genetics and Breeding, Swedish University of Agricultural Sciences in Svalöf.

References

Asmund, B. & Kristiansen, J. 1986. The genus *Mallomonas* (Chrysophyceae). *Opera Bot.* **85**: 1–128.
Cronberg, G. & Sandgren, C.D. 1986. A proposal for the development of standardized nomenclature and terminology for chrysophycean statospores. In: Kristiansen, J. & Andersen, R.A. (eds.) *Chrysophytes: Aspects and Problems.* Cambridge University Press, Cambridge, pp. 317–28.
Eloranta, P. & Palomäki, A. 1986. Phytoplankton in Lake Konnevesi with special reference to eutrophication of the lake by fish farming. *Aqua Fenn.* **16**(1): 37–45.
Hällfors, G. & Hällfors, S. 1988. Records of chrysophytes with siliceous scales (Mallomonadaceae and Paraphysomonadaceae) from Finnish inland waters. *Hydrobiologia* **161**: 1–29.

15

Scale morphology and growth characteristics of clones of *Synura petersenii* (Synurophyceae) at different temperatures

BRIGITTE MARTIN-WAGENMANN &
ANTJE GUTOWSKI

Introduction

Synura petersenii has a worldwide distribution. It is often an important, and frequently the most common, *Synura* species of the phytoplankton (Kristiansen 1979; Siver 1987; Hällfors & Hällfors 1988; Hickel & Maass 1989). This taxon appears to be pH and temperature indifferent and often occurs under eutrophic conditions (Kristiansen 1975, 1986, 1988; Kies & Berndt 1984; Roijackers 1986).

The first monograph of the genus *Synura* by Korshikov (1929) was based on scale morphology. Balonov & Kuzmin (1974) later differentiated the genus into three sections: Synura, Petersenianae and Lapponicae. The section Petersenianae is an exception, as the species are distinguished by differences in the construction of elements of scales, whereas the species of the other sections exhibit explicitly different elements (Fott & Ludvik 1957). Several subspecific taxa within this section have been described on the basis of variations in the scales (Siver 1987, 1988); however, it is often difficult to distinguish them from the type for several reasons. First, there is variability among scales of the armour of the same cell (e.g. between body and caudal scales; Fig. 15.4(1, 2)). Second, some taxa are distinguished even though not all scales of the cell armour are significantly different from the type. Third, continuous transitions among subspecific taxa often occur (Siver 1988).

The above is especially true for *S. petersenii* var. *glabra* (Takahashi 1967; Kristiansen 1979, 1986; Nicholls & Gerrath 1985; Siver 1988; Gutowski 1989), which was first described by Korshikov (1929) as a separate species and then converted by Huber-Pestalozzi to a variety in

1941. Siver (1987) proposed to describe it as a forma, because the differences did not seem to be genetically distinct. Hällfors & Hällfors (1988) accepted no taxonomic recognition at any level and described *S. petersenii* var. *glabra* as a 'weakly silicified ecophene or phenecotype of *S. petersenii* f. *petersenii*'. Nevertheless, to date there are no data on the range of variability of the characters, and the possible effects of external influences such as temperature, pH or silica concentration of the medium on scale morphology. Our aim was to study this variability and determine whether it could be influenced by culture temperature or age (i.e. length of time after isolation from nature).

Materials and methods

Specimens of *S. petersenii* were isolated from three different sites and established from single cells detached from the colonies by washing them in distilled water. Table 15.1 provides details regarding isolation. Clones are available from the personal culture collection of the authors. Stock cultures were held as unialgal batch cultures in chemically defined DY III medium (Lehman 1976) at 12.5 °C and about 80 µEin m^{-2} s^{-1} of cool white fluorescent light on a 10:14 h light:dark cycle. At every transfer into new medium, the old culture was fixed (90% alcohol, 5% formalin, 5% glycerin), enabling us to study the influence of culture duration on scale morphology. Experiments were carried out in triplicate Erlenmeyer flasks at 5–25 °C, in 5 deg steps, under continuous fluorescent illumination of about 60 µEin m^{-2} s^{-1} light intensity in a different culture chamber. The clones were first adapted to the experimental temperature for at least 3 weeks. As there was only one chamber available for these experiments, it was not possible to start all experiments simultaneously. The cultures were of different age and may have been taken from different phases of growth.

Table 15.1. *Clones used in the study*

Clone	Isolation information
I	Isolated from 'Riemeisterfenn', Berlin, boggy eutrophic area, 27 October 1988, 5.2 °C, 717 µS cm^{-1}, pH 7.4, by B. Martin-Wagenmann
II	Isolated from 'pond Lichterfelde', Botanical Garden, Berlin, eutrophic, 10 November 1988, 3.5 °C, 719 µS cm^{-1}, pH 7.8, by B. Martin-Wagenmann
III	Isolated from 'Nikolassee', Berlin, eutrophic small lake, 8 November 1988, 3.3 °C, 680 µS cm^{-1}, pH 8.0, by B. Martin-Wagenmann

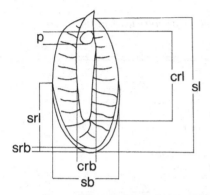

Figure 15.1. Measured characters. sl, scale length; sb, scale breadth; crl, central ridge length; crb, central ridge breadth; srl, scale rim length; srb, scale rim breadth; p, porus diameter.

For the determination of growth rates, an initial inoculum of 3–5 ml was transferred into 35–40 ml of new medium, and experiments were sampled every 2 days until the stationary phase was reached. Samples were fixed with Lugol's solution and the cell number determined in a haemocytometer (Fuchs-Rosenthal ruling) after breaking up the colonies ultrasonically. Growth rate was expressed as $\mu = (\ln N_2 - \ln N_1)/(t_2 - t_1)$, where N_2 and N_1 are the number of cells in the culture at times t_2 and t_1 respectively. The division time in units of days was calculated from the growth rate (Guillard 1973). These experiments were performed following Sandgren (1988).

To answer the question of the dependence of scale morphology variability on culture age, samples taken at the start of cloning and after 7 months in culture were studied using transmission electron microscopy (TEM). The question of the influence of temperature on scale variability was investigated using samples taken during the exponential phase of the experiments cultivated at 5 and 20 °C. Scales were washed in distilled water, air dried on Formvar-coated grids, and photographed at a magnification of × 12000 (corrected by the manufacturer to × 13 398) at 50 kV on filmsheets (8.3 × 10.2 cm) in a Zeiss EM109. Fifty scales from at least three grids (600 scales in total) were analyzed from the TEM negatives using a viewing table and a magnifying glass. Figure 15.1 and Table 15.2 show the characters studied.

Results and discussion

Classifying the scales of the different clones

Scales of *Synura petersenii* f. *petersenii* exhibit bilateral symmetry and are heteropolar. Each scale (Fig. 15.1) consists of an oval perforated basal

Table 15.2. *Qualitative scale characters used in this
study*

Strut length
Total numbers of struts
Number of struts reaching the scale edge
Bifurcation of struts (yes/no)
Longitudinal connections (yes/no)
Spine (yes/no)
Central ridge posteriorly curved (yes/no)
Pore size of the basal plate (fine/coarse)
Degree of silicification

plate with a median, hollow cylinder (central ridge) which ends in an
oblique, pointed cone called the spine. At the base of the spine is a large
circular porus. The upturned posterior rim extends at best to two-thirds
of the scale length. The basal plate is ornamented with a series of parallel
struts radiating from the central ridge to the scale rim.

Scales of *S. petersenii* var. *glabra* show several characters different from
the above type. Scales are smaller, and are circular to oval. The central
ridge is not as developed and is often curved in the posterior. The struts
do not touch the rim of the scale or are often completely missing. The
spine is also either missing or is not very pronounced, whereas the rim is
considerably developed. In general, the scales are not as strongly silicified
as scales of f. *petersenii* (Korshikov, 1929; Petersen & Hansen 1956; Fott
& Ludvik 1957; Takahashi 1967; Kies & Berndt 1984; Starmach 1985;
Kristiansen 1986).

Table 15.3 gives data for some scale characteristics measured at the
start of cloning. The largest scales are from clone I, whereas scales from
clones II and III are significantly smaller. Scale form is expressed by
calculating the quotient sl/sb (for definition see Fig. 15.1) Clone I differs
significantly from the others. It has more oblong-oval scales, whereas II
and III possess more circular scales (Fig. 15.4(1, 2, 9, 18, 19)). Clone I
also exhibits a pronounced central ridge, in contrast to the other clones
that do not differ significantly from each other with regard to this feature.
The rim is not strongly accentuated within the clones with more circular
scales, but there are pronounced differences in the development of struts.
As shown in Fig. 15.2 (I 1, II 1, III 1), 98% of scales of clone I have struts,
whereas only 30% of scales of clone II and 10% of scales of clone III have
struts, but most of these are longer than two-thirds of the space between

Table 15.3. *Mean values and standard deviations of scale characters at the start of cloning*

	Clone I 1	Clone II 1	Clone III 1
sl (μm)	4.08 ± 0.33	3.27 ± 0.33	2.90 ± 0.31
sb (μm)	2.06 ± 0.44	1.98 ± 0.34	1.88 ± 0.21
sl/sb	1.85 ± 0.21	1.69 ± 0.27	1.55 ± 0.21
crl (μm)	2.74 ± 0.26	1.96 ± 0.34	1.75 ± 0.27
sl/crl	1.59 ± 0.14 (63%)[a]	1.70 ± 0.21 (59%)[a]	1.68 ± 0.17 (60%)[a]
srl (μm)	2.61 ± 0.30	2.27 ± 0.23	1.96 ± 0.33
sl/srl	1.49 ± 0.17 (67%)[a]	1.47 ± 0.12 (68%)[a]	1.50 ± 0.18 (67%)[a]

[a] Quotient converted into percent.
For abbreviations see Fig. 15.1. Roman numerals refer to clones; arabic numerals refer to sample series (1 = start of cloning).

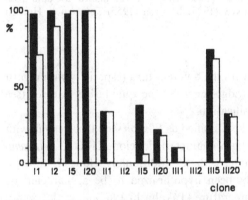

Figure 15.2. Percentage of scales with struts (black bars), and percentage of struts exceeding two-thirds of the space between the central ridge and rim (open bars), in scales from different sample series (arabic numerals) of clones (roman numerals). 1, start of cloning; 2, start of experiments; 5, 5 °C; 20, 20 °C.

the central ridge and rim (i.e. they reach or nearly reach the scale edge). This differentiating character is thus problematic. Scales of clone I are more heavily silicified than scales of the other clones as evidenced by the extent of scale ornamentation. Figure 15.3 gives some data from the literature for comparison; few such data exist, and often their significance is difficult to evaluate (number of measurements, maximum values, mean values, etc.). Generally, scales of clone I are markedly different from those of the other clones, where clones II and III show some overlap. Nevertheless,

Chrysophyte algae

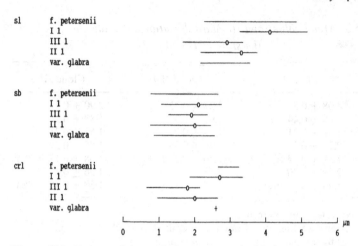

Figure 15.3. Ranges of size variation, compared with data from the literature. For abbreviations, see Fig. 15.1. Open circles, mean value; lines, variability range; +, single data point from the literature. 'f. *Petersenii*' and 'var. *glabra*' use data from the following references: Korshikov (1929) Manton (1955) Petersen & Hansen (1956) and Takahashi (1967).

the clones can be associated with the different taxa (especially with regard to scale length and central ridge length). The qualitative features (see Table 15.2) support the classification.

In conclusion, clone I can be classified as *S. petersenii* f. *petersenii*, with the exception of the strut length characteristic, clone II as *S. petersenii* var. *glabra*, and clone III is intermediate. Additionally, clone III shows a coarse basal plate which has been hypothesized to be a characteristic developed in response to temperature (Takahashi 1967).

Growth experiments

On the basis of floristic studies *S. petersenii* var. *glabra* was first characterized as oligothermal and later classified as eurythermal, as was f. *petersenii* (Gutowski 1989). Only a few culture experiments deal with chrysophytes, and those available generally deal with statospore production (Sandgren 1981, 1983, 1988; Sandgren & Flanagin 1986), silicon and nutrient demands (Klaveness & Guillard 1975; Lehman 1976; Sandgren & Barlow 1989; see Leadbeater and Barker, this volume), and light and temperature tolerances (Healey 1983; Sandgren 1988). For comparison, only data for *S. petersenii* cultures were considered here.

Table 15.4 gives the maximum cell density for the experimental temperatures. All clones showed a pronounced decrease in density at 25 °C. Maximum cell density remains more or less stable for the f. *petersenii* clone (I) from 5 to 20 °C. *S. petersenii* var. *glabra* (II) exhibits increasing cell density with decreasing temperature for this range, whereas the intermediate clone (III) reaches maximum densities at 10 °C. These data agree with values in the literature, although cultures may reach

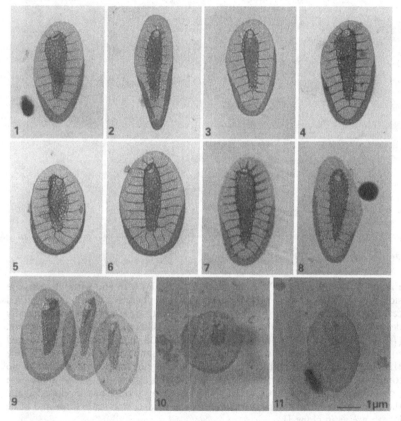

Figure 15.4. Scales from different sample series. (*1, 2*) Clone I, body and caudal scale at the start of cloning. (*3, 4*) Clone I, after 7 months in culture. Note the different degree of silicification. (*5, 6*) Clone I, 5 °C. Scales are rounded to circular. (*7, 8*) Clone I, 20 °C. Scales are oblong to oval. Note the different degree of silicification and bifurcation of struts. (*9*) Clone II, at the start of cloning. Note the different degree of silicification; scales with and without struts are side by side. (*10, 11*) Clone II, after 7 months in culture. Scales have become rounded, the central ridge has diminished, and there is a very low degree of silicification.

(*continued*)

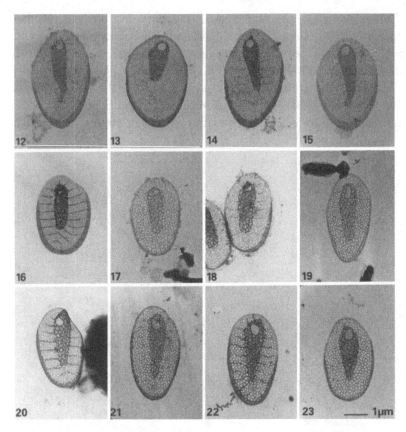

Figure 15.4 (*continued*) (*12–14*) Clone II, 5 °C. Scales with and without struts are side by side; the central ridge may develop again; there is a pronounced curved central ridge and rim. (*15, 16*) Clone II, 20 °C. Struts are sometimes pronouncedly developed; scale form is characteristic. Note the different degree of silicification. (*17, 18*) Clone III, at the start of cloning. There is a coarse basal plate, rounded to oval, different degree of silicification, scales with and without struts. (*19*) Clone III, after 7 months in culture. There is uniformity of scales, and struts can no longer be found. (*20, 21*) Clone III, 5 °C. Struts occur again and there is a pronounced spine. (*22, 23*) Clone III, 20 °C. The central ridge has developed again, there is a pronounced spine, no curved central ridge, and scale form and scale length are as in clone II.

higher densities with higher concentrations of silicon in the medium (Klaveness & Guillard 1975; Sandgren & Flanagin 1986).

Table 15.5 shows the calculated growth rates and division times. Clones used for the present study were isolated from cold water conditions (Table 15.1). Nevertheless, all clones exhibit a pronounced rise in growth rates at

Table 15.4. *Maximum cell densities (10^3 cells ml^{-1}) of clones at different temperatures (triplicate cultures)*

Temperature (°C)	I	II	III
5	310	390	320
10	300	320	360
15	280	310	260
20	280	200	270
25	70	97	117

Table 15.5. *Growth rates and division times of the clones at different temperatures (triplicate cultures)*

	Growth rate, μ			Division time, t_g [days]		
Temperature (°C)	I	II	III	I	II	III
5	0.25 ± 0.04	0.17 ± 0.01	0.15 ± 0.01	2.8	4.1	4.6
10	0.26 ± 0.01	0.15 ± 0.01	0.25 ± 0.03	2.6	4.7	2.7
15	0.25 ± 0.04	0.18 ± 0.01	0.22 ± 0.01	2.8	3.9	3.1
20	0.34 ± 0.03	0.25 ± 0.08	0.28 ± 0.01	2.1	2.8	2.5
25	0.16 ± 0.03	0.15 ± 0.02	0.22 ± 0.06	4.4	4.6	3.2

20 °C and then a drop at 25 °C. Clone f. *petersenii* (I) and the var. *glabra* clone (II) show no marked changes in growth rates at 5–15 °C, whereas the intermediate clone (III) exhibits a somewhat linear relation between growth rate and temperature. *S. petersenii* f. *petersenii* had the highest growth rates of all clones, var. *glabra* the lowest, and the third clone was intermediate. Compared with the literature data the measured growth rates and division times are very low, although clone-to-clone variability exists (Klaveness & Guillard 1975; Sandgren & Flanagin 1986; Sandgren 1988). The proposed linear relation for chrysophyte growth at temperatures between 5 and 20 °C was observed only with the intermediate clone (Sandgren 1988). Nevertheless, maximum rates at 20 °C agree well.

Changes in scale morphology

Mean values and variation ranges of the most important scale characters are listed in Tables 15.6 and 15.7, with respect to differences in culture

Table 15.6. *Mean values and variation ranges (v.r.) of scale characters*

Clone	sl (μm)	sb (μm)	crl (μm)	srl (μm)
I 1	4.08	2.06	2.74	2.61
I 2	3.81	2.07	2.43	2.57
I 5	3.89	2.11	2.52	2.32
I 20	4.02	1.86	2.59	2.43
I$_{total}$	3.95	2.02	2.57	2.48
v.r.	2.75–5.06	1.06–2.75	1.15–3.55	1.46–3.46
II 1	3.27	1.98	1.96	2.27
II 2	2.45	1.60	0.70	1.71
II 5	3.38	2.21	1.87	2.32
II 20	3.26	2.12	1.92	2.42
II$_{total}$	3.09	1.98	1.61	2.18
v.r.	1.60–3.90	0.80–2.66	0.27–2.57	0.89–3.28
III 1	2.90	1.88	1.75	1.96
III 2	3.28	1.75	1.98	2.16
III 5	3.19	1.70	1.93	2.05
III 20	3.22	1.84	1.96	2.38
III$_{total}$	3.15	1.79	1.91	2.14
v.r.	1.69–4.17	1.06–2.31	0.71–2.66	0.89–2.9

For abbreviations, see Fig. 15.1. Roman numerals refer to clones; arabic numerals refer to sample series (1 = start of cloning; 2 = after 7 months; 5, 20 = cultivated at 5 and 20 °C respectively).

Table 15.7. *Significances of changes in mean values*

	I		II		III	
	A	B	A	B	A	B
sl	** ↓	–	*** ↓	* ↓	*** ↑	–
sb	–	*** ↓	*** ↓	–	* ↓	** ↑
crl	*** ↓	–	*** ↓	–	*** ↑	–
crb	* ↓	–	*** ↓	*** ↑	*** ↓	*** ↑
srl	–	–	*** ↓	–	*** ↑	*** ↑
srb	*** ↓	*** ↓	*** ↓	*** ↓	*** ↑	*** ↑
p	–	–	–	–	*** ↑	–

For abbreviations see Fig. 15.1. Roman numerals refer to clones; A, changes with culture age; B, changes with temperature; –, no significance; * weakly significant; ** significant; *** strongly significant; ↑ ↓ increase or decrease in the mean value.

age (A) and temperature (B). It is worth mentioning that variation of the length characteristics sl, crl and srl (for definition see Fig. 15.1) are found to be only 10.6–15.8%, whereas those of the breadth characteristics sb, crb and srb and of the porus diameter p range from 16.6% to 29.8%, making the latter characteristics less reliable for differentiation.

Variation with culture age

Scales of f. *petersenii* and var. *glabra* clones shorten with increasing culture age (4.08 to 3.81 µm, 3.27 to 2.45 µm), whereas scales of the intermediate clone lengthen to values of the var. *glabra* clone (2.90 to 3.28 µm). The central ridge changes similarly (2.74 to 2.43 µm, 1.96 to 0.70 µm, 1.875 to 1.98 µm). A decrease in scale breadth is recorded for scales of the var. *glabra* and intermediate cones, but scales of the f. *petersenii* clone remain stable. Scale form changes from oblong-oval to more rounded oval scales in the f. *petersenii* clone, whereas the scales of the other clones reach a more oval shape starting from a circular form. Scale rim length is variable in the var. *glabra* and intermediate clones, but remains stable within the f. *petersenii* clone. With regard to all other measured characters, marked variability is exhibited, especially in scales of the var. *glabra* and inter-mediate clones. Generally, scales of the f. *petersenii* clone exhibit less changes in characters and differ significantly from the other two clones, whereas the intermediate clone approaches values of the var. *glabra* clone.

Variation with temperature

Concerning scale length, there are no significant changes with temperature in the f. *petersenii* and intermediate clones, whereas scales of the var. *glabra* clones shorten. Much variability can be recognized with regard to scale breadth. Scales of the f. *petersenii* clone develop a more oblong oval form at higher temperature, whereas the other clones remain rounded to circular in shape. No temperature effect can be recognized with regard to the central ridge in any of the clones. Scale rim length remains constant for the f. *petersenii* and the var. *glabra* type clones, but it lengthens in the intermediate clone. The clones thus exhibit only a few changes in the measured characters in scale morphology due to temperature. Only the breadth of scales of the f. *petersenii* clone appears to be significantly influenced.

Table 15.8. *Correlation coefficients of selected pairs of scale characters*

	I				II				III			
	1	2	5	20	1	2	5	20	1	2	5	20
sl/sb	−0.26	0.46	−0.32	−0.11	0.77	0.42	0.50	0.70	0.34	−0.44	−0.19	−0.11
sl/crl	0.77	0.84	0.91	0.83	0.93	0.35	0.67	0.81	0.88	0.76	0.91	0.69
sb/crb	0.87	0.53	0.71	0.92	0.80	0.01	0.85	0.84	0.67	0.88	0.73	0.76
crl/crb	0.14	0.52	−0.05	0.17	0.51	0.88	0.56	0.60	0.52	0.06	−0.03	0.36
sl/crl	−0.22	0.37	0.33	0.51	0.42	0.39	0.42	0.73	0.26	0.24	0.35	0.31

For abbreviations, see Fig. 15.1. Roman numerals refer to clones; arabic numerals refer to sample series (1 = start of cloning; 2 = after 7 months; 5, 20 = cultivated at 5 and 20 °C, respectively).

Changes in qualitative characters as a result of temperature and culture age

As shown in Fig. 15.2, temperature and culture age appear to have no influence on the struts of the f. *petersenii* clone. They may disappear with cultivation time on the scales of the other clones, but can reappear with cultivation at different temperature (a high percentage was found especially for the intermediate clone). The mean abundance of struts for the different clones is significantly different (I, 19; II, 13; III, 10) and there is some variability in the length of struts in the f. *petersenii* clone. It is interesting that the lowest strut abundance of this clone overlaps with the range of the intermediate clone, but is always distinctly different from the var. *glabra* clone. Bifurcation of struts, which is mentioned as a differentiating character of f. *petersenii*, does not occur in the var. *glabra* and intermediate clones but was observed in the f. *petersenii* clone (12%). A comparatively stable character for the var. *glabra* clone is the curved central ridge, and the intermediate clone is distinctly different with regard to this feature. The curved ridge almost never occurs in the f. *petersenii* clone (I, 1%; III, 4%; II, 42%).

A pronounced difference could be recognized with regard to silicification of scales. The var. *glabra* and intermediate clones exhibited very weakly silicified scales. Scales of f. *petersenii* were also weakly silicified, but nevertheless distinctly different from the others. A pronounced spine, which is thought to be a differentiating character for f. *petersenii*, was never important in our study in comparison with the other features. In conclusion, all clones exhibit more changes as a result of culture age than temperature.

Construction principles

Construction principles may be evaluated by showing correlations (Table 15.8). The results show the highest number of correlating character pairs for the var. *glabra* clone: sl/sb, sl/crl, sb/crb, crl/crb, sl/crl. The other clones have high correlation coefficients for scale length versus central ridge length (sl/crl), and scale breadth versus central ridge breadth (sb/crb) only. However, these two correlations show high values for all clones. The only exception is within the var. *glabra* clone (II 2), where only a short or no central ridge was developed. Scale rim breadth varies within the clones, but a rim is always present. This may be because it is required for scale case construction (Leadbeater 1986, 1990). The abundance of struts seems to be fixed within the clones, as no correlation between this value and scale length could be found.

Summary

Three clones of *Synura petersenii* Korshikov were cultured at temperatures of 5–25 °C, and growth rates were calculated. To analyze scale morphology, the differentiating characters were statistically evaluated for the experiments at 5 and 20 °C as well as for the influence of culture age after isolation. The first clone exhibited f. *petersenii*-type scales, the second var. *glabra* (Korsh.) Huber-Pestalozzi-like scales, and the third one had scales of intermediate character. Although overlaps occur with regard to some features, the clones investigated were always distinct from each other, and these differences would not be manipulated. As a result, the influence of culture age on scale morphology was much stronger than any temperature effect. To take this variation range into account, several scale features need to be studied. The most promising differentiating characters were scale size, a pronounced central ridge, and the number of struts for f. *petersenii*; strut length is not a definitive character. The curved central ridge and the low degree of silicification are distinct for var. *glabra*. The intermediate clone is distinctly different from the f. *petersenii* clone on the basis of the above features. The variation range of var. *glabra* must, therefore, be expanded. As the evaluation of these three clones does not allow us to draw definitive taxonomic conclusions, additional studies should be pursued, especially with regard to other environmental factors, such as silica and other nutrients. Also, additional studies of field material would be most helpful. Clumping the different taxa together would preclude the possibility of determining the reasons for the observed scale variation.

Acknowledgments

The authors wish to thank Prof. Dr. U. Geissler for discussions, and in particular Mrs U. Gaul for her continuous and attentive technical assistance.

References

Balanov, I.M. & Kuzmin, G.V. 1974. Species of the genus *Synura* Ehr. (Chrysophyta) in water reservoirs of the Volga Cascade. *Bot. Zurn.* **59**: 1675–86.

Fott, B. & Ludvik, J. 1957 Die submikroskopische Struktur der Kieselschuppen bei *Synura* und ihre Bedeutung für die Taxonomie der Gattung. *Preslia* **29**: 5–16.

Guillard, R.R.L. 1973. Division rates. In: Stein, J.R. (Ed.) *Handbook of Phycological Methods*. Cambridge University Press, Cambridge. pp. 289–312.

Gutowski, A. 1989. Seasonal succession of scaled chrysophytes in a small lake in Berlin. In: Kristiansen, J., Cronberg, C. & Geissler, U. (eds.) *Chrysophytes: Developments and Perspectives. Beih. Nova Hedwigia* **95**: 131–58.

Hällfors, G. & Hällfors, S. 1988. Records of chrysophytes with silicious scales (Mallomonadaceae and Paraphysomonadaceae) from Finnish inland waters. *Hydrobiologia* **161**: 1–29.

Healey, F.P. 1983. Effect of temperature and light intensity on the growth rate of *Synura sphagnicola. J. Plankton Res.* **5**: 767–74.

Hickel, B. & Maass, I. 1989. Scaled Chrysophytes, including heterotrophic nanoflagellates from the lake district in Holstein, northern Germany, In: Kristiansen, J., Cronberg, C. & Geissler, U. (eds.) *Chrysophytes: Developments and Perspectives. Beih. Nova Hedwigia* **95**: 233–57.

Kies, L. & Berndt, M. 1984. Die *Synura*-Arten (Chrysophyceae) Hamburgs und seiner nordöstlichen Umgebung. *Mitt. Inst. Allg. Bot. Hamburg* **19**: 99–122.

Klaveness, D. & Guillard, R.R.L. 1975. The requirement for silicon in *Synura petersenii* (Chrysophyceae). *J. Phycol.* **11**: 349–55.

Korshikov, A.A. 1929. Studies on the chrysomonads. I. *Arch. Protistenkd.* **67**: 253–88.

Kristiansen, J. 1975. On the occurrence of the species of *Synura* (Chrysophyceae). *Ver. Int. Ver. Limnol.* **19**: 2709–15.

Kristiansen, J. 1979. Problems in classification and identification of Synuraceae (Chrysophyceae). *Schweiz, Z. Hydrobiol.* **40**: 310–19.

Kristiansen, J. 1986. Silica-scale bearing chrysophytes as environmental indicators. *Br. Phycol. J.* **21**: 425–36.

Kristiansen, J. 1988. Seasonal occurrence of silica-scaled chrysophytes under eutrophic conditions. *Hydrobiologia* **161**: 171–84.

Leadbeater, B.S.C. 1986. Scale case construction in *Synura petersenii* Korsch. (Chrysophyceae). In: Kristiansen, J. & Andersen, R.A. (eds.) *Chrysophytes: Aspects and Problems*. Cambridge University Press, Cambridge. pp. 121–31.

Leadbeater, B.S.C. 1990. Ultrastructure and assembly of the scale case in *Synura* (Synurophyceae Andersen). *Br. Phycol. J.* **25**: 117–32.

Lehman, J.T. 1976. Ecological and nutritional studies on *Dinobryon* Ehrenb.: seasonal periodicity and the phosphate toxicity problem. *Limnol. Oceanogr.* **21**: 646–58.

Manton, I. 1955. Observations with the electron microscopy on *Synura caroliniana* Whitford. *Proc. Leeds Philos. Soc. Sci. Sect.* **6**: 306–16.

Nicholls, K.H. & Gerrath, J.F. 1985. The taxonomy of *Synura* (Chrysophyceae) in Ontario with special reference to taste and odour in water supplies. *Can. J. Bot.* **63**: 1482–93.

Petersen, J.B. & Hansen, J.B. 1956. On the scales of some *Synura* species. *Biol. Medd. Dan. Vid. Selsk.* **23**: 1–27.

Roijackers, R.M.M. 1986. Development and succession of scale-bearing Chrysophyceae in two shallow freshwater bodies near Nijmegen, The Netherlands. In: Kristiansen, J. & Andersen, R.A. (eds.) *Chrysophytes: Aspects and Problems*. Cambridge University Press, Cambridge. pp. 241–58.

360   *Chrysophyte algae*

Sandgren, C.D. 1981. Characteristics of sexual and asexual resting cyst
(statospore) formation in *Dinobryon cylindricum* Imhof (Chrysophyta).
J. Phycol. **17**: 199–210.
Sandgren, C.D. 1983. Morphological variability in populations of
chrysophycean resting cysts. I. Genetic (interclonal) and encystment
temperature effects on morphology. *J. Phycol.* **19**: 64–70.
Sandgren, C.D. 1988. The ecology of chrysophyte flagellates: their growth and
perennation strategies as freshwater phytoplankton. In: Sandgren, C.D.
(ed.) *Growth and Reproductive Strategies of Freshwater Phytoplankton.*
Cambridge University Press, Cambridge. pp. 9–104.
Sandgren, C.D. & Barlow, S.B. 1989. Siliceous scale production in chrysophyte
algae. II. SEM observations regarding the effects of metabolic inhibitors on
scale regeneration in laboratory population of scale free *Synura petersenii*
cells. In: Kristiansen, J., Cronberg, C. & Geissler, U. (eds.) *Chrysophytes:
Developments and Perspectives. Beih. Nova Hedwigia* **95**: 27–44.
Sandgren, C.D. & Flanagin, J. 1986. Heterothallic sexuality and density
dependent encystment in the chrysophycean alga *Synura petersenii* Korsh.
J. Phycol. **22**: 206–16.
Siver, P.A. 1987. The distribution and variation of *Synura* species
(Chrysophyceae) in Connecticut, USA. *Nord. J. Bot.* **7**: 107–16.
Siver, P.A. 1988. A new forma of the common chrysophycean alga *Synura
petersenii. Trans. Am. Microsc. Soc.* **107**: 380–5.
Starmach, K. 1985. Chrysophyceae und Haptophyceae. In: Ettl, H., Gerloff, J.,
Heynig, H. & Mollenhauer, D. (Eds.) *Süsswasserflora von Mitteleuropa,*
vol. 1. G. Fischer, Stuttgart.
Takahashi, E. 1959. Studies on genera *Mallomonas, Synura,* and other plankton
in freshwater by electron microscope: I. *Bull. Yamagata Univ., Agric. Sci.*
3: 117–51.
Takahashi, E. 1967. Studies on genera *Mallomonas, Synura,* and other plankton
in fresh-water by electron microscope. IV. Morphological and ecological
observations on genus *Synura* in ponds and lakes in Yamagata Prefecture.
Bull. Yamagata Univ., Agric. Sci. **5**: 199–218.

16

Status of the Chrysamoebales (Chrysophyceae): observations on *Chrysamoeba pyrenoidifera*, *Rhizochromulina marina* and *Lagynion delicatulum*

CHARLES J. O'KELLY & DANIEL E. WUJEK

Introduction

Until fairly recently, most concepts of higher taxa in the Chrysophyceae have been based on some combination of vegetative morphological features and flagellar number, position and behavior. Application of ultrastructural and molecular techniques to chrysophyte systematics research in the last few years has led to major taxonomic changes (Andersen *et al.* 1993; Moestrup, this volume; Preisig, this volume). However, the small number and limited diversity of species examined critically to date hampers further progress.

One concept that has been particularly difficult to evaluate, for lack of evidence, is that of the order Chrysamoebales, erected to accommodate chrysophytes that have ameboid ('rhizopodial') vegetative cells 'during the greater part of [the] life history' (Kristiansen 1990: pp. 439, 441). There have been few accounts of general ultrastructure in ameboid chrysophytes (Hibberd 1971, 1976; Hibberd & Chretiennot-Dinet 1979; Grell *et al.* 1990), and none on the detailed architecture of their flagellar apparatus.

Isolation into unialgal culture of two ameboid chrysophytes from New Zealand freshwaters – *Chrysamoeba pyrenoidifera* Korshikov and *Lagynion delicatulum* Skuja – prompted us to investigate their ultrastructure in detail and, in addition, to re-examine the marine species *Rhizochromulina marina* Hibberd et Chretiennot-Dinet. This paper summarizes the results of these investigations and thereby assesses the validity and status of the Chrysamoebales. Comprehensive accounts of the structure and reproduction of *C. pyrenoidifera* and *L. delicatulum* will appear elsewhere.

Materials and methods

Chrysamoeba pyrenoidifera was isolated from *Sphagnum* gametophores growing among potted insectivorous plants in the glasshouse of the Department of Plant Biology and Biotechnology, Massey University, Palmerston North, New Zealand. It was grown either in an infusion medium (OI) prepared from oxygenweed (*Elodea canadensis* Michx.) or in a 1:1 mixture of OI and Bold's 3N Bristol's medium with added soil-water extract (BBM; Starr & Zeikus 1987). *Lagynion delicatulum* was isolated from Lake Namunamu, North Island, New Zealand (Wujek & O'Kelly 1992), where it occurred as an epiphyte on *Elodea canadensis*. It was maintained in BBM. *Rhizochromulina marina* (CCAP 950/1) was obtained from the Culture Centre for Algae and Protozoa (CCAP; Thompson *et al.* 1988) and maintained in Provasoli's enriched seawater medium (PES; Starr & Zeikus 1987). All cultures were unialgal but not axenic.

Fixation of cells for transmission electron microscopy (TEM) was by a mixture of 2.5% glutaraldehyde and 0.1–0.5% osmium tetroxide in culture medium for *c.* 30 min at room temperature (15–20 °C). Vegetative cells of *C. pyrenoidifera* and *R. marina* were fixed by pouring the fixative onto cells growing in a plastic Petri dish. The cells were then collected on cellulose acetate filters, dehydrated in an acetone series, embedded in Polybed-Araldite or Spurr resin, serially sectioned with a diamond knife, stained with alcoholic uranyl acetate and lead citrate, and examined with a Philips 201c TEM.

Results

Chrysamoeba pyrenoidifera

In most features visible in the light and electron microscopes, *C. pyrenoidifera* closely resembles *C. radians* Klebs, as studied by Hibberd (1971). The structure of the pyrenoid, which in *C. pyrenoidifera* is visible in the light microscope, is completely surrounded by chloroplast thylakoids, is not appressed to the nucleus and is not penetrated or traversed by membranes (Figs. 16.1, 16.2), separates the two species.

Hibberd (1971) did not observe microtubules in the rhizopodia of *C. radians*. Rhizopodia of *C. pyrenoidifera*, however, contain microtubules in small numbers not organized into well-structured axonemes (Fig. 16.3). In serial sections, at least some of these microtubules could be traced back to the flagellar apparatus, specifically to the secondary cytoskeletal microtubules emanating from microtubular root R_1 of the flagellar apparatus (Fig. 16.8).

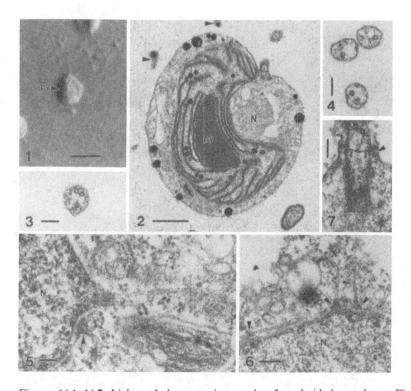

Figures 16.1–16.7. Light and electron micrographs of ameboid chrysophytes. Figures 16.1–16.3. *Chrysamoeba pyrenoidifera*. Figure 16.1. Light micrograph (differential interference contrast optics) of vegetative cells showing essential features, especially the pyrenoid (py). Scale bar represents 10 μm. Figure 16.2. TEM micrograph of a vegetative cell showing the pyrenoid. Note the absence of membranes penetrating the pyrenoid matrix and the presence of chloroplast lamellae between the pyrenoid and the nucleus. Neither a girdle lamella nor a ring nucleoid is present in the chloroplast. Arrowheads indicate rhizopodia. Scale bar represents 1.0 μm. Figure 16.3. TEM cross section of a rhizopodium showing microtubules. The number of microtubules is variable (usually one to five are present), and they are typically located near the periphery of the rhizopodium, as shown here. There is no evidence for the organization of the microtubules into axonemes. Scale bar represents 0.1 μm. Figures 16.4–16.7. *Rhizochromulina marina*, TEMs. Scale bars represent 0.1 μm. Figure 16.4. Cross sections of vegetative cell rhizopodia showing microtubules. The number of microtubules is variable (usually one to seven are present), and they may be located centrally, as here, or peripherally, especially in larger rhizopodia into which organelles such as mitochondria have protruded. There is no evidence for the organization of the microtubules into axonemes. Figure 16.5. Termination of two rhizopodial microtubules (confirmed by serial sections) at a common point on the nuclear envelope. Arrowhead shows attachment point of an additional microtubule that projects out of the plane of the section. N, nucleus. Figure 16.6. Termination of two microtubules (contributing to a single rhizopodium) at right angles to the nuclear envelope, and a third that departs from the nuclear envelope at a small angle. Note attachment points (arrowheads) often associated with electron-dense conical structures. Figure 16.7. Longitudinal section of the functional basal body in a zoospore, showing the two-gyred transitional helix proximal to the basal plate. Arrowhead indicates position of (partial) ring structure outside the basal body microtubules.

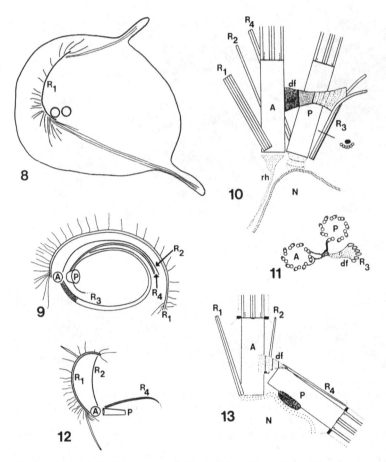

Figures 16.8–16.13. Diagrammatic illustrations of cytoskeletal features in ameboid chrysophytes. Not to scale. Abbreviations: A, anterior (hair-bearing) flagellum/ basal body; df, distal connecting fiber; N, nucleus; P, posterior (non-hair-bearing) flagellum/basal body; py, pyrenoid; R_n, microtubular root (numbering scheme according to Andersen 1987); rh, rhizoplast. Figures 16.8–16.11. *Chrysamoeba pyrenoidifera*. Figure 16.8. Origin of rhizopodial microtubules as secondary cytoskeletal microtubules emanating from root R_1. Figure 16.9. The flagellar apparatus as seen from the anterior end of the cell, showing the paths of the microtubular rootlets and array of secondary microtubules emanating from root R_1. Secondary microtubules from root R_2 omitted for clarity. Figure 16.10. The basal body complex in ventral view (as defined by e.g. Barr & Allan 1985; O'Kelly 1989; compare Owen *et al.* 1990a); a slightly exploded diagrammatic view showing major components. Secondary cytoskeletal microtubules and the distal terminus of root R_3 omitted for clarity. Figure 16.11. Cross-sectional view through the basal bodies (tip-to-base orientation, ventral side down), showing the shape of the distal fiber and its connection points to the basal bodies and to root R_3. Figures 16.12, 16.13. *Lagynion delicatulum*. Figure 16.12. The flagellar apparatus as seen from the anterior end of the cell, showing the paths of the microtubular root. Figure 16.13. The basal body complex in ventral view, showing major components. Secondary cytoskeletal microtubules have been omitted for clarity.

Vegetative cells and zoospores of *C. pyrenoidifera* have essentially identical flagellar apparatus features. Two (or, in premitotic cells, four) basal bodies, four microtubular roots and a rhizoplast connecting the anterior (younger?) basal body to the nucleus, are present (Figs. 16.9, 16.10). The basal bodies describe an angle of *c*. 20°, and they are connected by a complex distal fiber that also is attached to root R_3 (Figs. 16.10, 16.11). Transitional helices lie distal to the basal plate. The four microtubular rootlets form a loop around the posterior flagellar groove; roots R_1, R_2 and R_4 terminate at the far extremity of the loop, but two microtubules of root R_3 extend back towards the basal bodies, terminating at or near the chloroplast envelope just below the basal bodies.

The basal body angle in *C. radians* is similar to that in *C. pyrenoidifera*, and Hibberd (1971) observed a three- or four-stranded microtubular root that probably is equivalent to root R_1 of *C. pyrenoidifera*. Hibberd (1971, 1979) did not observe transitional helices in *C. radians*.

Rhizochromulina marina

Many of the light and electron microscopic features of *R. marina* closely resemble those of *Chrysamoeba* species, but Hibberd & Chretiennot-Dinet (1979) noted several differences, particularly in basal body/centriole organization and in the placement of the Golgi apparatus in zoospores (parabasal in *Chrysamoeba*, posterior to the nucleus in *R. marina*).

Hibberd & Chretiennot-Dinet (1979) did not observe rhizopodia in their TEM preparations of *R. marina*. The rhizopodia observed here contain small numbers of microtubules, as do those of *C. pyrenoidifera* (Fig. 16.4). However, these microtubules terminate individually or in small groups at the nuclear envelope (Figs. 16.5, 16.6). Not all the microtubules present at the nuclear envelope extended into pseudopodia (not shown).

As Hibberd & Chretiennot-Dinet (1979) observed, vegetative cells of *R. marina* contain centrioles, while zoospores contain one functional basal body and one probasal body. So far as conventional TEM can determine, roots are absent. Hibberd & Chretiennot-Dinet (1979) did not observe a transitional helix in the functional basal body. A two-gyred helix was found in the present study; remarkably, it lies proximal to the basal plate (Fig. 16.7). A few micrographs also show the presence of a single (partial?) ring-like structure in the same plane as the transitional helix but outside the basal body microtubules (Fig. 16.7). No flagellar rod or other inclusion has been found in the flagellum.

Lagynion delicatulum

The only prior examination with TEM of a member of the genus *Lagynion* was by Hibberd (1976), who looked only at vegetative cells of an unidentified species. No other member of the family Stylococcaceae, to which *Lagynion* belongs (Meyer 1986), has been studied.

The vegetative cells of *L. delicatulum* produce a single unbranched rhizopodium that extends through the aperture of the lorica. Settled zoospores produce several more robust rhizopodia prior to and during lorica formation. None of these rhizopodia contain microtubules or any other discernible cytoskeletal elements. The flagellar apparatus in zoospores – which are biflagellate, not uniflagellate as previously has been thought to be typical of the Stylococcaceae (Meyer 1986) – possesses two fully developed basal bodies and three microtubular roots (Figs. 16.12, 16.13). The basal body angle is *c.* 130°, and transitional helices, of three gyres in the anterior and 0 or one gyre in the posterior basal body, lie distal to the basal plate. The distal connecting fiber is of simple construction. None of the microtubular roots forms a loop, and no root corresponding to root R_3 of other chrysophytes is present. Vegetative cells possess fully developed basal bodies, but without attached flagella and with only a template for root R_1 present. The included angle is usually *c.* 30°.

Discussion

The organism here identified as *Chrysamoeba pyrenoidifera* is clearly a close relative of *C. radians*. The apparent discrepancies between the two in rhizopodial structure and the presence or absence of the transitional helix are unlikely to be significant. Hibberd (1971) fixed *C. radians* cells in osmium tetroxide at 0 °C, which may have removed any microtubules present in the rhizopodia. Several examples, particularly among pedinellids (Patterson 1986) and mallomonads (Zhang *et al.* 1990), are known of closely related chrysophytes that may or may not have transitional helices.

The *Chrysamoeba* species are also clearly closely related to chromulinalean (*sensu* Preisig, this volume) chrysophytes. The small basal body angle in *Chrysamoeba* aside, the configuration and arrangement of flagellar apparatus components, especially the roots and the basal body connecting fibers, are essentially identical to those reported in detail for *Poterioochromonas stipitata* Scherffel (Schnepf *et al.* 1977), *Chrysosphaerella brevispina* Korshikov (Andersen 1990) and *Dinobryon cylindricum* Imhof (Owen *et al.* 1990*a*). In fact, some organisms presently considered to be members of Chromulinales, such as *Uroglena americana* Calkins (Owen *et al.* 1990*b*) and *Chrysonephele palustris* Pipes *et al.* (Pipes *et al.* 1990), have flagellar apparatus features more unlike those of *Poterioochromonas, Chrysosphaerella* and *Dinobryon* than are those of *Chrysamoeba.*

Although superficially similar to *Chrysamoeba*, the ultrastructural features of *Rhizochromulina marina* rule out any possibility of a close relationship between the two genera; Hibberd (1986), in fact, considered *R. marina* 'aberrant'. Several features observed here suggest that *R. marina* belongs to a group that also includes the pedinellids (Zimmermann *et al.* 1984; Preisig *et al.* 1991), the silicoflagellates (Moestrup & Thomsen 1990) and the pelagomonads (Andersen *et al.* 1993). These features of *R. marina* include: (1) the microtubule-containing rhizopodia with the microtubules anchored on the nuclear envelope, commonly found among pedinellids and silicoflagellates; (2) the Golgi body located posterior to the flagellate-cell nucleus, as found in pedinellids; (3) the transitional helix proximal to the basal plate in the functional basal body of zoospores, a feature found elsewhere only in some pedinellid species, including *Apedinella radians* (Lohmann) Campbell (figures 11, 17, 35 and 36 in Koutoulis *et al.* 1988), *Ciliophrys infusionum* Cienkowsky (Davidson 1982, as *C. marina* Caullery), and *Pteridomonas danica* Patterson et Fenchel (Patterson & Fenchel 1985), and in the pelagomonad *Pelagomonas calceolate* Andersen et Saunders (Andersen *et al.* 1993); (4) the ring-like structure in the same plane as the transitional helix but outside the basal body microtubules, known in some pedinellids and in the silicoflagellate *Dictyocha speculum* Ehrenberg (Moestrup & Thomsen 1990). The rhizopodia of *R. marina* contain a variable number of non-coalescent microtubules, and therefore are more similar to those of the silicoflagellates (Moestrup & Thomsen 1990) than to those of pedinellids, which bear a fixed number of microtubules organized into discrete axonemes (Davidson 1982; Zimmermann *et al.* 1984).

Since *R. marina* has: (1) both vegetative cells and zoospores; (2) vegetative cells that are dorsiventral with rhizopodia that are not uniform in diameter or length; and (3) microtubules in the rhizopodia that are irregular in number and not organized into axonemes, the monospecific genus *Rhizochromulina* cannot be accommodated within the Pedinellales, the Dictyochales, or the Pelagophyceae. Therefore the order Rhizochromulinales and family Rhizochromulinaceae are proposed. If the class Dictyochophyceae (botanical nomenclature) = Silicoflagellata (zoological nomenclature) is circumscribed to include both Dictyochales and Pedinellales, as suggested by Moestrup & Thomsen (1990) and Moestrup (this volume), then the Rhizochromulinales would represent a third order in this class. Andersen *et al.* (1993) have suggested that the Pedinellales (as Pedinellophyceae), Dictyochales (as Dictyochophyceae) and Rhizochromulinales form a clade that also includes the Pelagomonadales (Pelagophyceae) and the diatoms (Bacillariophyceae).

The features of *Lagynion delicatulum* are quite different from those of either *Chrysamoeba* or *Rhizochromulina*, noticeable particularly in the absence of microtubules from the rhizopodia and in the architecture of the flagellar apparatus. The greatest similarity is with *Hibberdia magna* (Belcher) Andersen, a chrysophaeralean species which Andersen (1989) placed in its own order, the Hibberdiales. Similarities appear in the pronounced obtuse angle between the basal bodies, the simple distal fiber construction and, in particular, the absence of root loops and of root R_3 in zoospore flagellar apparatuses. There are also some significant differences. For example, *H. magna* zoospore flagellar apparatuses have rhizoplasts, whereas none is apparent in *L. delicatulum*. Also, *H. magna* zoospores have eyespots but *L. delicatulum* zoospores do not, and the structures of the posterior flagella are markedly different. Nevertheless, assuming that the Stylococcaceae as conceived by Meyer (1986) is monophyletic, and that *L. delicatulum* is representative, the family should be transferred to the previously monotypic Hibberdiales.

From the examination of these three ameboid chrysophytes it is clear that the Chrysamoebales as conceived by, for example, Kristiansen (1990), is paraphyletic or polyphyletic, and that its type family and genus* can be subsumed within the Chromulinales. There is therefore neither a nomenclatural nor a conceptual basis for the retention of the order Chrysamoebales. As has been shown for many other morphological types among the chrysophytes, the ameboid level of vegetative organization is a poor indicator of taxonomic and evolutionary relationships (Andersen *et al.* 1993; Preisig, this volume). Studies on many more chrysophytes are required if we are to find out just what these relationships really are.

---

* For obvious reasons, we are treating *Rhizochrysis* Pascher and Rhizochrysidaceae as separate from *Chrysamoeba* and Chrysamoebaceae (see Hibberd 1971; Preisig, this volume), and as *sedis mutabilis*, pending critical examination of entities referable to *Rhizochrysis*.

Appendix: Latin diagnoses
Rhizochromulinales ord. nov. (botanical nomenclature)
Rhizochromulinida ord. nov. (zoological nomenclature)

Protista straminopila plerumque unicellularia cellulis vegetativis/trophicis dorsi-ventralibus, saltem interdum rhizopodialibus, nullis flagellis praeditis, vulgo chloroplastos aureos foventibus. Rhizopodia microtubulos varii numeri exhibentia, involucro nucleari affixos, non in axonemata dispositos. Reproductio zoosporis effecta rhizopodiis carentibus. Exosceleton siliceum per totam vitam nullum.

Predominantly unicellular stramenopile protists with dorsiventrally organized, aflagellate, at least sometimes rhizopodial vegetative/trophic cells, usually with golden chloroplasts. Rhizopodia containing irregular numbers of microtubules, which are anchored on the nuclear envelope but not organized into axonemes. Reproduction by zoospores without rhizopodia. Siliceous exoskeleton lacking from all life history stages.

Type family: Rhizochromulinaceae (botanical nomenclature)
Rhizochromulinidae (zoological nomenclature)

Cellulae vegetativae/trophicae plerumque amoebas unicellulares nudas constituentes, rarius plasmodia multinuclearia formantes. Zoosporae nudae, lageniformes, flagellis emergentibus singulis. Praeterea characteres ordinis exhibentes.

Vegetative/trophic cells naked unicellular amebae, more rarely multinucleate plasmodia. Zoospores naked, lageniform, with a single emergent flagellum. Otherwise as for the order.

Type genus:
Rhizochromulina Hibberd et Chretiennot-Dinet, *J. Mar. Biol. Assoc. U.K.* 59: 189, 1979.

Summary

Ultrastructural features of three ameboid chrysophytes – *Chrysamoeba pyrenoidifera* Korshikov (Chrysamoebaceae), *Rhizochromulina marina* Hibberd et Chretiennot-Dinet (Rhizochromulinaceae fam. nov.) and *Lagynion delicatulum* Skuja (Stylococcaceae) – were examined to assess the validity of the order Chrysamoebales. Rhizopodia of *C. pyrenoidifera* contain microtubules, many of which originate from flagellar apparatus root R_1. Flagellar apparatus architecture, essentially identical in vegetative cells and zoospores, closely resembles that found in ochromonadalean chrysophytes. Rhizopodia of *R. marina* also contain microtubules, but these originate from the nuclear envelope. Rootlets are absent from the zoospore flagellar apparatus, and the transition region of the single emergent flagellum has a transitional helix proximal to the basal plate.

Rhizopodia of *L. delicatulum* apparently lack discrete cytoskeletal elements. Zoospore flagellar apparatus architecture closely resembles that of *Hibberdia magna*. On this: (1) the Chrysamoebaceae is referable to the Chromulinales (Chrysophyceae), (2) the Rhizochromulinales ord. nov. and Rhizochromulinaceae fam. nov. are referable to a position alongside the Pedinellales and Dictyochales (Dictyochophyceae), and (3) the Stylococcaceae are referable to the Hibberdiales (Chrysophyceae). There is neither a nomenclatural nor a conceptual basis for retention of the Chrysamoebales.

Acknowledgments

This work began during a sabbatical visit by D.E.W. to Massey University. The Massey University Council facilitated D.E.W.'s visit. The National Science Foundation (USA), the University Grants Committee (NZ), the Massey University Research Fund and the FRCE Committee of Central Michigan University provided partial funding. E.M. Nickless maintained the cultures, D.P. Barrett prepared much of the material for TEM, and D.H. Hopcroft and R.J. Bennett (Electronic Microscopy Unit, Hort. Research, Palmerston North) assisted with the electron microscopy and the preparation of electron micrographs. Tyge Christensen provided the Latin diagnoses. Our thanks to these persons and organizations.

References

Andersen, R.A. 1987. Synurophyceae classis nov., a new class of algae. *Am. J. Bot.* **74**: 337–53.

Andersen, R.A. 1989. Absolute orientation of the flagellar apparatus of *Hibberdia magna* comb. nov. (Chrysophyceae). *Nord. J. Bot.* **8**: 653–69.

Andersen, R.A. 1990. The three-dimensional structure of the flagellar apparatus of *Chrysosphaerella brevispina* (Chrysophyceae) as viewed by high voltage electron microscopy stereo pairs. *Phycologia* 29: 86–97.

Anderson, R.A. 1993. The chrysophytes: a review. *Protoplasma* (in press).

Andersen, R.A. 1991. The cytoskeleton of chromophyte algae. *Protoplasma* **164**: 143–59.

Andersen, R.A., Saunders, G.W., Paskind, M.P. & Sexton, J.P. 1993. Ultrastructure and 185 rRNA gene sequence for *Pelagomonas calceolata* gen. et sp. nov. and the description of a new algal class, the Pelagophyceae classis nov. *J. Phycol.* **29**: 701–15.

Barr, D.J.S. & Allan, P.M.E. 1985. A comparison of the flagellar apparatus in *Phytophthora*, *Saprolegnia*, *Thraustochytrium* and *Rhizidiomyces*. *Can. J. Bot.* **63**: 138–54.

Davidson, L.M. 1982. Ultrastructure, behavior, and algal flagellate affinities of the helioflagellate *Ciliophrys marina*, and the classification of the helioflagellates (Protista, Actinopoda, Heliozoea). *J. Protozool.* **29**: 19–29.

Grell, K.G., Heini, A. & Schüller, S. 1990. The ultrastructure of *Reticulosphaera socialis* Grell (Heterokontophyta). *Eur. J. Protistol.* **26**: 37–54.

Hibberd, D.J. 1971. Observations on the cytology and ultrastructure of *Chrysamoeba radians* Klebs (Chrysophyceae). *Br. Phycol. J.* **6**: 207–23.

Hibberd, D.J. 1976. The ultrastructure and taxonomy of the Chrysophyceae and Prymnesiophyceae (Haptophyceae): a survey with some new observations on the ultrastructure of the Chrysophyceae. *Biol. J. Linn. Soc.* **72**: 55–80

Hibberd, D.J. 1979. The structure and phylogenetic significance of the flagellar transition region in the chlorophyll c-containing algae. *BioSystems* **11**: 243–61.

Hibberd, D.J. 1986. Ultrastructure of the Chrysophyceae: phylogenetic implications and taxonomy. In: Kristiansen, J.& Andersen, R.A. (eds.) *Chrysophytes: Aspects and Problems.* Cambridge University Press, Cambridge. pp. 23–36.

Hibberd, D.J. & Chretiennot-Dinet, M.J. 1979. The ultrastructure and taxonomy of *Rhizochromulina marina* gen. et sp. nov., an amoeboid marine chrysophyte. *J. Mar. Biol. Assoc. U.K.* **59**: 179–93.

Koutoulis, A., McFadden, G.I. & Wetherbee, R. 1988. Spine-scale reorientation in *Apedinella radians* (Pedinellales, Chrysophyceae): the microarchitecture and immunocytochemistry of the associated cytoskeleton. *Protoplasma* **147**: 25–41.

Kristiansen, J. 1990. Phylum Chrysophyta. In: Margulis, L., Corliss, J. O., Melkonian, M. & Chapman, D.J. (eds.) *Handbook of Protoctista.* Jones and Bartlett Publishing Co., Boston. pp. 438–53.

Meyer, R.L. 1986. A proposed phylogenetic sequence for the loricate rhizopodial chrysophyceae. In: Kristiansen, J. & Andersen, R.A. (eds.) *Chrysophytes: Aspects and Problems.* Cambridge University Press, Cambridge. pp. 75–85.

Moestrup, Ø. & Thomsen, H.A. 1990. *Dictyocha speculum* (Silicoflagellata, Dictyochophyceae): studies on armoured and unarmoured stages. *Biol. Skr. Dan. Vid. Selsk.* **37**: 1–57.

O'Kelly, C.J. 1989. The evolutionary origin of brown algae: information from studies of motile cell ultrastructure. In: Green, J.C., Leadbeater, B.S.C. & Diver, W.L. (eds.) *The Chromophyte Algae: Problems and Perspectives.* Clarendon Press, Oxford. pp. 255–78.

Owen, H.A., Mattox, K.R. & Stewart, K.D. 1990*a*. Fine structure of the flagellar apparatus of *Dinobryon cylindricum* (Chrysophyceae). *J. Phycol.* **26**: 131–41.

Owen, H.A., Stewart, K.D. & Mattox, K.R. 1990*b*. Fine structure of the flagellar apparatus of *Uroglena americana* (Chrysophyceae). *J. Phycol.* **26**: 142–9.

Patterson, D.J. 1986. The actinophryid heliozoa (Sarcodina, Actinopoda) as chromophytes. In: Kristiansen, J. & Andersen, R.A. (eds.) *Chrysophytes: Aspects and Problems.* Cambridge University Press, Cambridge. pp. 75–85.

Patterson, D.J. & Fenchel, T. 1985. Insights into the evolution of heliozoa (Protozoa, Sarcodina) as provided by ultrastructural studies on a new species of flagellate from the genus *Pteridomonas. Biol. J. Linn. Soc. (Lond)* **34**: 381–403.

Pipes, L.D., Tyler, P.A. & Leedale, G.F. 1989. *Chrysonephele palustris* gen. et sp. nov. (Chrysophyceae), a new colonial chrysophyte from Tasmania. *Beih. Nova Hedwigia* **95**: 81–97.

Preisig, H.R., Vørs, N. & Hällfors, G. 1991. Diversity of heterotrophic heterokont flagellates. In: Patterson, D.J. & Larsen, J. (eds.) *The Biology of Free-living Heterotrophic Flagellates*. Clarendon Press, Oxford. pp. 361–99.

Schnepf, E., Deichgräber, G., Röderer, G. & Herth, W. 1977. The flagellar root apparatus, the microtubular system and associated organelles in the chrysophycean flagellate, *Poterioochromonas malhamensis* Peterfi (syn. *Poterioochromonas stipitata* Scherffel and *Ochromonas malhamensis* Pringsheim). *Protoplasma* **92**: 87–107.

Starr, R.C. & Zeikus, J.A. 1987. UTEX: the culture collection of algae at the University of Texas at Austin. *J. Phycol.* **23**(Suppl.): 1–47.

Thompson, A.S., Rhodes, J.C. & Pettman, I. (eds.) 1988. *Culture Collection of Algae and Protozoa. Catalogue of Strains* 1988. Culture Collection of Algae and Protozoa, Ambleside, UK.

Wujek, D.E. & O'Kelly, C.J. 1992. Silica-scaled Chrysophyceae (Mallomonadaceae and Paraphysomonadaceae) from New Zealand freshwaters. *N. Z. J. Bot.* **30**: 405–14.

Zhang, X., Inouye, I. & Chihara, M. 1990. Taxonomy and ultrastructure of a freshwater scaly flagellate *Mallomonas tonsurata* var. *etortisetifera* var. nov. (Synurophyceae, Chromophyta). *Phycologia* **29**: 65–73.

Zimmermann, B., Moestrup, Ø. & Hällfors, G. 1984. Chrysophyte or heliozoon? Ultrastructural studies on a cultured species of *Pseudopedinella* (Pedinellales ord. nov.) with comments on species taxonomy. *Protistologica* **20**: 591–612.

17

The genus *Paraphysomonas* from Indian rivers, lakes, ponds and tanks

DANIEL E. WUJEK & LEELA C. SAHA

Introduction

The colorless chrysophyte genus *Paraphysomonas* de Saedeleer includes about 50 taxa. All have typical heterokont flagellation and silicified scales covering the cell body. Only a very few of the larger spine-scales are visible in the light microscope. Electron microscopy is necessary for the identification of most species and subspecies.

Examinations of the freshwater algal flora of the tropics are numerous, and in recent years have started to include the chrysophytes. Cronberg (1989) has summarized all of the literature on scaled chrysophytes from the tropics. However, there has been only one study using electron microscopy from India and two from countries adjacent to India: Bangladesh (Takahashi & Hayakawa 1979) and Sri Lanka (Dürrschmidt & Cronberg 1989).

During a study of the biota of silica-scaled protists in Indian freshwaters, a number of species not previously reported were observed (Saha & Wujek 1990). This paper focusses not only on *Paraphysomonas* species not previously recorded, adding to the knowledge of their occurrence and distribution, but also presents descriptions of two new species.

Materials and methods

The sites that contained *Paraphysomonas* taxa are listed in Table 17.1 Sample fixation and preparation were as described in Saha & Wujek (1990); or the specimens were shadowed with platinum/palladium.

Results

Over 150 sampling sites were examined for *Paraphysomonas* taxa representing a wide geographic area of India (Fig. 17.1). Thirteen species (Table

Table 17.1. *Location of sample sites for* Paraphysomonas *in India, February–April 1990, and available indications of their ecological conditions*

Site no.	Temperature (°C)	pH	City/village/district
18 February, Bihar			
2. Tilha Pond	22	7.3	Bhagalpur
3. River Ganga	21	7.0	Bhagalpur
5. Rani Talab Pond	22	7.3	Bhagalpur
6. Tiwari Pond	20	7.7	Bhagalpur
19 February, Bihar			
7. Mukhra Pond	21.5	7.5	Bhagalpur
8. TNB College Pond	22	8.0	Bhagalpur
9. Bhag. U. Zoology Pond	24	7.0	Bhagalpur
21 February, Bihar			
10. Champa Nala Nathnagar	24	7.2	Bhagalpur
11. Sultangani Pond	22	7.4	Sultangani
13. Bochahi Pond	23	7.2	Munger
14. Sita Kund	54	7.1	Munger
15. Shatrughan Kund	24	7.0	Munger
16. Jal Mandir	24	6.8	Munger
17. Fisheries Dept. Pond	27	6.8	Jamalpur
18. North Pond, Jamalpur	24	7.2	Jamalpur
19. Lower Pond, Jamalpur	25	6.8	Jamalpur
23. Jori Talab, Haveli-Kharagpur	25	6.7	Jamalpur
22 February, Bihar			
26. Kharagpur Pond	23	6.7	Munger
29. Middle Bridge Pond	24	6.8	Sultanganj
30. Sultanganj Pond	25	7.4	Sultanganj
23 February, Bihar			
34. River Koshi	23	7.5	Purnea
36. Harda Bridge Pond	24	8.0	Purnea
39. Kaptanpul River	25	7.0	Purnea
40. Forbesganj Mor Pond	26	6.8	Purnea
41. Kasba Pond	25	7.3	Purnea
43. Tatma Pond	21	6.6	Purnea
26 February, Bihar			
45. Central Jail Pond	26	8.2	Gaya
46. Katari Pond	26.5	8.1	Gaya
47. Vishar Pond	25	7.0	Gaya
49. Pita Maheswar	26	7.0	Gaya
50. Ram Sagar	27.5	7.0	Gaya
51. Baitarni Pond	25	7.0	Gaya
52. Gugri Tard Pond	29	7.0	Bodh-Gaya
53. High School Pond	26	9.0	Bodh-Gaya

Table 17.1 (*continued*)

Site no.	Temperature (°C)	pH	City/village/district
1 March, Bihar			
55. Mandu Pond	26	7.3	Bodh-Gaya
2 March, Bihar			
56. Line Tank	22	7.2	Ranchi
57. Jail Pond	22.5	7.1	Ranchi
58. Karam Toli Pond	21	7.0	Ranchi
59. Hatania Pond	23	7.5	Ranchi
60. Kanke Dam	23	7.5	Ranchi
61. Serpentine Lake	23.5	6.9	Ranchi
63. Saheed Park Pond	26	7.5	Ranchi
64. Dhurwa Dam	22	6.8	Ranchi
4 March, West Bengal			
65. Tarabag Pond	26	7.3	Burdwan
66. Krishna Sayar	26.5	7.2	Burdwan
67. Gargi Hostel Pond	26	7.2	Burdwan
69. Einstein Hostel Pond	28	7.1	Burdwan
70. Kamal Sayar Pond	27	7.4	Burdwan
71. Medical College Women's Hostel Pond	27	7.5	Burdwan
72. Midda Pond	27	7.5	Burdwan
73. Shyam Sayar Pond	28	7.6	Burdwan
74. Burdwan Univ. Guest House Pond	28	7.3	Burdwan
6 March, Orissa			
77. Nakaur Patra, CIFA	24	6.8	Kausalyaganga
78. Reservoir Pond no. 2, CIFA	27	7.2	Kausalyaganga
79. DBT-II, CIFA	27	7.2	Kausalyaganga
80. Reservoir Pond no. 1, CIFA	27	7.2	Kausalyaganga
7 March, Orissa			
81. Daudamukuudapur Pond	23	6.8	Near Puri
82. Birapratappur Pond	23	6.5	Near Puri
83. Malatipatapur Pond	23	6.8	Near Puri
84. Harekrishnapur Pond	28	7.5	Near Puri
85. Temple Tank I	28	7.5	Puri
87. Temple Tank III	31	7.3	Puri
8 March, Orissa			
88. Sardu Pond	32	6.8	Near Konarak
89. Vegunia Pond	28	6.8	Near Konarak
14 March, Tamil Nadu			
91. Kayambedu	28.2	7.2	Madras
92. Sembarambakkam	30	7.2	Madras

(*continued*)

Table 17.1 (*continued*)

Site no.	Temperature (°C)	pH	City/village/district
93. Thimperumputhur	31	6.9	Madras
94. Chengalpaton Canal	32	7.0	Madras
95. Vedanthangal	38	7.3	Madras
16 March, Karnataka			
97. Hebbal Tank	27	7.0	Bangalore
100. Thippegondanahally	28	6.8	Bangalore
101. Tavarekere Tank	28	6.9	Bangalore
17 March, Karnataka			
102. Hejjala Tank	28	7.1	Near Mysore
103. Channapatna Tank	28	7.0	Near Mysore
105. Guththalu Kere	29	7.0	Mandya
106. CDS Canal	29	7.0	Mysore
110. Devaraya Kere	28	7.3	Mysore
111. Kurrarahally Tank	28	7.5	Mysore
112. Bilikere Lake	26	7.5	Mysore
113. River Cauvery	28	6.0	At Balamuri spill-way
18 March, Karnataka			
115. Lalbag Lake	24	6.9	Bangalore
116. Indira Pond	28	7.0	Hyderabad
117. Hussain Sagar Lake	26	7.2	Hyderabad
120. Himayath Sagar Lake	32	8.0	Hyderabad
121. Osman Sagar Lake	30	8.0	Hyderabad
122. Lake Banjara	31	7.5	Hyderabad
27 March, Utar Pradesh			
129. Durgakund Pond	22	7.0	Varanasi
29 March, Utar Pradesh			
135. Paschim Sareera Pond	25	7.0	SW of Allahabad
136. Paschim Sareera Canal	24	7.2	SW of Allahabad
3 April, Utar Pradesh			
147. Talab Sheesh Mahal	25.5	7.5	Lucknow
148. Telibagh	26.5	7.0	Lucknow
149. Sharda Canal	20.5	7.2	Lucknow
10 April, Haryana			
152. Badkhal Lake	19.5	7.1	Delhi
153. Suraj Kund	24.5	8.1	Delhi
155. Okhla Reservoir	24.5	6.9	Yamuna River, Delhi

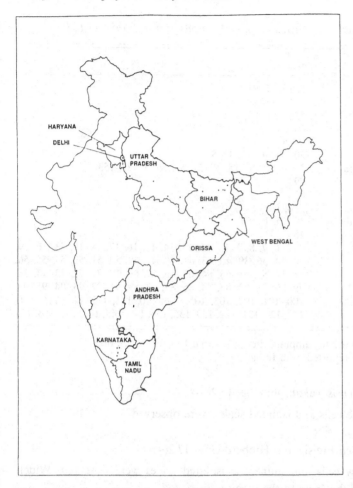

Figure 17.1. General location of sample sites in India.

17.2), 12 of which are new to India, and including two new to science, were observed. Species are arranged in alphabetical order; those not previously recorded in India are marked with an asterisk. Distribution records for all previously described taxa can be found in Preisig & Hibberd (1982*a, b*) Dürrschmidt & Cronberg (1989), Vørs *et al.* (1990) and Wujek & Bland (1991).

*Paraphysomonas acantholepis Preisig and Hibberd (Fig. 17.2(2))

The only reports for this species are from Europe; this is the first record for the tropics.

Table 17.2. *Distribution of* Paraphysomonas *taxa in India*

Taxon	Location[a]
P. acantholepis	5, 10, 17, 91
P. bandaiensis	40, 59, 87
P. caelifrica	5, 41, 85, 87, 91
P. coronata	10, 85
P. eiffelii	87, 91
P. gladiata	30, 39, 40, 41, 43, 85
P. imperforata	2, 6, 7, 9, 10, 18, 40, 51, 53, 56, 59, 83, 153
P. planus	79
P. porosa	85
P. runcinifera	5, 6, 16
P. runciniferopsis	10, 15, 16, 59
P. undulata	153
P. vestita[b]	3, 5, 6, 7, 8, 9, 10, 11, 13, 14, 15, 16, 17, 18, 19, 23, 26, 29, 30, 34, 36, 39, 40, 43, 45, 46, 47, 49, 50, 51, 52, 53, 55, 56, 58, 59, 60, 61, 63, 64, 65, 66, 67, 69, 70, 71, 72, 73, 74, 78, 79, 80, 81, 82, 83, 84, 85, 87, 88, 89, 91, 92, 93, 94, 95, 97, 100, 101, 102, 103, 105, 106, 110, 111, 112, 113, 115, 116, 117, 120, 121, 122, 129, 135, 136, 147, 148, 149, 152, 153, 155

[a] Numbers refer to sampling sites; see Table 17.1 for description.
[b] Previously reported from India.

**P. bandaiensis* Takahashi (Fig. 17.2(3))

Both whole cells and isolated scales were observed.

**P. caelifrica* Preisig and Hibberd (Fig. 17.2(4))

Only a few cells and numerous isolated scales were observed. Widely distributed, but new to the tropics.

**P. coronata* Moestrup and Zimmerman (Fig. 17.2(5))

Whole cells were not observed.

**P. eiffelli* Thomsen (Fig. 17.2(6))

This species, characterized by its tower-like spines-scales, has also been observed in nearby Sri Lanka (Dürrschmidt & Cronberg 1989).

**P. gladiata* Preisig and Hibberd (Fig. 17.2(7))

This is the first report of *P. gladiata* from the tropics. Its most recent report is from China (Kristiansen & Tong 1989).

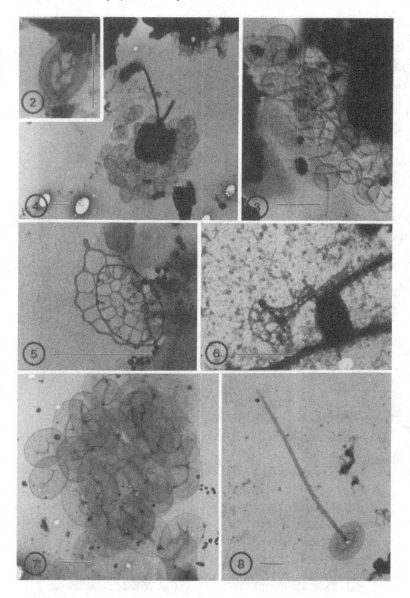

Figure 17.2. *Paraphysomonas.* (2) *P. acantholepis*; (3) *P. bandaiensis*; (4) *P. caelifrica*; (5) *P. coronata*; (6) *P. eiffelii*; (7) *P. gladiata*, (8) *P. imperforata*. Scale bars represents 1 μm.

(*continued*)

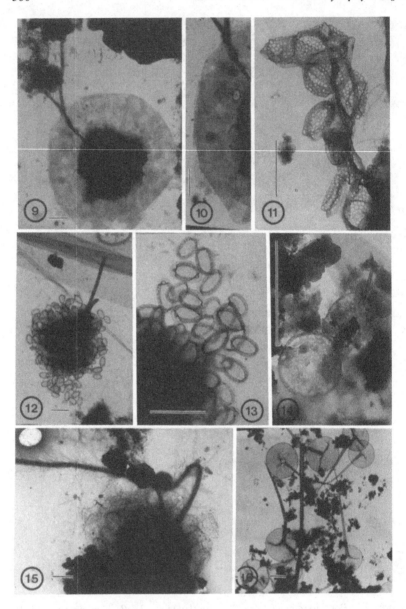

Figure 17.2 (*continued*). *Paraphysomonas*. (9, 10) *P. planus*; (11) *P. runcinifera*;
(12, 13) *P. runciniferopsis*; (14) *P. porosa*; (15) *P. undulata*; (16) *P. vestita*. Scale bar
represents 1 μm.

*P. imperforata Lucas (Fig. 17.2(8))

This species is distributed world-wide, including marine localities.

*Paraphysomonas planus sp. nov. (Figs. 17.2(9), 17.2(10))

Diagnosis Cellulae sphaericae vel subsphaericae, *c.* 3.5 µm in diametro, incoloratae. Flagella dua inaequalia, alterum longius pilisque armatum (*c.* 12 µm), alterum glabrum (3.5 µm). Cellulae squamis numerosis, uniformibus, indutae. Squamae sphaericae vel ellipticae (0.8–1.0 × 1.0–1.2 µm), planae. Cysta ignota.

Type locality DBT-II, CIFA, Kausalyaganga (Orissa), India, 6 March 1990.

Typus speciei Fig. 17.2(9)

The colorless cells are subspherical to spherical, motile, approximately 3.5 µm in diameter (measurements based on whole mounts). Two unequal flagella; one *c.* 12 µm long with two opposite rows of hairs, the other smooth and *c.* 3.5 µm long. Cell body covered with plate scales of one basic form. Each scale is flat, circular to oval.

Type material was collected in one of the experimental ponds, DBT-II, at the Central Institute of Fisheries and Aquaculture (CIFA), Kausalyaganga, in the state of Orissa, India, 6 March 1990.

The scales of *P. planus* resemble those of *P. amphiplana* Wujek (Wee & Wujek 1986) and to a lesser degree *P. corynephora* Preisig and Hibberd. The latter, however, has two types of scales (flat and spine-scales) while *P. planus* has only one type (flat plate scales). The scales of *P. amphiplana* are all elliptical and the flimmer flagellum is nearly twice as long as in *P. planus*.

*P. porosa Dürrschmidt and Cronberg (Fig. 17.2(14))

Scales identical to those from its type locality, Tissa Wewa, Sri Lanka, where it was found in a small farmland pond on the road to Puri.

*P. runcinifera Preisig and Hibberd (Fig. 17.2(11))

Its most recent report is from China (Kristiansen & Tong 1989).

*Paraphysomonas runciniferopsis sp. nov. (Figs. 17.2(12) and 17.2(13))

Diagnosis Cellulae incolores, circiter globosae, 2.0–3.0 µm diametro. Flagella dua, inaequalia; alterum 2.8–3.5 µm longum, mastigonematibus tubulosis instructum, alterum 0.8–1.0 µm longum, laeve. Cellulae squamis maculiformis uniformibus obtectae. Squama quaeque basi plana, lateribus manifeste concavis; basis plana proximalis. Basis circiter elliptica 0.25–0.4 × 0.4–0.8 µm, plana.

Type locality   Shatrughan Kund (Bihar), India, 21 February 1990.

Typus speciei   Fig. 17.2(12)

Cells colorless, approximately spherical, 2.0–3.0 µm in diameter. Two unequal flagella; one 2.8–3.5 µm long, bearing tubular mastigonemes, the other 0.8–1.0 µm long, smooth. Cells covered with one type of meshwork scale. Each scale with a flat base and concave sides; flat base is proximal. Base more or less elliptical, 0.25–0.4 × 0.4–0.8 µm, smooth and flat with no apertures.

Type material collected on 21 February 1990 in the state of Bihar, at a pond called Shatrughan Kund.

The scales of *P. runciniferopsis* superficially resemble those of *P. runcinifera* Preisig and Hibberd. However, the bases of the scales in the new species are smooth, sometimes even hollow, and contain no apertures as are present in *P. runcinifera.*

**P. undulata* Preisig and Hibberd (Fig. 17.2(15))

Reported world-wide from five other localities. This is the second report from a tropical/subtropical locality.

P. vestita (Stokes) de Saedeller (Fig. 17.2(16))

This species is distributed world-wide, including in the tropics (Dürrschmidt & Cronberg 1989) and a previous report from India (Saha & Wujek 1990).

Discussion

The present study revealed that *Paraphysomonas* species are unexpectedly abundant in tropical waters. Intensive sampling in similar habitats will probably reveal similar populations of *Paraphysomonas* irrespective of geographical location. Of the 152 different bodies of water sampled, 97 (64%) contained one or more *Paraphysomonas* species. *Paraphysomonas vestita* was present in 91 of the samples containing *Paraphysomonas* taxa. The majority of the species have been recorded frequently from Europe, North America and other parts of the world where the silica-scaled chrysophyte flora has been investigated. Most of the species are eury-thermal, tolerating a wide range of water temperatures.

Only recently, two species in our study (*P. porosa* and *P. vestita*) have been reported for neighboring Sri Lanka (Dürrschmidt & Cronberg 1989). As this and other studies on tropical chrysophyte floras are published, a geographically well-defined and characteristic tropical flora for *Paraphysomonas* may emerge.

Summary

Electron microscopic examination of 152 samples from Indian rivers, lakes, ponds and tanks has resulted in the identification of 13 taxa of *Paraphysomonas* (Chrysophyceae). Ten of these taxa have not previously been recorded from India (*P. acantholepis, P. bandaiensis, P. caelifrica, P. coronata, P. eiffelii, P. gladiata, P. imperforata, P. porosa, P. runcinifera* and *P. undulata*). Two further taxa are described as new (*P. planus* and *P. runciniferopsis*).

Acknowledgments

We would like to thank Paul Elsner for preparing the coated grids and shadowing, and the following individuals who assisted D.E.W. in the field and in making local travel and housing arrangements: Drs S.P. Adhirary, J. Ahmad, K.S. Bilgrami, B.R. Chaudhary, S.V.S. Chauhan, R.N. Das, S.K. Goyal, H.D. Kumar, R.K. Mehrota, P.K. Misra, D.D. Nautiyal, H. Polasa, R. Rao, V.N.R. Rao, C.P. Sharma, N.K. Shastri, R.K. Somashekar, G.L. Tiwari and V. Venkateswarlu. D.E.W. was supported by a CMU Research and Creative Endeavors grant and a Fulbright Fellowship while in India.

References

Cronberg, G. 1989. Biogeographical studies, scaled chrysophytes from the tropics. *Beih. Nova Hedwigia* **95**: 191–232.
Dürrschmidt, M. & Cronberg, G. 1989. Contribution to the knowledge of tropical chrysophytes: Mallomonadaceae and Paraphysomonadaceae from Sri Lanka. *Arch. Hydrobiol. Suppl.* **82**: 15–37.
Kristiansen, J. & Tong, D. 1989. Studies on silica-scaled chrysophytes from Wuhan, Hangzhou and Beijing, P.R. China. *Nova Hedwigia* **49**: 183–202.
Preisig, H.R. & Hibberd D.J. 1982a. Ultrastructure and taxonomy of *Paraphysomonas* (Chrysophyceae) and related genera: I. *Nord. J. Bot.* **2**: 397–420.

Chrysophyte algae

Preisig, H.R. & Hibberd, D.J. 1982b. Ultrastructure and taxonomy of
    Paraphysomonas (Chrysophyceae) and related genera: II. *Nord. J. Bot.* **2**:
    601–38.
Saha, L.C. & Wujek, D.E. 1990. Scaled chrysophytes from northeast India.
    Nord. J. Bot. **10**: 343–57.
Takahashi, E. & Hayakawa, T. 1979. The Synuraceae (Chrysophyceae) in
    Bangladesh. *Phykos* **18**: 129–47.
Vørs, N., Johansen, B. & Havskum, H. 1990. Electron microscopical
    observations on some species of *Paraphysomonas* (Chrysophyceae) from
    Danish lakes and ponds. *Nova Hedwigia* **50**: 337–54.
Wee, J.L. & Wujek, D.E. 1986. Two new species of Paraphysomonadaceae
    (Chrysophyceae). *Nova Hedwigia* **43**: 81–6.
Wujek, D.E. & Bland, R.G. 1991. Chrysophyceae (Mallomonadaceae and
    Paraphysomonadaceae) from Florida. III. Additions to the flora. *Florida
    Sci.* **54**: 41–8.

Index of scientific names

Subject index

benthic phase, 59
biflagellate, *see* flagella, number
biogeography, *see* distribution
biological indicators, *see* indicator species
biomanipulation, 291–2
biomass, 221–5, 229, 233, 240, 242, 246, 248–9, 270, 272, 274
  chrysophyte with phosphorus availability, 281–3
  zooplankton with chrysophyte correlation, 280, 281, 292–3, 295
  zooplankton with phytoplankton correlation, 280–1
biomineralization, 141–64
  definition, 141
biomonitoring, 303, 309–10
biotic interaction, 294; *see also* biomass
biotin
  deprivation, 99
  requirement, 129
birth rate, chrysophyte, 290; *see also* growth rates
body scales, *see* scales
bootstrap statistical test, 29, 33
boundary layer, 103, 105–6, 120
bristles (on scales), 145, 153, 157, 166, 168–9, 172, 279, 286, 291, 294, 304, 319, 338–9, 342–4
brown tide, *see* algal blooms
buoyancy, passive, 204

calibration sets, 293, 306–8, 312–13
canonical correspondence analysis, 244, 261, 307
capsoid morphology, 60–1
carbon
  acquisition, 99, 106, 111–12
  biochemistry, 96, 97–100, 101
  concentration, 130
  fixation, 99, 119
  inorganic, 236–7
  ratios, 106–7
  requirement, 128
  transfer, 120
  uptake, 237
carotenoid, 60; *see also* pigments
Cascade Lakes experiments, 292
cell
  coverings, 270; loricas, 270, 366; naked, 270, 366
  division, 57, 147, 151; vegetative, 60
  hairs, 57
  membrane, 100
  morphology, 337
  scales, *see* scales
  ultrastructure, 53, 56–7, 64, 67
  vegetative division, 57, 60, 147, 151

wall, 56–7, 59, 65–6, 105, 108; *see also specific organelles*
cellulose, 97
chemoorganotrophy, 96–9, 120
  obligate, 110–11
  phagotrophy, 80, 96–7, 99, 106–112, 120, 122–4, 127, 129, 130–1, 136, 197, 270; ancestral mode, 134; facultative, 128–9; obligate, 126, 128; trade-offs, 133
  saprotrophy, 96, 99, 106, 109
chlorophyll, *see* pigments
chloroplasts
  absence of, 84–5, 88
  color, 65
  function, loss of, 135
  girdle lamella, 76, 79, 82
  morphology of, 55–6, 65
  number, 76, 79, 82
  origins of, 84, 87, 121
  pyrenoids, 62, 64, 76, 79, 80, 82
Chrysophyte Lakes Database, 276, 280
citric acid, 184
cladocerans, 273–4, 295; *see also* zooplankton
classification, 9, 12, 14, 46–74, 78, 81, 88, 95, 143, 144
  Agardh, 4
  Andersen, 144
  biochemical criteria, 26, 67
  Bourrelly, 18, 19, 46, 143
  Christensen, 25–6
  Chromophyta, 26
  Chrysamoebales, 361–72
  Copeland, 18
  Dujardin, 7
  Ehrenberg, 6, 7
  importance of good, 262
  Kristiansen, 46, 48, 50
  Lameere, 18
  Linnean, 3
  methods, 19
  Moestrup, 50
  morphological, 15, 26, 46
  Pascher, 13, 15, 18, 50
  polyphyletic, 81
  Starmach, 46–7
  Stein, 8
climate studies, 254, 258
  changes, 318–19
coccoid morphology, 57, 60, 65
colchicine, 155
color, water, 224, 249
competitive ability, 241, 270
  zooplankton, 274, 292–4
conductivity, *see* specific conductance gradient
Connecticut flora, *see* American flora
contractile vacuoles, *see* vacuoles

scales (*cont.*)
  cell, 52–4, 77, 80, 83, 152
  diversity, 165
  flagellar, *see* flagella
  fossilized, 304
  'glue', 172–3
  for herbivore avoidance, 286, 291, 294
  identification, 315–17; *see also*
    biomonitoring
  immunological studies, 165–78
  insertion, 160, 165, 168
  lack of, 279, 286; *see also* cell, coverings
  mineralized, 141
  morphogenesis, 157, 166, 172–4, 176
  morphology, 144–5, 165–6, 319, 337,
    343–4, 345–58
  movement, 167–8, 172
  silicified, 142, 155, 160–1, 214, 279, 373
  spine, 168–9, 170–1
  variability, 346–7, 350, 352–3; with
    culture age, 355, 357–8; with
    temperature, 347, 351–2, 355, 357, 358
Scandinavia, 269
seasonal monitoring, 229
seasonality, *see* distribution
sediments
  analysis, 310, 320
  surface, 308–10, 312–13, 315–16
  'seed' population, 185, 269
selective feeding, *see* feeding biology
sexual reproduction, 78, 81
sheath, 55
silica, 141
  absorption of, 142
  crystalline polymers of, 141
  cycling, impact of chrysophytes on,
    159–60
  deposition vesicles (SDVs), 142, 152–3,
    155, 157, 161, 168–9
  dissolution of, 142, 157–9
  fossils, 306
  limitation, 279
silicon, 96
  cell content, 148–9
  growth requirement for, 148, 150, 160–1
  replenishment, 150–2
skeleton, silicified, 83, 85; *see also* scales
species diversity, 222, 224, 229, 232–5, 242,
    245–6
specific conductance gradient, 238–40,
    253–4, 255, 258–9, 334
specific growth rate, *see* growth rates
spindle microtubules, *see* microtubules
spine scales, *see* scales
spontaneous generation, 4, 6
state transitions, 98
statistics, *see* bootstrap statistical test,
    canonical correspondence analysis *and*
    principal component analysis

statospores, *see* cysts
sterol, 63, 97
stomatocysts, *see* cysts
stratification, algal, 192–3, 196–7, 204, 237;
    see also distribution *and* metalimnetic
    zone
stratigraphy, *see* depth–time profile
submetalimnetic zone, 248
succession, 244, 247, 311
sulphur emission, *see* acid rain
surface
  area, lake, *see* lake
  microlayer, 199–200
  sediment approach, *see* sediments
swarmers, *see* zoospores
swimming
  cell rotation during, 107–8
  energy costs, 106
symbiosis, parasitic, 203; *see also*
    endosymbiont theory
systematics, 1–91

taxonomic scale, 233
taxonomic significance, 57
taxonomy, *see* classification
temperature, 334
  algal blooms, 192, 197
  changes, 244, 253, 343
  cold water flora, 248, 250
  cool water flora, 248
  effects on growth, *see* growth rates
  effects on scales, *see* scales
  gradients, 204, 235, 243–7, 255, 259
  morphological changes, 319
  optimum, 192
  physiological limits, 249
  warm water flora, 250
tension-resistant matrix, 105
tentacles, 78, 81, 83, 84
thallus, morphology of, 61, 64; *see also*
    organization
thiamine deprivation, 99
thylakoid reactions, 98
tolerances, ecological, 232; *see also*
    temperature, light *and* nutrients
toxicity, *see* algal blooms; *see also*
    volatiles
trace element requirements, 197, 200, 203
transitional helix, 77, 80, 82–3
trophic gradient, 240–3, 254, 274–5, 281
trophic level, 119, 225, 229, 258
trophic status, 311

ultrastructure, *see* organelle
uniflagellates, *see* flagella, number
urea, 100, 102

Printed in the United States
By Bookmasters